Against Relativism

For Reg Coates,
especially Chapters 9 and 11

Against Relativism

Philosophy of Science, Deconstruction and Critical Theory

Christopher Norris

First published 1997

2 4 6 8 10 9 7 5 3 1

Blackwell Publishers Ltd
108 Cowley Road
Oxford OX4 1JF
UK

Blackwell Publishers Inc.
350 Main Street
Malden, Massachusetts 02148
USA

British Libray Cataloguing-in-Publication Data

A CIP catalogue record for this book is available from the British Library.

Library of Congress Cataloging-in-Publication Data

Norris, Christopher.
Against relativism: philosophy of science, deconstruction, and critical theory / Christopher Norris.
p. cm.
Includes bibliographical references and index.
ISBN 0-631-19864-4 (alk. paper). – ISBN 0-631-19865-2 (pbk.)
1. Science – Philosophy. 2. Deconstruction. 3. Realism.
I. Title.
Q175.N594 1997
501–dc21 97-4300 CIP

Typeset in $10\frac{1}{2}$ on 12pt Garamond 3 by
Best-set Typesetter Ltd, Hong Kong
Printed and bound in Great Britain by
Hartnolls Limited, Bodmin, Cornwall

This book is printed on acid-free paper

Contents

Preface

The past few years have witnessed some heated debate about various forms of cultural-relativist (or social-constructivist) thinking that would appear to undermine all accepted standards of truth, objectivity and reputable method in the natural sciences. On one side are ranged those defenders of truth – scientists mostly – who reject such ideas as merely a species of irrational mystery-mongering. (Thus Richard Dawkins: 'show me a cultural relativist at thirty thousand feet and I will show you a hypocrite. Airplanes built according to scientific principles work.') Alongside them are realist philosophers of science who insist that what makes our theories and beliefs true or false is the way things stand in reality and not just the way reality is 'constructed' in accordance with this or that cultural world-view, descriptive scheme, communal belief-system or whatever. On the other side are ranged the anti-realists and cultural relativists for whom such an argument is both nonsensical – involving as it does the appeal to truths outside or beyond our best current knowledge – and also (most often) an instrument of power for silencing dissident views.

I should say straight off that my own position is that of a causal realist who believes – as a matter of inference to the best explanation – that science has achieved genuine advances in our knowledge of a (largely) mind-independent and language-independent reality. That is to say, I reject any version of the claim that truth *just is* whatever we make of it under this or that currently favoured description. I shall therefore offer some detailed criticism of those various movements of thought – in philosophy, sociology, cultural theory and 'science studies' – which make a full-scale programme of reducing truth to what is currently and contingently 'good in the way of belief'. This argument takes various forms, from the 'strong programme' in sociology of knowledge to the neo-pragmatism of philosophers like Richard Rorty and the relativist doctrine – derived from Thomas Kuhn – that scientists on either side of a major paradigm change quite literally 'live in different worlds', so that no comparison is possible between or across such 'incommensurable' theories. Yet more extreme are the statements of postmodernist thinkers, such as Jean-François Lyotard, who maintain that science has now moved on into an epoch when 'performativity' (not truth) is the operative criterion, and when the guiding rule is to maximize 'dissensus' – to multiply the range of noncommensurable theories and paradigms – rather than aim for some illusory idea of truth at the end of enquiry. Then again there are those, Paul Feyerabend among them, who recommend that we abandon all notions of scientific 'method' and henceforth adopt an anarchic pluralist approach unbeholden to accepted (authoritarian) standards of truth, consistency and rigour.

I shall criticize these arguments at some length since they seem to me both philosophically confused and a source of much confusion elsewhere, especially among cultural and literary theorists keen to view science ('postmodern' science) in the image of their own concerns. Hence the vague allusions to fashionable *topoi* – chaos theory, fractals, quantum nonlocality, Heisenberg's uncertainty principle and so forth – often to be found in the writings of Lyotard and like-minded commentators. However, I shall also be concerned to develop an alternative (realist) approach drawing on the work of various philosophers in the Anglo-American and Continental traditions of thought. These claims are backed up by reference to particular episodes in the history of science which strengthen the case for causal realism as the only theory capable of explaining our knowledge of the growth of knowledge. On the 'Continental' side I shall trace various lines of descent from the programme of critical epistemology initiated by Gaston Bachelard and Georges Canguilhem. Bachelard's theory of 'epistemological breaks' is of particular interest in this context since it has given rise to two sharply divergent schools of thought. Thus on the one hand its influence can clearly be seen in Foucault's sceptical-relativist 'archaeology' of the natural and human sciences. On the other – I shall argue – that influence extends to Jacques Derrida's far more rigorous deconstructive account of the relation between concept and metaphor in the texts of Western philosophy, including those of Bachelard.

Anti-realism derives much of its current appeal from the widespread turn toward language-based or hermeneutic approaches, coupled with an equally widespread scepticism regarding traditional ('foundationalist') modes of philosophical enquiry. The result has been a confluence of ideas from many sources – Heidegger, late Wittgenstein, Quine, Kuhnian philosophy of science, post-structuralism, postmodernism, neo-pragmatism, the 'strong' programme in sociology of knowledge – all of which are taken as signalling the end of that old, truth-fixated, epistemological paradigm. However, I contend, there is nothing to be gained (and a great deal to be lost) by this over-reactive retreat from standards of truth and rational accountability. For there exists a whole range of more promising alternative resources – in history and philosophy of science, epistemology, ontology and philosophical semantics – which point a way forward from this sceptical-relativist impasse. They include not only causal-realist approaches as developed by philosophers in the broadly 'analytic' (Anglo-American) tradition but also that mode of epistemo-critical reflection on the discourse of the natural sciences practised by thinkers like Bachelard. Indeed, I shall put the case that these two (supposedly non-communicating) schools of thought – 'analytic' and 'Continental' philosophy – are much better viewed as variant responses to a number of shared preoccupations. This will involve some detailed commentary on texts by Derrida and Paul de Man which raise crucial issues concerning the status of scientific truth-claims and (as in Derrida's essay 'White Mythology') the relation between concept and metaphor in the process of scientific theory construction. I shall also emphasize the close kinship that exists between developments in so-called 'post-analytic' philosophy – deriving very often from Quine's ontological-relativist arguments and Kuhn's doctrine of radical meaning-variance – and the kinds of ultra-nominalist argument to be found in post-structuralism and its various source texts, Foucault's among them.

In each case, I suggest, these ideas can be traced back to one or another version of the 'linguistic turn' whose effect is to promote a generalized scepticism with regard to scientific knowledge. What is needed, therefore, is a clearer view of the differences *within* those traditions that have lately given rise to a realist counter-movement in Anglophone philosophy of science, and which also distinguish Derrida's work from the wider currency of postmodernist/post-structuralist thought. For deconstruction harbours critical resources that are not to be had from the kind of blanket anti-realist approach which has dominated recent 'radical' critiques of science and philosophy of science. I shall thus propose some far-reaching realignments of interest and a conception of critical realism that cuts across conventional accounts of recent intellectual history. I shall also devote some extended commentary to a number of individual thinkers – Davidson, Hanson, Kuhn, Quine, Rorty, van Fraassen and others – whose work has been central to recent debate in this area. Through examples from various fields of science (including rival interpretations of quantum mechanics) I seek to demonstrate the incoherence of anti-realist doctrines and the way that their exponents always have recourse at some point to a covert realist ontology.

There are also chapters on the science of aerodynamics – a particular interest of mine – taken as a test-case for cultural-relativist or social-constructivist doctrines, and again, as a challenge to more moderate forms of anti-realism (such as van Fraassen's) which find no room for 'laws of nature' as a means of causal explanation. One weakness of recent work in this field – i.e., mainstream philosophy of science – has been its failure to engage with studies of applied technological development such as Edward Constant's path-breaking book *The Turbojet Revolution*. Most often these issues have been treated in a relatively abstract fashion or with insufficient regard to the various real-world constraints and possibilities which in practice decide what shall count as a viable theory, paradigm, research programme, etc. I therefore include some fairly detailed discussion of work in this field, along with a critique of Heidegger's thinking on the 'question of technology', since his ideas have had a large (and I think a predominantly harmful) influence on current attitudes to science among philosophers, sociologists and cultural theorists.

Elsewhere I offer a critical account of the 'strong' sociological approach, typified by works like Andrew Pickering's *Constructing Quarks* and Shapin and Schaffer's *Leviathan and the Air-Pump*. This latter has become a classic of the genre and is often cited in support of the idea that experimental science is just one 'language-game' or cultural 'life-form' among others, a discourse whose nature can best be understood with reference to its social and ideological conditions of emergence. What this argument reveals – I argue – is a regular confusion between the context of discovery and the context of justification, or a failure to acknowledge the distinctive standards (of truth, warrant, evidential reasoning or inference to the best explanation) which properly characterize these different contexts. Another main factor is the sceptical argument – deriving from Hume and revived by present-day philosophers such as Nelson Goodman – that inductive explanations can always be shown to rest upon circular (inductive) premises or to presuppose what they seek to establish, i.e., the existence of law-governed regularities in nature. Whence comes the sceptical meta-induction by which it is maintained that we can never have adequate grounds for believing any scientific theory to be true or for thinking that science has shown some progress toward truth at the end of enquiry. However, this argument works more convincingly the other way around. That is to say, it lends strong support to the case that science has attained a progressively more adequate (depth-explanatory) knowledge of just those real-world causal attributes, structures, dispositions or properties which themselves serve to justify the process of inductive reasoning.

I would hope that this book will be of interest to critical theorists, to philosophers and historians of science, and to sociologists of knowledge willing to question their own as well as other people's disciplinary investments. In particular, I shall challenge the idea that there exists an unbridgeable gulf – or, in Lyotard's postmodernist parlance, a 'differend' of strictly incommensurable phrase-genres – between the truth-claims of science and the value sphere of ethical and sociopolitical debate. On the contrary: such debate very often raises issues which cannot be discussed to any purpose without an adequate (scientifically informed) and thus a properly realist assessment of our practical options. What is required, in short, is an ethics responsive to the range of material factors – both enabling and restrictive – that bear upon our thinking in matters of real-world social and moral choice. For otherwise the way is wide open to those varieties of wholesale anti-realist doctrine (Richard Rorty's prominent among them) which offer us the merely notional 'freedom' to redescribe reality however we choose according to this or that preferred language game or set of cultural values. Such ideas have given rise to much confused thinking in the social and human sciences. Thus my chief aim in what follows is to offer a critical but constructive challenge to these currently widespread forms of anti-realist and cultural-relativist thinking.

Acknowledgements

Once again I should like to thank my colleagues and postgraduate students in the Philosophy Section at Cardiff for offering ideas, comments and criticism in response to the various lines of argument developed here. 'There is the bit where you say it,' as J. L. Austin famously remarked, 'and the bit where you take it back.' Probably I have not taken back enough, or taken back all the wrong bits, or put in some other bits that will make them wish they had tactfully let things stand. Anyway, the book would have been a much poorer thing without their many constructive suggestions and constant supply of intellectual stimulus over the past three years. So thanks to Robin Attfield, Andrew Belsey, Michael Durrant, Andrew Edgar, Stephen Moller, Peter Sedgwick, Alessandra Tanesini, and Barry Wilkins; also to Gideon Calder, John Doherty, Stephen Horton (whose Ph.D. thesis raised all the right issues), Christa Knellwolf (who commented very usefully on an early draft), Carol Jones, Marianna Papastephanou, David Roden and Daniele Procida. I had a welcome chance to air some of these ideas in my two undergraduate seminars (on Philosophy of Science and Deconstruction) during the academic year 1996/7. I should like to thank the students for their lively interest and for contributing more than they knew to my work in progress. Some chapters of this book have previously appeared in the journals *Inquiry, Metaphilosophy, Philosophy and Social Criticism, Paragraph* and *SubStance.* I am grateful to the editors and publishers for their permission to reprint this material in a slightly modified from.

Once again the chief burdens were borne by Alison, Clare and Jenny, who kept me pretty much on an even keel through the stormier passages. 'Another damned thick square book? Scribble, scribble, scribble,' as somebody once said to somebody else.

Christopher Norris
Cardiff
December, 1996

1

Metaphor, Concept and Theory Change: Deconstruction as Critical Ontology

I

I shall here contend that the cultural-relativist approach to philosophy of science is misguided both in theory and in the kinds of ethical or sociopolitical judgement to which it very often gives rise. By equating 'science' with its own worst abuses – i.e., with its exploitative, its purely instrumental or technocratic forms – this approach fails to recognize that the quest for truth carries its own ethical imperative, that is to say, an obligation (on the part of scientists and philosophers or sociologists of science) to get things right so far as possible and not to be swayed by the pressures of conformist ideology or consensus belief. For, as William Empson remarks, the human mind very often has to labour against unknown odds of prejudice, unreason and deep-laid resistance to heterodox ideas.[1] From which it follows that any new discovery – any 'paradigm shift' or challenge to established (consensus-based) norms – will encounter such resistance not only in the wider institutional context but also from those forces that are ranged against it in the scientist's own pre-existent habits of thought. Where science breaks new ground it is most often through what Empson describes as a 'complicated churning' of facts and theories, a process which may look decidedly messy (mixed up with all sorts of extraneous private or social motivation) if treated from a viewpoint primarily focused on the original context of discovery. But whatever the historical or psychobiographical interest of such enquiries, it is in the scientific context of justification that any truth-claims must finally be assessed.[2] And this applies equally to the ethical issue of how far a scientist may be justified in proposing theories at variance with the ideological self-images of the age. For such conflicts could simply not arise if 'truth' were indeed – as the cultural relativists would have it – just a product of short-term localized consensus belief.[3] And conversely, the moral argument would lack all force if the thinker's heterodox view of things were shown to rest on nothing more substantial than an attachment to some idiosyncratic quirk of method or technique, whatever its long-run pragmatic or instrumental yield.

No doubt such cases of serendipitous discovery have occurred fairly often in the history of science and provide welcome grist to the mill of relativists, sceptics and anti-methodologists like Feyerabend.[4] But to view them as *representative* cases – typical examples of how science works through a mixture of adhockery, cranky obsessions, fake methodology and sheer opportunism – is to fall into a version of that widespread error which confuses the context of discovery with the context of justification. Altogether more significant are those paradigm instances – like Priestley versus Lavoisier on the process of combustion – where two rival theories could indeed for a while claim warrant from the evidence to hand, but where one proved wanting while the other was borne out through a mode of explanation better equipped to cope with that evidence.[5] On the Feyerabend view we might just as well still be working with the phlogiston theory, since any arguments against it could only have been a matter of contingent sociocultural interests or peer-group pressure. And the same goes for other well-documented cases – like Boyle's controversy with Hobbes over the air-pump – where competing ideologies or political beliefs can be shown to have exerted an influence that effectively decided the issue on each side.[6] Thus justice requires, in Lyotard's postmodernist parlance, that we respect the narrative 'differend' between them and not set up as arbiters of a truth that would wrong at least one (perhaps both) of the parties in dispute by overriding their specific interests, motives, criteria of judgement, etc.[7] Least of all should we suppose – in the wisdom of hindsight – that when Boyle sought to verify the vacuum hypothesis by laboriously constructing his air-pump, he was thereby espousing the values of free, open-minded, empirical enquiry, whereas Hobbes's advocacy of the plenum thesis came down to nothing more than a displaced expression of his conservative or social-legitimist views. On the contrary, so this argument runs: Boyle's commitment to those values was just as 'ideological', just as much a reflex of his own self-image as spokesman for a newly emergent (bourgeois and empiricist) scientific culture. Its establishment – contra the Hobbesian upholders of state and monarchical power – required precisely this promotion of a hands-on experimental method, linked to the legitimizing rhetoric of science as a quest for knowledge not beholden to any such external source of authority. Thus there is simply no deciding the issue between them in point of 'scientific' method, validity or truth.[8]

I shall argue here that this whole line of thinking is radically misconceived. On the one hand it presents a false (distorted and reductive) understanding of science as compared with alternative, more informed varieties of historical, cultural and critical enquiry. On the other it deflects the force of any criticism that would engage with science on reasoned and principled terms, rather than resorting to polemics or routine gestures of dismissal. What perhaps started out as a justified protest against arrogant technocratic reason has now become, in many quarters, a pretext for the crudest, most wholesale forms of cultural-relativist dogma. Even in its milder variants this approach can give rise to troubling ethical as well as epistemological confusions. Empson makes the point with regard to E. A. Burtt's claim that the Copernican theory was 'conceived as a charming piece of mathematical fitness, not scientifically, as a thing empirically true', a claim supposedly borne out by the evidence that 'people ignored the apparent absence of stellar parallaxes.'[9] Here Burtt anticipates Feyerabend's point about the often far from perfect match between theory and experimental data, along with Quine's ontological-relativist argument that observations are always theory-laden and theories underdetermined by the observational evidence.[10] Moreover, according to Burtt, 'Aristotle's view of space prevented him from conceiving the heavenly bodies as bodies, and as subject to geometry, so that astronomy had to wait for a revival of Platonism.' But, as Empson remarks, 'Aristotle himself produced the argument about parallaxes, and Copernicus overrode it; I find this a pleasing historical fact because it shows that both these great men were more intelligent (less at the mercy of their own notions) than Mr Burtt wishes to think them.'[11] Indeed, one objection to modish talk of 'paradigms', 'discourses', 'interpretive communities' and the like is that it tends to promote a placid assurance that change comes about for no better reason – and at no greater cost of strenuous disavowal – than mere boredom with the old paradigm and desire for some substitute discourse whereby to enliven the ongoing cultural conversation. This is at any rate the view of neo-pragmatists like Richard Rorty and Stanley Fish, convinced as they are that persuasion (or rhetoric) is indeed the bottom line of enquiry, and that truth-talk or kindred high-toned appeals to fact, evidence, theory or principle are just so many ploys for enlisting assent among members of the relevant peer-group community.[12]

That these notions currently exert such widespread appeal is a symptom not only of the postmodern retreat from values of truth and reason but also of the ethical bankruptcy entailed by this slide into an outlook of *laissez-faire* relativism. The point is well made in the following passage from Empson's review of Burtt.

> This assumption, that the metaphysics of science might have been quite different, makes Mr Burtt tease Newton about absolute space as if it was a wild and unnecessary fancy; Newton ought to have been sensible, and believed in a sort of halfpenny-press relativity. But this (apart from being ungenerous) is untrue; something has to be thought of as settled; the point of that ill-named theory is not that everything became relative but that a new thing (not space but the velocity of light) was found which could be treated as absolute. Before (say) the Michelson-Morley experiments it was necessary to fix space, if not really for physics, then habitually for the imagination. The fact that Newton gives so much attention to the metaphysical idea of a fixed space does not show him as an inconsistent empiricist, but as a man with an extraordinary grasp of the assumptions of his own thought, an extraordinary foreknowledge of the lines on which it must be modified. Mr Burtt is unfair again about a passage where Newton says that 'the *variety* of motion is

> always decreasing and therefore there must be some principle of conservation, perhaps something directly in the hands of God, to stop the universe from running down like a clock.' Mr Burtt wrongly repeats this as 'motion was on the decay' and treats it as a theological misunderstanding of the principle of the Conservation of Energy. Of course it is an extraordinary forecast of the second law of Thermodynamics; it is precisely where Newton's mind must be regarded with something hardly less than reverence that Mr Burtt accuses him of prescientific modes of thought.[13]

I have quoted this passage at length for two main reasons. First, it offers an eloquent defence of science as a truth-seeking, intellectually strenuous, and at times morally courageous activity. Second, it stands in sharp contrast to the cynicism displayed by sceptics like Feyerabend when they treat those values as really nothing more than a careerist ploy to gain 'scientific' credence, or a strategy to hoodwink the wider (non-specialist) community of lay believers.

One could extend this argument to a point-for-point comparison between Empson's reading of the Newton case and Feyerabend's attitude of prosecuting zeal towards Galileo in his battle with the forces of entrenched religious dogma and prejudice.[14] Thus what Empson remarks about Burtt's imputations of theologico-metaphysical bias – that 'this (apart from being ungenerous) is untrue' – would apply just as aptly to Feyerabend's claim that truth is neither here nor there in such matters, that Galileo was in any case fudging his results or bending the evidence to fit the theory, and hence that he ought to have shown more respect for the wisdom of the Church authorities. That he runs these last three arguments together without, so far as one can tell, perceiving their logical incompatibility is evidence enough of the strains imposed – or the confusions engendered – by this ultra-relativist line. And there is a further sense in which it offers support to Empson's criticism of Burtt. That is, it shows how easily relativism moves across from (1) investigating the pre-scientific origins of scientific truth-claims, via (2) treating those claims as on a par with other (e.g., metaphysical or theological) items of belief, to (3) the full-blown Feyerabend doctrine that holds the scientist *intellectually and morally at fault* for venturing truths that happen to conflict with existing (past or present) ideas of the social good.

II

Post-structuralism is one influential source of this present-day fashion for cultural-relativist and kindred anti-realist doctrines. However, the influence can also be seen to work the other way around. That is, post-structuralists tend to suppose that the 'linguistic turn' is a veritable *fait accompli*, that even science (or philosophy of science) must henceforth take it on board, and therefore that values such as 'truth', 'reason' and 'reality' can have no further role in any discourse that aspires to play by the current rules of the game.[15] This conclusion is typically arrived at through a selective reading of Saussure, backed up – as they suppose – by the Nietzschean–Derridean sceptical assault on truth-values, logic, validity conditions and suchlike deplorable residues of a Western logocentrist 'metaphysics of presence'. Other commentators – Maudemarie Clark among them – have done a good job in showing how far this misrepresents Nietzsche's attitude towards the natural sciences of his day.[16] In Derrida's case (as I shall argue here) it is likewise a product of fixed preconceptions allied to a manifest failure to read his work beyond the handful of oft-cited passages which, taken out of context, appear to support the view of deconstruction as a kind of all-purpose rhetorical licence for cutting 'philosophy' down to size. Yet nothing could be further from Derrida's position as argued with the utmost conceptual rigour – that is to say, through close reading of particular texts, but also through the meticulous analysis of truth-claims, categorical judgements, orders of logical priority, etc. – in essays such as 'White Mythology' and 'The Supplement of Copula.'[17] What these writings reveal above all is the distance that separates Derrida's work from the attitude of *de rigueur* epistemological scepticism which has become such a hallmark of recent post-structuralist thought.

I hardly need to stress the irony of this situation, given the *soi-disant* 'radical' claims that very often go along with the new orthodoxy. Post-structuralism indeed started out as a vigorous protest against the kind of lazy-minded 'commonsense' thinking that passed itself off, in departments of literature, as an adequate response to the absurd pretensions of all that newfangled 'theoretical' stuff. More specifically, it offered a worthwhile critique of the truth-claims vested in a certain moralizing discourse, an uncritical appeal to supposedly self-evident values (or standards of 'mature' judgement) with punitive sanctions attached for anyone who dared to question its sacrosanct articles of faith. Where the project miscarried – or so I would claim – was in pushing this argument way beyond the confines of a particular (intradisciplinary) dispute about culture-specific habits of thought and judgement that claimed special access to some privileged realm of timeless, transcendent, self-validating truth. Such scepticism was altogether justified in so far as it sought to reveal those claims as impositions of an authoritarian rhetoric upon texts and also upon readers – especially students – who were not well placed, for fairly obvious reasons, to contest their presumptive warrant. But it is a different matter when these arguments are extended from literature (or literary criticism) to the entire range of humanistic disciplines and then – since there is nothing left to obstruct the all-encompassing horizon of 'language', 'text' or 'discourse' – to the natural sciences also. For what drops completely out of sight at this stage is the distinction between *on the one hand* those varieties of discourse (scientific, philosophic, theoretical, historical, sociological, etc.) where truth-claims are always in question, and where the relevant criteria include – for instance – observational warrant, factual evidence, logical consistency and inference to the best (most adequate) explanation, and *on the other hand* language as deployed in the creation of fictive, poetic or imaginary worlds where such conditions are understood not to apply, or if so, then at a certain implicit remove from the normative sphere of veridical utterance.[18] It is the difference, in short, between contexts of enquiry where statements may be judged true or false, and contexts – like that of literary interpretation – where the assignment of determinate truth-values is rarely (if ever) possible.

This distinction is everywhere under attack in the more 'advanced' quarters of postmodern cultural debate. Thus sceptical historiographers like Hayden White see no valid grounds for distinguishing fictive from avowedly factual or truth-telling discourse, since both exhibit a similar range of tropological devices, descriptive techniques, modes of narrative emplotment and so forth.[19] Comparative ethnographers – Clifford Geertz among them – also take the view that theirs is a discipline primarily characterized by its status as 'writing', that is to say, by its success in producing 'thick descriptions' whose value lies not so much in their truth to the observed detail of this or that cultural life-form, but rather in the process of textual composition – or narrative *mise-en-scène* – that evokes some strange (sublime or fabulous) moment of encounter exceeding the utmost powers of accurate recall.[20] For their part the New Historicists – much influenced by Geertz – make a full-scale programme of conflating these realms, to the point where 'historical' and 'literary' texts are treated as wholly on a par with respect to both their value as social-documentary sources and whatever quality or interest they possess as products of narrative contrivance.[21] Then again there is Rorty's idea of philosophy as likewise just another 'kind of writing', a kind all too often given over to talk of 'truth', 'reason', '*a priori* concepts', 'sense-data', 'conditions of possibility' and the like. Far better – Rorty thinks – if we came around to treating philosophy as an open multiplicity of language games, a range of optional vocabularies whose interest and value consist in their providing new metaphors we can live by or styles of narrative 'self-description', new ways in which to characterize the history of philosophy and our own relationship to it.[22]

In each of these cases – and others besides – it is taken pretty much for granted that the entire disciplinary map has been altered (indeed radically transformed) by the turn towards language, discourse, textuality or representation as the limit-point of intelligibility. But it is also clear that in each case the proclaimed liberation from old disciplinary constraints goes along with a whole new set of orthodox bans on any talk of 'reality' or 'truth', or any questions concerning the conceptual adequacy of these various textualist paradigms. The issue is posed most sharply with regard to the status of historical truth-claims and the idea (propounded by White) that these can amount to no more than a choice among the various rhetorical strategies – metaphor, metonymy, synecdoche, irony – and the corresponding range of generic options which between them map out the terrain of discursive or narrative representation.[23] I have written at length elsewhere – as have other critics of the textualist line, among them Perry Anderson – about the failure of this model to account for what is specific to the kinds and modalities of historical discourse, i.e., those features that distinguish history from other (fictive or literary) narrative genres.[24] They include, in brief, the historian's concern with factual evidence, causal explanation, chronological consistency, the use of counter-factual hypotheses with a view to eliminating rival (inconsistent) explanatory accounts, attribution of motives to agents individual and collective, constraints upon their freedom of action imposed by exigencies of real-world practical choice, and so forth. What is so striking about the current ultra-relativist or textualist turn, here as in philosophy of science, is the attitude of resolute indifference it displays towards any such criteria of valid argument that would place certain limits on the range of admissible or good-faith interpretative options.

Nor can it be argued – unless through a species of massive ontological confusion – that all of the above-mentioned factors are likewise operative in the writing and reading of fictional texts. For such a viewpoint can only be maintained if one takes it as a matter of sheer self-evidence (or maybe with the usual obligatory nod towards Derrida) that there is indeed nothing 'outside the text', or nothing that could possibly serve to demarcate the fictive and historical domains. But this is, I shall argue, so blatant a misreading of Derrida's texts as to place the burden of proof very squarely back where it belongs, on the side of those who would regard history (along with science and our other deluded questings for truth) as a wholly textual or rhetorical construct, a 'kind of writing' without the least claim to get things right according to its own specific disciplinary criteria. Again it seems fair to conjecture that this situation has come about very largely as a consequence of certain modish ideas – the utopian 'free play' of the signifier portending revolutionary change, *écriture* (and more especially *écriture féminine*) as likewise aimed towards the overthrow of the transcendental (or patriarchal) signified, the 'classic bourgeois realist text' as intrinsically in league with the forces of injustice and oppression – which started life among post-structuralist literary theorists and have now gained a hold upon other disciplines, among them historiography and philosophy of science.[25] In the literary field these ideas could do little harm, apart (that is) from their role in creating a new 'radical' orthodoxy with its own kinds of firmly dissuasive technique for heading off awkward questions. But there was much more at stake in their subsequent passage – via writings like Barthes's influential essay 'The Discourse of History' – from literary criticism to those other disciplines where questions of truth, validity and method were now dismissed as so many remnants of an obsolete 'positivist' paradigm with its own covert ideological agenda.[26] Whatever their express political alignment, such notions fell in all too readily with the wider trend towards revisionist (often right-wing revisionist) views on the relation between historical 'truth' and present-day social values.[27] At any rate, they could offer no resistance to such views – no reasoned or principled resistance – except by evoking the Nietzschean–Foucauldian idea of history as the site of an ongoing struggle over meanings and values where 'truth' once again had no role to play save that of a strategic rhetorical move in the agonistic game of power/knowledge interests.[28] And it is not hard to see why this message should be welcomed by strong revisionist ideologues whose hour was coming round at last with the swing towards highly conservative forms of hegemonic consensus-building.[29]

Of course it would be wrong – another species of vulgar reductionist error – to maintain that these ideas are philosophically ill-founded on just the guilt-by-association grounds mentioned above. Any adequate critique has to press much farther back and enquire into the conditions of validity for the entire structuralist and post-structuralist project, including its originary formulations in Saussure's *Cours de Linguistique générale.*[30] Valentine Cunningham has some good points here which merit a lengthy quotation from his recent book *In the Reading Gaol.* 'Preferring the line of non-referentiality,' he writes,

> is to opt for the truly dark side of Saussure's imagination, the vision that drove him to spend up to four years filling forty or so notebooks with his efforts to show that certain poetic texts, certain ancient and modern poems, had a secretive anagrammatical content – that is, cryptic, often self-referential meanings (authors' names, for example) seeded deep within them. . . . What the long search indicates is a high level of faith in the arbitrary, almost magical subsistence of meaning. And it was this idea of a kind of self-assertiveness of language that did not, it appears, allow him instantly to dismiss as ludicrous the Genevan spiritualist medium Mlle Hélène Smith when she was brought to him in his capacity as a Professor of Sanskrit to investigate her claim to be inspired by spirit guides to speak Sanskrit. She also, more unusually (for there was a lot of pseudo-Indian mysticism knocking about Europe at the end of the nineteenth century) claimed to speak Martian. Arrestingly, Saussure decided that she probably was uttering bits of Sanskrit, and probably some Martian as well.[31]

Cunningham makes a lot of this anecdote, and I think justifiably so. Thus he remarks that Saussure attended Mlle Smith's séances from 1895 to 1898, that he was working on the anagram notebooks between 1906 and 1909, and that his famous series of lectures at Geneva – transcribed by various students and eventually published as the *Cours générale* – took place from 1907 to 1911. Cunningham also draws attention to the quite extraordinary appeal exerted – mainly on critics of a high formalist or structuralist bent – by Saussure's distinctly wacky anagrammatical researches.[32]

However, his main point in digging up this odd chapter of episodes is that it leads us to cast a more quizzical eye on those notions (like the 'arbitrary' nature of the sign) which have since become a high point of post-structuralist doctrine. Thus:

> whether he was questing for hidden, possibly random, and unintentional names, words, meanings in a kind of super-structuralist (or cabbalistic) search for the deviously internalized senses of poems, or taking the medium's glossolalic capacities in Sanskrit and Martian seriously enough to delve into her manifestly non-referential utterances, Saussure was only doing, as Tzvetan Todorov has suggested, what he tried to systematize in parts of the *Cours*, namely treating language and text as arbitrary, non-referential, non-symbolic.[33]

This is not just a piece of knockabout polemics at Saussure's expense, or a merely *ad hominem* argument that fails to address the deeper theoretical issues. It has to do precisely with the motivated character – the nexus of idiosyncratic needs and desires – that went into producing a theory of language one of whose central precepts was the non-motivated ('arbitrary') relation between signifier and signified. Thus it points to a deep-laid contradiction in Saussure's work and, beyond that, a sense in which the structuralist and post-structuralist enterprise necessarily partakes of what Derrida would call a constitutive aporia in its own theoretical project. (Indeed it is worth remarking, since Cunningham makes no mention of the fact, that Derrida arrives at some strikingly similar conclusions through his own deconstructive readings of Saussure in *Of Grammatology* and elsewhere.)[34] All the same, the case as Cunningham presents it is more concerned with the context of discovery – with the background interests and obscure psychological promptings that led to this set of claims about language – than with the context of justification wherein those claims should properly be assessed for their consistency, validity or truth. The latter would require a demonstrative argument to the effect that Saussure's theory is inadequate to account for certain strictly ineliminable features of language, among them its referential capacity and its expressive (that is to say, its motivated) aspect as a bearer of speaker's intentions.

This is just the kind of argument that Derrida provides in his early essays exploring the interface – the points of mutual interrogative exchange – between structuralism and phenomenology.[35] The latter project is one which seeks to do justice to language in its expressive dimension (i.e., those aspects that cannot be reduced to a purely structural account) but which always at a certain point comes up against the need to postulate a pre-given structural economy that makes language and communication possible. The former, conversely, stakes its claim on the absolute priority of structure over meaning, or of *langue* over everything pertaining to the realm of *parole*, only to discover – or unwittingly to reveal through symptomatic blind spots in its own discourse – that language involves a signifying surplus which cannot be accounted for in purely structural terms. Both projects are sufficiently rigorous or self-critical to bring these aporias to light *despite and against* their express methodological commitments. Thus 'i[t] is always something like an *opening* which will frustrate the structuralist project. . . . What I can never understand, in a structure, is that by means of which it is not closed.'[36]

On the other hand, phenomenology has to acknowledge that meaning (or expression) cannot come about except in so far as there already exists a domain of pre-constituted signifying structures – Saussure's *la langue* – which alone makes it possible for speech-acts to possess intelligible meaning or force. It is for this reason, as Derrida writes, that 'a certain structuralism has always been philosophy's most spontaneous gesture.'[37] Only by holding itself open to both these requirements – to phenomenology in its concern with that which exceeds any purely structural account, and to structuralism in its rigorous attempt to theorize the conditions of possibility for language in general – can thinking avoid what Derrida calls 'the Scylla and Charybdis of logicizing structuralism and psychologistic geneticism'.[38]

In short, there is no escaping the difficulties that arise from any serious engagement with these issues in philosophy of mind and language. They are ineluctable problems in precisely the way that Kant discovered in his dealing with the various antinomies of pure reason, among them (most crucially) the freewill/determinism issue.[39] But they do show clearly that Cunningham is right – albeit basing his argument more on anecdotal than philosophic grounds – when he claims that the purebred structuralist approach is de facto predisposed to ignore certain features of language that would figure necessarily in any account which made due allowance for its other (i.e., expressive and referential) aspects. The same applies to his point about Saussure's motives in treating the sign as a two-term relation between signifier and signified, and in thus seeking to exclude – or bracket out – all concern with language in its referential aspect. Here again there is good warrant for Cunningham's suspicion that this move had to do not only with Saussure's somewhat specialized disciplinary interests (that is, his desire to establish the science of structural-synchronic linguistics) but also with certain ulterior commitments of a less than scientific kind. But if so, then the argument needs to go further than pointing out the various possible or likely relationships that hold between Saussure's doctrine of the 'arbitrary' nature of the sign and his obsessive quest for anagrammatical structures of sound and sense in poetry (leaving aside Mlle Smith and the Martians). That is to say, it has to raise – among other things – the Fregean issue about 'sense' and 'reference', and the question whether *any* theory of language can do without an argued and viable account of how those aspects are logically related.[40] From here one might go on to enquire (again following Frege) why it is that Saussure and his post-structuralist disciples treat the 'sign' as their basic unit of discourse, rather than the sentence or the proposition. For such a shift of focus might enable them to see that words acquire meaning (and meanings acquire truth-values) only at this higher stage of logico-semantic grasp where reference necessarily plays a role – that is, in distinguishing veridical from false, empty, non-referring or fictive expressions – and where nothing therefore requires us to think of language as paradigmatically a network of differential signs 'without positive terms'. All this would tend to bear out the substance of Cunningham's claims but would do so by way of a closely argued philosophical critique that yielded no hostages to the charge of reductive psycho-biographical treatment.

III

In philosophy of science it is also the case that postmodern scepticism thrives on the lack of alternative (more adequate) conceptual resources, or rather on the attitude of willed indifference that counts them a priori irrelevant. Such is at any rate the view canvassed by postmodernists and other proponents of the view that scientific truth-claims in the end come down to so many discourses, language games, paradigms or instrumental fictions whose warrant is simply a function of their answering to present-day social and cultural needs.[41] Here again one often finds Derrida invoked – on the basis, one suspects, of a cursory or second-hand acquaintance with his work – as having shown beyond doubt (taking a cue from Nietzsche) that concepts are just a species of sublimated metaphor, products of the 'arbitrary' will-to-truth within language, and hence quite devoid of veridical content or explanatory power. From which it follows that scientific conceptions of 'reality' are in fact nothing more than matters of descriptive convenience, metaphors (or operative fictions) adopted for no better reason than their happening to fit with some currently prestigious research programme, ontological scheme, conceptual paradigm or whatever. Thus only the most naïve of scientists or philosophers could nowadays persist in the absurd belief that science has anything at all to do with questions of reality and truth.

One source of these ideas, I suggest, is the line of philosophico-scientific enquiry descending from the work of Gaston Bachelard and Georges Canguilhem.[42] It is mostly encountered – at least by present-day cultural theorists – through its subsequent transformation into a doctrine of full-blown epistemological scepticism. This process began with Althusser's attempt to construct an account of Marxist 'theoretical practice' that would break altogether with naïve (empiricist) conceptions of reality and truth, while at the same time holding the line against forms of idealist mystification.[43] Thus:

> if the process of knowledge does not transform the real object, but only transforms its perception into concepts and then into a thought-concrete, and if all this process takes place, as Marx repeatedly points out, '*in thought*', *and not in the real object*, this means that, with regard to the real object, in order to know it, 'thought' operates on the transitional forms which designate the real object in the process of transformation in order finally to produce a concept of it, the thought-concrete. . . . But neither can the thought-concrete which is finally produced be confused with the real.[44]

In which case the standard of theoretical truth can only be a standard specific or internal to the discourse within which that standard is itself proposed, defined or elaborated. But this has the unfortunate upshot – much seized upon by Althusser's critics – of relativizing 'truth' to the immanent criteria of this or that 'theoretical practice', which latter phrase (so these critics maintain) amounts to no more than a face-saving substitute for 'discourse', 'paradigm', 'language game', etc.[45] That Althusser was indeed uncomfortably aware of these problems is evident enough from his tortuous efforts to explain how reality (the 'object of thought') must at all costs not be confused with its representation (the 'concrete-in-thought') arrived at through precisely such a rigorous practice of theoretical elaboration.[46] But this leaves the question wide open as to how, on Althusser's account, we could ever gain knowledge of those real-world objects, processes or events which constitute the proper domain of Marxist theoretical enquiry.

So it was – to cut a short story shorter still – that the critics could view him as self-impaled on the horns of a familiar post-Kantian dilemma, on the one hand committed to criteria of truth and to the existence of an 'outside' (mind- and language-independent) reality, while on the other conceding both 'reality' and 'truth' to be concepts specific to the process of thought wherein they figured as operative terms. All that remained, in order for Althusser's dislodgement to be effected with minimum fuss, was that obsolete (epistemological) notions like 'process of thought' be replaced with the new-minted idiom of 'discourse' taken over from Foucault and other exponents of the linguistic/textualist turn.[47] And there seemed no reason to stop with Althusser since his, after all, had been only the latest, most desperate attempt to prop up that entire philosophical enterprise – from Descartes to Kant, Marx *et au-délà* – whose overweening claims went along with the idea of truth as something more than a product of discourse or a figment of the various intra-discursive 'subject-positions' that periodically laid claim to such truth. Thus it is no coincidence that the vaunted 'eclipse' of Althusserian Marxism occurred just in time to vacate the high ground of cultural theory for a range of alternative projects – post-structuralism, neo-pragmatism, discourse theory, Foucauldian 'genealogy' and the like – that were quick to exploit the opportunity thus opened up. Henceforth it could scarcely be doubted that 'discourse' was the absolute limit-point of critical enquiry and 'truth' nothing more than a mirage created by the Lacanian 'subject-presumed-to-know', or what Foucault – in his blithely dismissive reading of Kant – was pleased to name 'that strange empirical-transcendental doublet'.[48] On a wider front this period was also marked by the post-1968 retreat from political positions which now appeared hopelessly naïve in maintaining that values like reality, truth and critique (let alone 'theoretical practice') might actually have something useful to contribute to the project of human emancipation.

My purpose in offering this thumbnail sketch of the trajectory from Bachelard, via Althusser to Foucault has been to point up some of the consequences – ethical and political – that ensued from the drift towards a conventionalist (or cultural-relativist) stance on issues in epistemology and philosophy of science. Bachelard figures importantly here since his writings on the role of metaphor in scientific thought were an unacknowledged source for much that now passes as received post-structuralist wisdom. In Althusser's case this debt was of course quite explicit. It had to do with his notion of a clearly marked 'epistemological break' in the development of Marx's thinking, a moment of radical *coupure* when the 'early' (i.e., Hegelian or humanist) writings gave way to the 'mature' (adequately theorized) concepts and categories that emerged in the *Grundrisse* and *Das Kapital*.[49] This argument was 'structuralist' in the specific sense that it involved a symptomatic reading of the Marxian texts alert to the various orders of structural relationship – in Saussurean terms, the 'synchronic' disposition of elements – which characterized the break with an earlier, less rigorously theorized approach to questions of ideology, class-consciousness, practical agency, relations between economic 'base' and socio-cultural 'superstructure', etc. There was also an evident affinity between Althusser's theory of ideological 'recruitment' or 'interpellation' and Lacan's psychoanalytic understanding of this process in terms of the subject's Oedipal passage from the 'imaginary' to the 'symbolic' realm.[50] And if further evidence were needed of Althusser's structuralist leanings, then one could instance his clear indebtedness to Benveniste on the role of subject-positions in discourse and the way that those positions are constructed – or distributed – through the differential system of personal pronouns.[51]

All of this lends some credence to the claim that Althusser was indeed a 'structural Marxist', and that the fortunes of his project were inexorably tied to the rise and fall of structuralist thinking in its 'classic' theoretical form. But we can now see plainly – so the argument goes – that this project was always doomed to miscarry, compelled to push the linguistic turn to its ultimate relativist conclusion. That is to say, it ran up against the painful recognition – consistent with its own premises – that all truth-claims, scientific paradigms, subject-positions, ideas of Marxist 'theoretical practice' and so forth were relative to, or constructed in, some particular localized or culture-specific discourse. It was therefore disqualified from laying any claim to articulate either the conditions of possibility for its own 'scientific' (theoretically warranted) status or the ways in which other such discourses fell into modes of 'ideological' self-misrecognition. And so the way was open for critics like Hindess and Hirst – along with the disciples of Foucault – to proclaim the demise of Althusserian Marxism and the passage to a post-Marxist, post-scientific, post-epistemological paradigm that would guard against these errors by taking 'discourse' as its bottom-line term of analysis.

We can best see what is wrong with all this by returning to Bachelard's philosophy of science and the few pages of densely argued commentary that Derrida devotes to Bachelard in his essay 'White Mythology'.[52] For the point that most urgently needs making in this context is that neither thinker subscribes to the view – the simplistic postmodernist view – that 'all concepts are metaphors,' that reality is ultimately a linguistic or discursive construct, or that science deals only in images and tropes whose 'truth' is a function of their instrumental yield for short-term pragmatic purposes. In this regard, Derrida writes,

> Bachelard is faithful to tradition: metaphor does not appear to him either simply or necessarily to constitute an obstacle to scientific or philosophical knowledge. It can work for the critical rectification of a concept, reveal a concept as a bad metaphor, or finally 'illustrate' a new concept. In the process of scientific knowledge the 'verbal obstacle' often has the form of metaphor ('metaphoric contrivance', 'generalized image', 'deficient metaphorical character of the explanation' etc.), doubtless. And doubtless the domain of metaphor is extended even beyond language, taken in the strict sense of verbal 'expression': 'metaphors seduce reason'. But, on the other hand, the psychoanalysis of objective knowledge must above all denounce 'immediate metaphors' ('The danger of immediate metaphors in the formation of the scientific spirit is that they are not always passing images; they push towards an autonomous kind of thought; they tend to completion and fulfilment in the domain of the image'); as we shall see, it is the *system* of metaphors that interests Bachelard initially: and on the other hand, a nonimmediate, constructed metaphor is useful when it comes to 'illustrate' knowledge wrested from bad metaphor. (WM, p. 259)

Of course this passage cannot be read as a straightforward approving exposition of Bachelard's views or as a paraphrase intended to signal agreement on all the main points at issue. Indeed, there is much evidence elsewhere in 'White Mythology' that Derrida thinks it strictly impossible to attain that degree of control over metaphor – that 'rational' command of its expressive powers and resources – which would enable science (as Bachelard thinks) to distinguish 'bad' from 'good' examples of the kind. For there will always be at least one metaphor too many, one trope belonging to the network of interwoven images and figures that make up the discourse of science itself, and likewise the discourse of philosophy, despite its best endeavours – from Aristotle down – to comprehend metaphor on its own conceptual terms. Thus:

> there is no properly philosophical category to qualify a certain number of tropes that have conditioned the so-called 'fundamental', 'structuring', 'original' philosophical oppositions: they are so many 'metaphors' that would constitute the rubrics of such a tropology, the words 'turn' or 'trope' or 'metaphor' being no exception to the rule. To permit oneself to overlook this vigil of philosophy, one would have to posit that the sense aimed at through these figures is an essence rigorously independent of that which transports it, which is an already philosophical thesis, one might even say philosophy's *unique* thesis, the thesis which constitutes the concept of metaphor, the opposition of the proper and the nonproper, of

essence and accident, of intuition and discourse, of thought and language, of the intelligible and the sensible. (WM, p. 229)

So there can be no question of Derrida's simply concurring with Bachelard in the view that science – or philosophy of science – has the means at its disposal to conceptualize metaphor and thereby to discipline its more unruly or troublesome effects. For one would then need to draw a rigid line of demarcation between good and bad, 'proper' and 'improper', or 'intelligible' and merely 'sensible' (i.e., intuitive or inadequately theorized) figures of thought. And this claim would always be open to the objection – brought against it at numerous points in 'White Mythology' – that despite such hygienic precautionary measures philosophy remains dependent upon metaphor for its most 'fundamental', 'structuring' or 'originary' concepts.

Hence, no doubt, the version of 'White Mythology' that enjoys widespread credence among cultural and literary theorists. On this account, Derrida's sole purpose is to deconstruct the concept/metaphor distinction to the point where philosophy becomes just a species of rhetoric (or 'kind of writing'), and where science can be shown to consist in nothing more than a range of 'pre-scientific' tropes and images masquerading as hypothesis, proof, inductive warrant, law-governed causal explanation, etc. But this raises an awkward question as to what could possibly be meant by the term 'pre-scientific' if *all* science – every last criterion of scientific method and truth – is thus reduced to the level of sublimated metaphor or 'white mythology'. It is the same question that Derrida poses when he remarks – in one of several key passages strategically ignored by most commentators – that '[t]he concept of metaphor, along with all the predicates that permit its ordered extension and comprehension, is a philosopheme' (WM, p. 228). That is to say, there is no possibility of addressing this issue of priority between 'concept' and 'metaphor' without engaging the *philosophic* discourse on metaphor that has defined the very terms of that debate from Aristotle to the present. To be sure it is the case, as Derrida writes, that once exposed to this suspicion of its etymopoeic or metaphoric origin 'the concept of the concept cannot retain the gesture of mastery, taking-and-maintaining-in-the-present, comprehending and grasping the thing as an object' (p. 224). But equally we can have no concept of metaphor – no resources for discussing or theorizing its nature, no criterion (even) for distinguishing 'metaphoric' from 'literal' sense – except by way of this necessary recourse to philosophical concepts and categories.

In philosophy of science likewise this constitutes a powerful objection to arguments in the postmodern-pragmatist or cultural-relativist mode. They include various assertions to the effect – in roughly descending order of intelligibility – that 'all scientific concepts are metaphors', 'all truth-claims a product of prescientific prejudice', or – in the Rorty version – 'science just the name for what counts as such according to the current conversational rules of the game'.[53] As I have said, there is a clearly marked line of descent that links these ideas with Bachelard's interest in the pre-scientific ('metaphorical') origin of scientific truth-claims and theories. But some crucial distinctions have somehow dropped out along the way, among them Bachelard's cardinal thesis – taken up by Althusser – that scientific progress comes about through the occurrence of 'epistemological breaks' that mark the passage from image-based thinking or intuitive sense-certainty to a discourse capable of adequate conceptual definition. And Derrida is far from rejecting this thesis, whatever his doubts as to the possibility of carrying such a programme through without metaphorical remainder. For it is a precondition of scientific thought – as it is for discussing the topic of metaphor in philosophically accountable terms – that one aims towards an order of conceptual adequacy that would at least strive to minimize the presence of naïve animistic or anthropomorphic residues.

Derrida's chief example here – taken from Georges Canguilhem – has to do with the cellular theory of biological growth and development, a theory over which (in Canguilhem's words) there 'hover, more or less closely, affective and social values of cooperation and association'. More specifically:

> This discovery of the thing and this invention of a word henceforth call for some comment. With the cell, we are in the presence of a biological object whose affective overdetermination is incontestable and considerable. . . . [W]ho knows whether, in consciously borrowing from the beehive the term cell in order to designate the element of the living organism, the human mind has not also borrowed from the hive, almost unconsciously, the notion of the cooperative work of which the honeycomb is the product?[54]

Thus Canguilhem, not so much rejecting the cellular theory as false or in any way 'unscientific', but asking – like Bachelard – what additional factors might have played some role in its initial formulation. Let me now quote a passage from Derrida's commentary since it makes the distinction yet clearer.

> This epistemological ambivalence of metaphor, which always provokes, retards, *follows* the movement of the concept, perhaps finds its chosen field in the life sciences, which demand that one adopt an unceasing critique of teleological judgment. In this field the animistic or (technical, social, cultural) analogy is *as* at home as possible. Where else might one be so tempted to *take the metaphor for the concept*? And what more urgent task for epistemology and for the critical history of the sciences than to distinguish between the word, the metaphoric vehicle, the thing and the concept? (WM, p. 261)

Again I should acknowledge that Derrida is here paraphrasing Canguilhem's argument, or writing in a mode of *oratio obliqua* that cannot safely be assumed to carry the force of assertoric statement. It is a problem that often arises with texts like 'White Mythology', where much of the argument is conducted through a mixture of critical commentary and direct or indirect quotation. Some critics of Derrida – J. Claude Evans among them – have seen this as evidence of his failure to abide by the ground-rules of responsible scholarship, that is to say, of his exploiting such speech-act 'undecidability' as a technique for never quite meaning what he says or unambiguously saying what he means.[55] But this charge will appear unfounded if one looks beyond isolated passages like those cited above to the wider context of argument wherein such passages acquire a more determinate meaning. For it then becomes clear – from what Derrida has to say not only about Bachelard and Canguilhem but also, more generally, about the role of metaphor *vis-à-vis* the discourses of science, epistemology and the 'great chain' of ontological relations in which (since Aristotle) metaphor has occupied a crucial place – that he cannot intelligibly be taken to argue for the wholesale reduction of truth-claims and concepts to the order of sublimated metaphor. Quite simply, that thesis fails to make sense on the threefold ground of its *incoherence* as a piece of philosophical doctrine, its *inadequacy* as an account of how advances come about in the field of scientific knowledge, and its *demonstrable falsehood* as purporting to derive from the arguments advanced by thinkers like Bachelard and Canguilhem.

By the same token, one will mistake the import of Derrida's texts if one takes them to be concerned solely with the rhetorical close reading of particular, isolated passages. For one will then fail to recognize that they also have to do with the conditions of possibility – 'transcendental' conditions in the strict (Kantian) sense of that term – which render such arguments valid or invalid. It is here that deconstruction differs most sharply from the outlook of blanket epistemological scepticism espoused by postmodernist culture-critics and philosophers of science like Feyerabend. For these latter any talk about validity conditions – whether with regard to scientific truth-claims or to questions in the philosophy of mind, knowledge or language – can always be reduced to contingent observations about the various beliefs that happen to prevail in this or that interpretive community. For Derrida, conversely, it is a matter of respecting the best, most rigorous standards of philosophic argument – of conceptual analysis, logical precision, alertness to non-sequiturs, category mistakes and so forth – even while remarking on those blind spots in the discourse of philosophy (especially in its dealing with metaphor) that emerge through a deconstructive reading.

These are also the standards that necessarily apply in science and philosophy of science. Here they involve what Derrida (following Bachelard) describes as a process of conceptual exegesis and critique, one that starts out from images, naturalized metaphors, intuitive sense-certainty and the like, but which then – through successive refinements and elaborations – achieves a more adequate theoretical grasp of the phenomena it seeks to describe or explain. Only thus can one avoid the strictly unintelligible (though currently widespread) idea that scientific truth *just is* whatever passes as such according to the dominant beliefs, metaphors or world hypotheses of some particular time and place. On this view there can be no past or present item of belief so utterly devoid of rational, evidential or explanatory warrant that we might feel confident in declaring it simply not a plausible candidate for truth. In short, postmodernism fails the most elementary test for philosophies of science, that which requires them to offer some convincing (historically and conceptually adequate) account of our knowledge of the growth of scientific knowledge.[56] For what the argument very often comes down to is a version of the etymopoeic fallacy, the idea that concepts can be reduced *without remainder* to their metaphoric origins or the fact of their deriving (at whatever genealogical remove) from some matrix of sublimated images or tropes.[57]

IV

Such is indeed the sort of argument that Derrida rehearses at length in the early part of 'White Mythology'. His text here is Anatole France's dialogue 'The Garden of Epicurus', where philosophers are portrayed as dupes of the idea, prisoners of abstraction, thinkers who substitute a lifeless currency of concepts for the original vitalizing sources of mythic or poetic inspiration.[58] 'By an odd fate,' as the speaker Polyphilos puts it, 'the very metaphysicians who think to escape the world of appearances are constrained to live perpetually in allegory. A sorry lot of poets, they dim the colours of the ancient fables, and are themselves but gatherers of fables. They produce white [blanched] mythology.'[59] It is a familiar argument, one that looks back to Nietzsche and forward to Heidegger, for both of whom the term 'metaphysics' denominates a process of epochal decline. On this account, the pre-Socratics figure as strong precursors with the courage of their own metaphorico-poetic convictions, whereas later thinkers – from Socrates down – betrayed those origins in the drive to subjugate metaphor to concept, or poetry to the overweening will-to-truth embodied in modes of merely abstract or dialectical reasoning.[60] Hence the well-known passage from Nietzsche's essay 'On Truth and Falsity in their Ultramoral Sense', a passage often cited out of context as the last word on Nietzschean 'genealogy' and Derridean 'deconstruction' alike. 'What then is truth?' Nietzsche rhetorically demands. Nothing more, it seems, than

> [a] mobile, marching army of metaphors, metonymies, anthropomorphisms: in short a sum of human relations which became poetically and rhetorically intensified, metamorphosed, adorned, and after long usage, seem to a nation fixed, canonic and binding; truths are illusions of which one has forgotten that they *are* illusions; worn out metaphors which have become powerless to affect the senses, coins which have their obverse (image) *effaced* and are now no longer of account as coins but merely as metal.[61]

On the face of it, this is the same line of argument put forward by Polyphilos in 'The Garden of Epicurus'. Philosophy (or 'metaphysics') amounts to no more than a species of 'blanched mythology', a decadent discourse chronically prone to forget or to efface its own originary, life-giving sources through the epistemic will that seeks to privilege truth over fiction, concept over metaphor, the 'intelligible' over the merely 'sensuous', and so forth.

From this point it is no great distance to Heidegger's brooding on the long-lost wisdom of the pre-Socratics and his identification of 'Western metaphysics' with an epochal forgetfulness of Being, one whose fateful (predestined) upshot is modern science and its wholesale elevation of the technocratic will-to-power.[62] Thus Polyphilos again, in markedly Nietzschean–Heideggerian vein:

> All these words, whether defaced by usage, or polished smooth, or even coined expressly in view of constructing some intellectual concept, yet allow us to frame some idea to ourselves of what they originally represented. So chemists have reagents whereby they can make the effaced writing of a papyrus or a parchment visible again. . . . If an analogous process were applied to the writings of the metaphysicians, if the primitive and concrete meaning that lurks yet present under the abstract and new interpretations were brought to light, we should come upon some very curious and perhaps instructive ideas.[63]

In this case there is, I think, good warrant to suppose that we should *not* take Polyphilos' statements – or the Nietzsche-inspired reading of metaphor in the texts of philosophy from which they purportedly derive – as bearing the mark of implied authorial endorsement. What counts decisively here is not so much the sense of this or that passage taken out of context but the role of such statements when referred to the overall structure and logic of Derrida's argument in 'White Mythology.' Thus it is worth remarking that Polyphilos doesn't have quite the last word when he mocks the philosophers as a 'sorry lot of poets', producers of a 'blanched' or 'anaemic' mythology. That word goes to Aristos, his interlocutor, cast in the dialogue as a 'defender of metaphysics' and therefore as one who doubts – on argued philosophical grounds – that the discourse of truth, reason and concept can be thus reduced to a repertoire of 'philosophemes' issuing from a yet more primordial repertoire of tropes, images and metaphors. Not that he is able to trump Polyphilos with any argument that the latter could be brought to acknowledge as possessing superior validity or force. For, as Derrida notes, Aristos merely 'finishes by *leaving*, determined to break off dialogue with a cheater: "I leave unconvinced. If only you had reasoned by the rules, I could have rebutted your arguments quite easily"' (WM, p. 213).

One way of reading this episode is as an ironic reversal of Socrates' dealing with the sophists, that is to say, his series of dialectical rejoinders practised upon those who deployed mere rhetoric (or the arts of suasive utterance) against the truth-claims of philosophic reason. On this view, 'White Mythology' would likewise constitute a kind of belated comeuppance, an exercise in the deconstructive-textualist mode whose subtitle might just as well have been 'The Sophist's Revenge.' The entire essay could then be read as an elaborate demonstration on Derrida's part – siding with Polyphilos – that philosophy is incapable of countering the rhetorician's charge since all its argumentative resources, its 'foundational' concepts and categories derive from the opponent's armoury. And this reading might be thought to find ample warrant elsewhere in Derrida's text. Thus the very distinction between 'concept' and 'metaphor' is one that philosophy is anxious to maintain, for instance, by proposing a 'general metaphorology' which would set limits to the aberrant play of figural language and restore philosophy to its proper role as arbiter of truth and reason. Yet it is thereby 'enveloped', as Derrida writes, in 'the field . . . that it would seek to dominate.' More specifically:

> [m]etaphor has been issued from a network of philosophemes which themselves correspond to tropes or to figures, and these philosophemes are contemporaneous to or in systematic solidarity with these tropes or figures. . . . Therefore it [philosophy] gets 'carried away' each time that one of its products – here, the concept of metaphor – attempts in vain to include under its own law the totality of the field to which the product belongs. (WM, p. 219)

So one could claim at least prima facie warrant – evidence, that is, from the above passage and others to kindred effect – for a reading that treats the issue as having been settled decisively in favour of rhetoric or metaphor. And this reading would gain additional support from the way in which Aristos abruptly backs out of the dialogue, attempting to disguise his all too evident discomfiture by invoking the old Socratic charge that his opponent hasn't played by the proper (philosophically respectable) rules of the game. For those rules have by now been shown up – or so it might seem – as nothing more than a handy dialectical bag of tricks whereby to claim the argumentative and moral high ground.

All of which puts large obstacles in the way of anyone seeking to argue the case for 'White Mythology' as a defence of philosophical and scientific truth-claims as against the postmodern-sceptical-relativist view. Still there is no escaping this latter conclusion if one reads the essay entire with adequate attention to its logical structure, rather than focusing (as most commentators do) on the early portions and those passages which, when taken out of context, appear to support that view. One crucial indicator here is the fact that Derrida doesn't rest content with citing the best-known (proto-deconstructionist) sentences from Nietzsche's essay 'On Truth and Falsity.' He also quotes the following passage, one that – understandably – tends to be ignored by those in search of Nietzschean warrant for reducing all truth-claims to so many sublimated metaphors, or science to a product of the epistemic will-to-power vested in language. It is all the more striking in so far as it prefigures Canguilhem's example of the beehive (or honeycomb), taken as a metaphor for the cellular structure of organic tissue and the way that this smuggles in anthropomorphic notions of cooperative labour and purpose. 'As we say,' Nietzsche writes,

> it is *language* which has worked originally at the construction of ideas; in later times it is *science*. Just as the bee works at the same time at the cells and fills them with honey, thus science works irresistibly at the great columbarium of ideas, the cemetery of perceptions, builds ever newer and higher storeys; supports, purifies, renews the old cells, and endeavours above all to fill that gigantic framework and to arrange within it the whole of the empiric world, i.e., the anthropomorphic world. And as the man of action binds his life to reason and its ideas, in order to avoid being swept away and losing himself, so the seeker after truth builds his hut close to the towering edifice of science in order to collaborate with it and to find protection. And he needs protection. For there are awful powers which press continually upon him, and which hold out against the 'truth' of science, 'truths' fashioned in quite another way, bearing devices of the most heterogeneous character.[64]

What this passage brings out is Nietzsche's respect for science as a genuinely truth-seeking mode of enquiry, and the error of those postmodernist or quasi-deconstructive readings that take him to be simply denouncing 'truth' as a species of rhetorical imposture. It is also of interest in so far as it moves over – like 'White Mythology' – from what looks at the outset like a wholesale sceptical position to one that acknowledges precisely this distinction between truths arrived at through a disciplined (scientific) labour of thought and truths which merely pass themselves off as such owing to their suasive or rhetorical power. That Derrida indeed takes the point of Nietzsche's distinction is evident enough from the passage of commentary that follows. 'This new articulation,' he suggests, 'without importing all the metaphysics of the classical opposition [i.e., between concept and metaphor], should also account for the specific divisions that epistemology cannot overlook, the divisions between what it calls metaphoric effects and scientific effects' (WM, p. 263).

To be sure, 'White Mythology' complicates the issue for anyone wishing to argue, like Bachelard, that 'bad' metaphors can always be 'corrected' – purged of their naïve imagistic or anthropomorphic content – by more rigorous procedures of theorization. But to claim that such a process can never be carried through completely – that *at least one* metaphor, as Derrida puts it, will always exceed the grasp of conceptual definition – is not to conclude, with Rorty or Feyerabend, that this whole way of thinking is disqualified since scientific concepts come down to *nothing more* than so many optional language games or metaphors indifferently ranked with regard to truth and devoid of epistemological criteria. For it remains the case that science (and a fortiori philosophy of science) cannot make sense of its own history, its progressive elaboration of more adequate descriptive and explanatory resources, without presupposing the existence of precisely such criteria. To misunderstand Derrida on this point is rather like declaring mathematics henceforth an impossible or futile activity since Gödel's theorem has shown beyond doubt that for every sufficiently complex system of axioms and proof procedures there will always exist at least one indispensable axiom whose validity cannot be proved in terms of that same system.[65]

No doubt there is a risk of fuzzy thinking when cultural and literary theorists appeal to *topoi* in the philosophy of science or mathematics – themes like undecidability, complementarity, set-theory paradoxes and the like – by way of loose analogy with their own interests and concerns.[66] For one thing, theories (or what count as such) in the interpretive and humanistic disciplines are rarely if ever carried to the point of a full-scale axiomatization where such problems could be found to emerge from a rigorous analysis of their own validity conditions. So it is mostly wishful thinking – or an insecure grasp of the arguments concerned – that inspires these attempts to emulate science or mathematics in a different field of study. However, the example of Gödel's theorem is one that Derrida often invokes (and I think with good reason) when seeking to specify just what it is that remains strictly 'undecidable' as regards the issue of priority between concept and metaphor, logic and rhetoric, theoretical and intuitive (sensuous or image-based) modes of understanding, etc.[67]

The following passage from 'White Mythology' comes as close as any to stating the case in a rigorously formalized mode. 'On the one hand,' he writes,

> it is impossible to dominate philosophical metaphorics as such, *from the exterior*, by using a concept of metaphor which remains a philosophical product. Only philosophy would seem to wield any authority over its own metaphorical productions. But, on the other hand, for the same reason philosophy is deprived of what it provides itself. Its instruments belonging to its field, philosophy is incapable of dominating its general tropology and metaphorics. It could perceive its metaphoricity only around a blind spot or central deafness. The concept of metaphor would describe this contour, but it is not even certain that the concept thereby circumscribes an organizing center, and this formal law holds for every philosopheme. And this for two cumulative reasons. (1) The philosopher will never find in this concept anything but what he has put into it, or at least what he believes he has put into it as a philosopher. (2) The constitution of the fundamental oppositions of the metaphorology (*physis/tekhne*, *physis/nomos*, sensible/intelligible, space/time, signifier/signified, etc.) has occurred by means of the history of a metaphorical language, or rather by means of 'tropic' movements which, no longer capable of being called by a philosophical name – i.e. metaphors – nevertheless, and for the same reason, do not make up a 'proper' language. (WM, pp. 228–9)

Clearly it is a gross misreading of passages like this to treat them as just a roundabout (quasi-philosophical) way of inverting the privilege traditionally accorded to concept over metaphor. On the contrary, Derrida maintains: both are caught up in a process of mutual interrogative exchange which permits no such simple inversion of priorities. This process, moreover, can only come about through philosophy's continued attempt to comprehend *so far as possible* the conditions under which metaphor gives access to a knowledge that can neither dispense altogether with metaphor – the dream of numerous philosophers from Plato to Locke, Kant and the logical positivists – nor confess itself wholly given over to metaphor as the limiting condition of knowledge.[68] For on the one hand, as 'White Mythology' makes abundantly clear, there is no escaping the ubiquity of metaphor as a vital resource – in Derrida's Heideggerian phrase, a perpetual 'standing reserve' – for those speculative advances in science which as yet lack the means of adequate proof or rigorous conceptualization. Yet on the other, there can be no science (or indeed philosophy of science) in the absence of certain criteria and decision procedures for establishing the point at which speculative metaphors give way to adequate concepts.

The primary mode of argument in 'White Mythology' is one that works through textual exegesis – through the close reading of various philosophers, among them Aristotle, Nietzsche and Bachelard – in order to assert this double necessity against more reductive or one-sided views. But there is also a further (twofold and inseparable) set of claims: that Derrida's account gets those thinkers right on the main points at issue, and also that they themselves were right – their arguments borne out by logic and right reason – as concerns the problematic topos of metaphor and its role in the discourse of philosophy and science. Here again it is a question of the conditions of possibility for explaining how science makes sense as an enterprise aimed towards knowledge and truth, and how philosophy can in turn make sense of science as just that kind of truth-seeking endeavour. This is why Derrida's readings of Nietzsche and Bachelard go so markedly against the current postmodernist grain. His point – simply put – is that these issues require much more in the way of detailed and rigorous critical treatment than is offered by the kind of all-purpose reductive ploy that would treat such values as mere emanations of the will-to-power vested in forms of institutionalized scientific discourse.[69] What is distinctive about Derrida's approach is the way that 'White Mythology' marshals evidence from the texts of Nietzsche and Bachelard in order to challenge the widely-held view of these thinkers as precursors of the postmodern sceptical-relativist line. But in so doing it also casts doubt on that equally widespread view of deconstruction which takes it to be just another version of the same argument, that is to say, a style of rhetorical close reading that *always and inevitably* undercuts the discourse of philosophic reason and truth.

This in turn helps to clarify what Bachelard – and Althusser after him – meant by positing the necessity of an 'epistemological break' between pre-scientific and scientific modes of knowledge.[70] Of course it might be seen as merely begging the question to speak in such terms, since the very criteria adopted – i.e., those of ideational 'distinctness', conceptual clarity, theoretical insight, etc. – are themselves heavily mortgaged in advance to a certain (intra-philosophical) discourse which takes their validity for granted. More than that, they are indebted to the single most pervasive of 'logocentric' images or metaphors, that which associates veridical knowledge, via accurate perception, with the *lumen naturale* of intellect or reason, the faculty that allows us to attain such 'clear and distinct' ideas through the exercise of its own self-critical procedures for arriving at ultimate truth.[71] Hence – to repeat – Derrida's claim that there will always be at least one metaphor, one strictly irreducible trope, which eludes philosophy's most strenuous attempts to purge its language of all such naïve (imagistic or sensuous) residues. Nor does he have any difficulty in showing how philosophers from Plato and Aristotle to Descartes, Kant, Hegel and Husserl have constantly invoked a whole covert network of 'heliotropic' images and metaphors, even – or especially – at the moment of proclaiming some new-found 'epistemological break', some passage beyond this unfortunate dependence on metaphoric habits of thought.

Thus '[p]rior to every determined presence, to every representative idea, natural light constitutes the very ether of thought and of its proper discourse' (WM, p. 267). And again, with reference to Descartes: '[n]atural light, and all the axioms it brings into our field of vision, is never subjected to the most radical doubt. The latter unfolds *in light*: "for I cannot doubt that which the natural light causes me to believe to be true, as, for example, it has shown me that I am from the fact that I doubt"'.[72] Hence the very pointed double significance of Derrida's title 'White Mythology.' On the one hand it refers to that sensuous origin, that 'fabulous scene' or blind spot of occulted metaphor that is forever just beyond the reach of philosophy's conceptual or explanatory grasp. On the other it evokes – most explicitly in Hegel – a covert narrative dimension whereby philosophy (that is to say, *occidental* philosophy) recounts the history of its own deliverance from the realm of sensuous illusion. Thus the notion of concept as a species of 'sublimated' metaphor coexists quite easily with the idea of philosophy as that which enables thinking to advance through the progressive *Aufhebung* ('sublation') of image and metaphor into forms of adequate conceptual understanding. 'The sensory sun, which arises in the East, becomes interiorized, in the evening of its journey, in the eye and the heart of the Westerner. He summarizes, assumes, and achieves the essence of man, "illuminated by the true light"' (WM, p. 268).

Thus it would seem, once again, that 'White Mythology' goes as far as possible in deconstructing every last pretence of philosophy to control the field of its own metaphorics, or even to comprehend what is at stake in these questions beyond their restricted (intra-philosophical) pertinence. All the same, one should not assume too readily that this is where the argument ends, that is to say, at the point where philosophy is played off the field – shown up as just a product of the occidental will-to-power – by that one heliotropic metaphor too many. I shall now return to Derrida on Bachelard, having taken this detour via Descartes and Hegel in order to stress the genuine difficulties that arise with any attempt to explain how advances in scientific knowledge come about through the process of breaking with its own metaphorical or image-based sensory origins. The following passage is crucial here since the question it poses can only be resolved (if at all) with reference to the entire argumentative structure of Derrida's essay. 'Is rectification,' Derrida asks,

> henceforth the rectification of a metaphor by a concept? Are not all metaphors, strictly speaking, concepts, and is there *any sense* in setting metaphor against concept? Does not a scientific critique's rectification rather proceed from an inefficient tropic-concept that is poorly constructed, to an operative tropic-concept that is more refined and more powerful in a given field and at a determined phase of the scientific process? The criterion of this progress or mutation ('break', 'remodelling', and many other forms that should be distinguished from each other), has not been defined, certainly; but a double certainty now seems problematic: 1. That this criterion must necessarily put to work a rhetorical evaluation ('from metaphor to concept', for example); 2. That tropes must necessarily belong to a pre-scientific phase of knowledge. (WM, p. 264)

One will not get very far in thinking about science, philosophy, or the role of metaphor in either if one takes those opening questions to be merely rhetorical, that is, as declaring an end-point to enquiry because there is *simply no deciding* the issue of priority between concept and metaphor. Nor should the quotation marks around 'break' and 'remodelling' – and their implicit extension to kindred terms like 'rectification', 'more refined' and 'more powerful' – be taken as signalling an attitude of wholesale scepticism with regard to the idea that science might advance through precisely such forms of conceptual critique. For, as Derrida goes on to remark, 'there is also a *concept of metaphor*: it has a history, yields knowledge, demands from the epistemologist construction, rectification, critical rules of importation and exportation' (WM, p. 264). And in the case of scientific metaphors, such knowledge comes about through a more exact grasp of their explanatory powers and limits, along with a variety of relevant procedures – experimental, theoretical, hypothetico-deductive and so forth – for determining their validity conditions in any given case.

To be sure, the immediate context of this statement is such as to indicate that Derrida is paraphrasing Bachelard, making out the case for science (and philosophy) as a process of conceptual 'refinement', 'remodelling', 'rectification', etc., in order to test it against the counter-thesis that metaphor is ubiquitous – and hence uncontrollable – in these and all other modes of discourse. But there is, as I have argued, a larger context that encompasses not only the entire text of 'White Mythology' but also the conditions of possibility which that text establishes for any adequate (philosophically accountable) treatment of the issue between concept and metaphor. And at this level it is revealed as strictly *unintelligible* to suppose that science might indeed come down to just a choice between various metaphorical constructions, discursive paradigms, world hypotheses and so forth, each disposing of its own wholly immanent (discourse-specific) criteria, and hence beyond reach of challenge or criticism from the standpoint of some other (*ex hypothese* incommensurable) paradigm. For there could then be no question of scientific knowledge advancing through the practice of a more sustained, rigorous and critical reflection on the nature (and sometimes the hidden liabilities) of its own formative metaphors. Quite simply, we should lack the criteria for distinguishing 'good' from 'bad', valid from invalid, or efficacious (knowledge-productive) metaphors from those whose naïve (anthropomorphic) character constitutes a block to any further such advance.

V

This is not to espouse a pragmatist conception of scientific 'truth' as just the honorific title standardly bestowed on those language games (or metaphors) that best fit in with the beliefs, social policy interests, or ideological self-images of the age. Such is the cultural-relativist argument (at any rate in its 'strong' sociological form), and such also – on one widely credited account – the position entailed by Bachelard's view of metaphor and its role in scientific thinking. An example often cited in this context is that of the tetrahedral structure of carbon, discussed in some detail by Bachelard and by various of his critics in the realist camp, Roy Bhaskar prominent among them.[73] This looks like a plausible candidate for treatment as a purely speculative contruct, an artefact of theoretical discourse which at the time – i.e., when first introduced and given the limits of currently available technology – could by no means be verified (or falsified) through observation or experiment. From which the anti-realists conclude, supposedly with warrant from Bachelard, that this theory had the status of an operative metaphor – or an instrumental fiction – adopted for want of any better (more elegant) hypothesis. In short, the tetrahdral structure was posited as 'an object without a direct realistic value in ordinary experience . . . an object which has to be designated as a *secondary object*, . . . an object preceded by theories.'[74] It thus served to demonstrate, in Bachelard's words (disapprovingly cited by Bhaskar), that 'we have now reached a level of knowledge at which . . . scientific objects are what we make of them, no more and no less.'[75]

However, it is a misreading of Bachelard – one that enjoys wide currency among cultural theorists – which would take such arguments as weighing in favour of an anti-realist or strong-sociological approach. On this latter view scientific truth-claims – observation sentences included – are *always* so completely 'theory-laden' that nothing could count as an adequate criterion for distinguishing the various orders of concept and metaphor, fact and fiction, knowledge as arrived at through adequate procedures of experiment, inductive inference, theoretical elaboration, etc., and knowledge as a 'socially constructed' discourse wherein such claims function solely to legitimize some current set of power-knowledge interests.[76] Hence the idea that paradigms are radically 'incommensurable', that they can only be judged according to their own (intra-discursive) criteria, and therefore that scientific 'progress' is the merest of illusions, shown up as such by the sheer impossibility of translating across or between paradigms.[77] But this is all quite remote from Bachelard's philosophy of science, committed as he is to a very different view of the metaphor/concept relation, and also to a far stronger account of what is involved in the 'epistemological break' between pre-scientific and scientific modes of knowledge production.[78] Thus his point in the above passage is not that we make *what we like* of such 'objects', but rather that, in certain problematic cases, what we make of them is crucially dependent on theory and underdetermined by the evidence at present to hand. This is very different from claiming (nonsensically) that scientific truth *just is* the sum total of those quasi-objects – theories, fictions, speculative constructs, metaphors, myths and so forth – none of which can claim any privileged status in point of conceptual grasp or explanatory power. For there is still room for that ongoing process of 'refinement' and 'rectification', even in cases – like the tetrahedral structure of carbon – where science comes up against the limits imposed by its current (technologically determined) capacity for testing hypotheses through experiment and observation.

There are two main issues here, both well argued by Ian Hacking in his book *Representing and Intervening*.[79] First, it is always possible that some new technology will arrive, for instance, through the advent of faster particle accelerators or electron microsopes with higher powers of resolution. And second, there is no good reason to suppose that if it lacks such resources at present, then science is inevitably thrown back upon fictive or metaphorical quasi-objects that merely masquerade as genuine 'concepts' or 'theories.' On the contrary: some of its most striking advances – relativity theory and quantum mechanics among them – have come about precisely through a far-reaching exercise of disciplined speculative thought, such that, to begin with, its findings are not only 'underdetermined' by the evidence but viewed as both counter-intuitive and actually in conflict with the best observational data. Thus – to cite the *locus classicus* for this sort of case – it appeared early on that Einstein's conjectures were refuted by the Michelson–Morley experiments, the results of which (through erroneous measurement) appeared to be at variance with the predicted theoretical result.[80] That this error was eventually exposed – and the theory borne out – was partly the upshot of improved observational techniques or greater care with the measuring equipment. But it also goes to show that scientific advances, and indeed 'revolutions' in thought, may sometimes occur when theory runs so far ahead of its own technological support systems that it has to proceed (in Bachelard's terms) from a speculative starting-point – at this stage, quite possibly, a heuristic metaphor – which is then subjected to the process of conceptual 'rectification' or critique. In the case of relativity theory this involved an epistemological break not only with the orthodox scientific wisdom of the day but also with the sheer self-evidence of commonsense perception and – at least temporarily – with the canons of adequate proof procedure and observational warrant. It therefore provides an apt illustration of Bachelard's (and Derrida's) point: that any 'new articulation' of the concept/metaphor dualism must think its way beyond the facile stage of merely inverting their traditional order of priority. Such a project, 'without importing all the metaphysics of the classical opposition, should also account for the specific divisions that epistemology cannot overlook, the divisions between what it calls metaphoric effects and scientific effects' (WM, p. 263). For it is the case with science, as also with philosophy, that no critical insight can possibly accrue from the kind of *de rigueur* scepticism (or blanket antirealist approach) which holds simply that 'all concepts are metaphors', all truth-claims a species of rhetorical imposition, etc. What still needs explaining is the way that certain metaphors have proved themselves capable of refinement, elaboration and critique to the point where they attain validity as concepts in the discourse of science or philosophy.

Again there are pertinent lessons to be drawn from the case of relativity theory and its apparent lack of confirmation by those early experiments. For it is surely a perverse misreading of this episode that would take it as evidence that Einstein's conjecture was false, or its truth-value wholly undecidable, until the moment when the results came out right, and 'true' thereafter only in the sense that it managed to gain credence among members of the relevant scientific peer-group, 'interpretive community' or whatever.[81] Equally perverse (though consistent with the thinking of sceptics like Feyerabend) is the view that it remained just an operative fiction, one whose validity was wholly a product of the value placed upon it by those with the power – the cultural capital – to decide in such matters. Thus from Feyerabend's standpoint the Michelson–Morley episode would figure as just one example among many, from Galileo down, of the kind of evidence-bending that always goes on in the process of securing scientific truth-claims, but which is always tactfully concealed from view when that process gets written up in the form of reports, articles, requests for funding or authorized histories of science.[82] However, there is no good reason – postmodern fashion or anti-science prejudice apart – to extend this approach from the context of discovery to the context of justification. That is to say, the burden of proof rests entirely with the sceptics, given the sheer weight of evidence for science's capacity to achieve real advances through that process of 'rectification and critique' which for Bachelard constitutes the hallmark of scientific method.

This applies not only to unproblematical cases, the sort that Thomas Kuhn takes to characterize periods of 'normal' scientific activity.[83] At such times there exists a close (mutually supportive) relation between theory and practice, along with broad agreement on what should count as adequate standards of proof, theoretical consistency, observational warrant, inductive justification, and so forth. It also applies to those other, more difficult cases – like the early period of relativity theory – when science passes through a stage of Kuhnian 'revolutionary' ferment and when no such consensus obtains. Such instances naturally exert a strong appeal upon critics of science – or postmodern sceptics – anxious to 'deconstruct' the terms of Kuhn's distinction and demonstrate that science *always* proceeds through a mixture of guesswork, speculative metaphor, ad hoc improvisation, the positing of merely hypothetical 'quasi-objects', and (above all) sheer opportunist self-interest. But there is little, if anything, to commend such a view beyond its answering to certain deep-laid psychological and cultural needs among the present-day debunkers of scientific truth. For what emerges most impressively from all this – as also, at a different but related level, from Derrida's reading of Bachelard and Canguilhem – is the fact that knowledge may continue to accrue even during periods when theory must proceed in the absence of agreed proof procedures or standards of experimental warrant.

Of course this is not to deny that Derrida, like Bachelard, shows a special interest in problematic cases where the relationship between concept and metaphor – or science in its 'normal' and 'revolutionary' modes – is most visibly thrown into doubt. Nor is it to play down those many passages where he writes about the strictly *undecidable* order of priority that can be shown to operate whenever science (or philosophy) comes up against the limit of its powers to explain or conceptualize the workings of metaphor. But we should not be misled – like so many commentators – into taking this as a deconstructive *fait accompli*, or an all-purpose rhetoric of 'undecidability' which can then be applied pretty much across the board, without the kind of rigorous exegetical commentary that Derrida brings to his reading of Bachelard and Canguilhem. It is the same habit of premature generalization – often (one suspects) from a weak grasp of the issues involved – that leads cultural and literary theorists to invoke notions like 'indeterminacy', 'complementarity', 'undecidability' and the like as if these could serve as a knock-down argument against the appeal to obsolete criteria of truth, reason or logic.[84] From here it is a short distance to the Kuhn-derived notion that by adverting to the 'theory-laden' character of observation sentences one has shown that truth-claims are always underdetermined by the evidence, or – what purportedly follows from this – that science has entered a postmodern phase where 'performativity' is the name of the game, measured by the power to persuade fellow-scientists, get published in the most prestigious journals, or attract the biggest research grants.[85]

Several remarks are in order here by way of gathering my various lines of argument together. One is that anti-realist philosophies of science tend to be highly selective in the particular aspects, episodes and periods of scientific thought that they single out for privileged treatment. Thus it is scarcely surprising that the case for ontological relativity has been argued very often – by Quine among others – with reference to the more speculative branches of recent theoretical physics.[86] And it is against the same background of prolonged revolutionary 'crisis' that Kuhn's ideas about theory change, meaning variance, paradigm incommensurability and so forth have achieved such wide currency.[87] However, there is no reason to treat this as a *typical* situation, nor – for that matter – to discount the various arguments that critics have brought against anti-realism in the field of subatomic particle physics.[88] Least of all should one generalize too quickly from the example of (say) relativity theory during its early, speculative phase or from the kinds of deep-laid metaphysical, conceptual and ontological perplexity that have marked the discussion of quantum mechanics since the beginning of this century.[89] For it is the case here also – as Derrida shows to very pointed effect in 'White Mythology' – that vaguely drawn analogies from one to another domain cannot substitute for the patient analytic approach that examines each instance with a due regard to its contextual specificity, conditions of emergence and contribution to our knowledge of the growth of knowledge.

References

1 William Empson, review of E. A. Burtt, *The Metaphysical Foundations of Modern Physical Science*, in Empson, *Argufying: essays on literature and culture*, ed. John Haffenden (London: Chatto and Windus, 1987), pp. 530–3.

2 On this distinction between context of discovery and context of justification, see Hans Reichenbach, *Experience and Prediction* (Chicago: University of Chicago Press, 1938) and Carl G. Hempel, *Aspects of Scientific Explanation* (New York: Macmillan, 1965). It is attacked or (in the loose sense) 'deconstructed' by the various anti-realists, cultural relativists and 'strong' sociologists of knowledge whose works are represented in note 3 below.

3 See, for instance, Barry Barnes, *About Science* (Oxford: Blackwell, 1985); Harry Collins and Trevor Pinch, *The Golem: what everyone should know about science* (Cambridge: Cambridge University Press, 1993); Paul K. Feyerabend, *Science in a Free Society* (London: New Left Books, 1978); Steve Fuller, *Social Epistemology* (Bloomington: Indiana University Press, 1988); K. Knorr-Cetina and M. Mulkay (eds), *Science Observed* (London: Sage, 1983); Bruno Latour and Steve Woolgar, *Laboratory Life: the social construction of scientific facts* (London: Sage, 1979); Andrew Pickering, *The Mangle of Practice: time, agency, and science* (University of Chicago Press, 1995); Pickering (ed.), *Science as Practice and Culture* (Chicago: University of Chicago Press, 1992); Richard Rorty, *Objectivity, Relativism, and Truth* (Cambridge University Press, 1991); Andrew Ross, *Strange Weather: science and technology in the age of limits* (London: Verso, 1991); Steven Shapin, *A Social History of Truth: civility and science in*

seventeenth-century England (University of Chicago Press, 1994); Steve Woolgar, *Science: the very idea* (London: Tavistock, 1988).

4 Feyerabend, *Science in a Free Society* and *Against Method* (London: New Left Books, 1975).

5 For the best-known discussion of this episode from a 'strong' ontological-relativist standpoint, see Thomas S. Kuhn, *The Structure of Scientific Revolutions*, 2nd eds (Chicago: University of Chicago Press, 1970).

6 Steven Shapin and Simon Schaffer, *Leviathan and the Air-pump: Hobbes, Boyle, and the experimental life* (Princeton, NJ: Princeton University Press, 1985).

7 Jean-François Lyotard, *The Differend: phrases in dispute*, trans. Georges van den Abbeele (Manchester: Manchester University Press, 1988).

8 Shapin and Schaffer, *Leviathan and the Air-pump.*

9 Empson, review of Burtt (see note 1), p. 531.

10 Feyerabend, *Against Method*; W. V. Quine, 'Two Dogmas of Empiricism', in *From a Logical Point of View*, 2nd edn (Cambridge, MA: Harvard University Press, 1961), pp. 20–46.

11 Empson, review of Burtt, pp. 531–2.

12 See, for instance, Richard Rorty, *Consequences of Pragmatism* (Brighton: Harvester, 1982); *Contingency, Irony, and Solidarity* (Cambridge: Cambridge University Press, 1989); *Objectivity, Relativism, and Truth*; also Stanley Fish, *Is There a Text in This Class? the authority of interpretive communities* (Cambridge, MA: Harvard University Press, 1980) and *Doing What Comes Naturally: change, rhetoric and the practice of theory in literary and legal studies* (Oxford: Clarendon Press, 1989). For further discussion – including some attempts to outflank Fish on his own ground – see W. J. T. Mitchell (ed.), *Against Theory: literary theory and the new pragmatism* (Chicago: University of Chicago Press, 1985).

13 Empson, review of Burtt, pp. 530–1.

14 Feyerabend, *Against Method.*

15 See, for instance, Derek Attridge, Geoff Bennington, and Robert Young (eds), *Post-Structuralism and the Question of History* (Cambridge: Cambridge University Press, 1987); J. V. Harari (ed.), *Textual Strategies: perspectives in post-structuralist criticism* (London: Methuen, 1980); Philip Rice and Patricia Waugh (eds), *Modern Literary Theory: a reader*, 3rd edn (London: Arnold, 1996); Robert Young (ed.), *Untying the Text: a post-structuralist reader* (London: Routledge and Kegan Paul, 1981).

16 Maudemarie Clark, *Nietzsche on Truth and Philosophy* (Cambridge: Cambridge University Press, 1990). See also Robert C. Holub, *Nietzsche* (New York: Twayne Publishers, 1995), especially pp. 55–78.

17 Derrida, 'White Mythology: metaphor in the text of philosophy' and 'The Supplement of Copula: philosophy before linguistics', in *Margins of Philosophy*, trans. Alan Bass (Chicago: University of Chicago Press, 1982), pp. 207–71 and 175–205.

18 See, for instance – from a range of disciplinary standpoints – Alex Callinicos, *Theories and Narratives: reflections on the philosophy of history* (Cambridge: Polity Press, 1995); Richard Campbell, *Truth and Historicity* (Oxford: Oxford University Press, 1992); Dominic LaCapra, *History and Criticism* (Ithaca, NY: Cornell University Press, 1985); Peter Lamarque and Stein H. Olsen, *Truth, Fiction and Literature: a philosophical perspective* (Oxford: Clarendon Press, 1994); Maurice Mandelbaum, *Philosophy*,

History and the Sciences (Baltimore: Johns Hopkins University Press, 1984); Ruth Ronen, *Possible Worlds in Literary Theory* (Cambridge: Cambridge University Press, 1994).

19 Hayden White, *Metahistory: the historical imagination in nineteenth-century Europe* (Baltimore: Johns Hopkins University Press, 1973); *Tropics of Discourse* (Johns Hopkins University Press, 1978); *The Content of the Form* (Johns Hopkins Univeristy Press, 1988); also – in a kindred anti-realist or 'strong' textualist vein – Keith Jenkins, *Re-Thinking History* (London: Routledge, 1991).

20 Clifford Geertz, *The Interpretation of Cultures* (New York: Basic Books, 1973) and *Local Knowledge: further essays on interpretive authority* (Basic Books, 1983).

21 See for instance H. Aram Veeser (ed.), *The New Historicism Reader* (London: Routledge, 1994).

22 See note 12 above.

23 See note 19 above.

24 See Christopher Norris, *What's Wrong with Postmodernism: critical theory and the ends of philosophy* (Hemel Hempstead: Harvester-Wheatsheaf, 1990), *Uncritical Theory: postmodernism, intellectuals and the Gulf War* (London: Lawrence and Wishart, 1992), and *Truth and the Ethics of Criticism* (Manchester: Manchester University Press, 1994); also Perry Anderson, *A Zone of Engagement* (London: Verso, 1992) and entries under note 18 above.

25 See note 15 above; also Roland Barthes, *S/Z*, trans. Richard Miller (London: Jonathan Cape, 1975); Catherine Belsey, *Critical Practice* (London: Methuen, 1980); Rosalind Coward and John Ellis, *Language and Materialism: developments in semiology and the theory of the subject* (London: Routledge and Kegan Paul, 1977); Philippe Sollers, *Writing and the Experience of Limits*, ed. D. Hayman, trans. P. Barnard (New York: Columbia University Press, 1983); *Tel Quel: Théorie d'ensemble* (Paris: Editions de Seuil, 1968); Hayden White, *The Content of the Form*.

26 Roland Barthes, 'The Discourse of History', in *The Rustle of Language*, trans. Richard Howard (Oxford: Blackwell, 1986), pp. 127–40; also 'The Reality Effect' and 'Writing the Event', ibid., pp. 141–8 and 149–54.

27 See Anderson, *A Zone of Engagement* and *The Ends of History* (London: Verso, 1994).

28 See, for instance, Michel Foucault, *Language, Counter-Memory, Practice*, ed. D. F. Bouchard and S. Weber (Oxford: Blackwell, 1970) and *Power/Knowledge: selected interviews and other writings*, ed. C. Gordon (Brighton: Harvester, 1980).

29 See note 27 above; also Callinicos, *Theories and Narratives* and Christopher Norris, 'Postmodernizing History: right-wing revisionism and the uses of theory', *Southern Review* (Adelaide), 21/2 (July 1988), pp. 123–40.

30 Ferdinand de Saussure, *Course in General Linguistics*, trans. Wade Baskin (London: Fontana, 1974).

31 Valentine Cunningham, *In the Reading Gaol: postmodernity, texts and history* (Oxford: Blackwell, 1994), p. 33.

32 See Jean Starobinski, *Words Upon Words*, trans. Olivia Emmett (New Haven: Yale University Press, 1979); also Michel Riffaterre, *Text Production*, trans. Terese Lyons (New York: Columbia University Press, 1983) and Julia Kristeva, *Semeiotiké* (Paris: Seuil, 1969).

33 Cunningham, *In the Reading Gaol*, p. 33.

34 Jacques Derrida, *Of Grammatology*, trans. G. C. Spivak (Baltimore: Johns Hopkins University Press, 1976).
35 Derrida, 'Force and Signification' and 'Genesis and Structure', in *Writing and Difference*, trans. Alan Bass (London: Routledge and Kegan Paul, 1978), pp. 3–30 and 154–68.
36 Derrida, 'Genesis and Structure', p. 160.
37 Ibid., p. 159.
38 Ibid., p. 158.
39 Immanuel Kant, *Critique of Pure Reason*, trans. N. Kemp Smith (London: Macmillan, 1974).
40 Gottlob Frege, 'On Sense and Reference', in Max Black and P. T. Geach (eds), *Translations from the Philosophical Writings of Gottlob Frege* (Oxford: Blackwell, 1952), pp. 56–78.
41 See Notes 3 and 4 above.
42 See Gaston Bachelard, *La Formation de l'esprit scientifique* (Paris: Corti, 1938); *Le Matérialisme rationnel* (Paris: Presses Universitaires de France, 1953); *The Philosophy of No: a philosophy of the new scientific mind* (New York: Orion Press, 1968); *L'Activité rationaliste de la physique contemporaine* (Press Universitaires de France, 1975); *The New Scientifc Spirit* (Boston: Beacon Press, 1984); Georges Canguilhem, *Etudes d'histoire et de philosophie des sciences* (Paris: Vrin, 1968); *La connaissance de la vie*, 2nd edn (Vrin, 1969). See also Dominique Lecourt, *Bachelard, ou le jour et la nuit* (Paris, 1974); Lecourt, *Marxism and Epistemology: Bachelard, Canguilhem and Foucault* (London: New Left Books, 1975); Mary Tiles, *Bachelard: science and objectivity* (Cambridge: Cambridge University Press, 1984).
43 See, for instance, Louis Althusser, *For Marx*, trans. Ben Brewster (London: Allen Lane, 1969); *'Lenin and Philosophy' and Other Essays*, trans. B. Brewster (London: New Left Books, 1971); Louis Althusser and Etienne Balibar, *Reading Capital*, trans. Ben Brewster (New Left Books, 1970).
44 Louis Althusser, 'Is It Simple to be a Marxist in Philosophy?', in *'Philosophy and the Spontaneous Philosophy of the Scientists' and Other Essays*, ed. Gregory Elliott (London: Verso, 1990), pp. 203–40; pp. 227–8.
45 Among the critics of Althusserian Marxism on these and other grounds see, for instance, Barry Hindess, Paul Hirst, Anthony Cutler and Athar Hussein, *Marx's 'Capital' and Capitalism Today*, 2 vols (London: Routledge and Kegan Paul, 1977–8); E. P. Thompson, 'The Poverty of Theory', in *'The Poverty of Theory' and Other Essays* (London: Merlin, 1978); Ernesto Laclau and Chantal Mouffe, *Hegemony and Socialist Strategy* (London: Verso, 1985).
46 See entries under note 43 above; also Althusser, 'On Theoretical Work: difficulties and resources', in *'Philosophy and the Spontaneous Philosophy of the Scientists' and Other Essays*, pp. 43–67 and *Essays in Self-Criticism*, trans. Ben Brewster (London: New Left Books, 1976).
47 For a more detailed account of this history, see Ted Benton, *The Rise and Fall of Structural Marxism* (London: New Left Books, 1984) and Gregory Elliott, *Althusser: the detour of theory* (London: Verso, 1987).
48 Michel Foucault, *The Order of Things: an archaeology of the human sciences* (London: Tavistock, 1970).

49 Althusser, *For Marx*.
50 See Jacques Lacan, *Ecrits: a selection*, trans. A. Sheridan-Smith (London: Tavistock, 1977); Louis Althusser, 'Freud and Lacan', in *'Lenin and Philosophy' and Other Essays*, pp. 177–202.
51 Emile Benveniste, *Problems in General Linguistics*, trans. Mary E. Meek (Miami: Miami University Press, 1971); Althusser, 'Ideology and Ideological State Apparatuses', in *Lenin and Philosophy*.
52 See Derrida, 'White Mythology' and note 42 above. All further references indicated by 'WM' and page number in the text.
53 See note 12 above.
54 Canguilhem, *La Connaissance de la vie* (see note 42 above), pp. 48–9; cited by Derrida in 'White Mythology', p. 261.
55 J. Claude Evans, *Strategies of Deconstruction: Derrida and the myth of the voice* (Minneapolis: University of Minnesota Press, 1991).
56 See, for instance, D. W. Hamlyn, *Experience and the Growth of Understanding* (London: Routledge and Kegan Paul, 1978); Imre Lakatos and Alan Musgrave, *Criticism and the Growth of Knowledge* (Cambridge: Cambridge University Press, 1970); Peter Muntz, *Our Knowledge of the Growth of Knowledge* (London: Routledge, 1985); Peter J. Smith, *Realism and the Progress of Science* (Cambridge University Press, 1981); Nicholas Rescher, *Scientific Progress* (Oxford: Blackwell, 1979); also – for a more sceptical (social-constructivist) view – Barry Barnes, *Interests and the Growth of Knowledge* (London: Routledge, 1978).
57 See note 12 above; also Rorty, *Philosophy and the Mirror of Nature* (Oxford: Blackwell, 1980) and *Essays on Heidegger and Others* (Cambridge: Cambridge University Press, 1991).
58 Anatole France, *The Garden of Epicurus*, trans. Alfred Allinson (New York: Dodd, Mead, 1923).
59 Ibid., pp. 213–14.
60 See, for instance, Nietzsche, 'On Truth and Falsehood in their Ultramoral Sense', in *Complete Works of Friedrich Nietzsche*, ed. D. Levy (London and Edinburgh, 1911), vol. II; also Nietzsche, *The Will to Power*, trans. Walter Kaufmann and R. J. Hollingdale (New York: Random House, 1967); Sarah Kofman, *Nietzsche and Metaphor*, trans. Duncan Large (Stanford, CA: Stanford University Press, 1993); Martin Heidegger, *Introduction to Metaphysics*, trans. Ralph Mannheim (New Haven: Yale University Press, 1954); *Nietzsche*, 4 vols, trans. David F. Krell (New York: Harper and Row, 1971); *The End of Philosophy*, trans. Joan Stambaugh (Harper and Row, 1971); *Basic Writings*, ed. D. F. Krell (Harper and Row, 1977). Thus Heidegger: 'The truth of *physis, aletheia* as the unconcealment that is the essence of the emerging power, now becomes *homoiosis* and *mimesis*, assimilation and accommodation. . . . It becomes a correctness of vision, of apprehension as representation' (*Introduction to Metaphysics*, pp. 184–5).
61 Nietzsche, 'On Truth and Falsity', p. 180.
62 See note 60 above; also Heidegger, *The Question Concerning Technology and Other Essays*, trans. William Lovitt (New York: Harper and Row, 1977).
63 Anatole France, *The Garden of Epicurus*, pp. 201–2; cited in Derrida, 'White Mythology', p. 211.

64 Nietzsche, 'On Truth and Falsity', pp. 187–8.
65 Kurt Gödel, 'On Formally Undecidable Propositions of *Principia Mathematica* and Related Systems', trans. B. Meltzer (New York: Basic Books, 1962); also S. G. Shanker (ed.), *Gödel's Theorem in Focus* (London: Routledge, 1987).
66 See David Wayne Thomas, 'Gödel's Theorem and postmodern theory', *Publications of the Modern Language Association of America*, 110/2 (March 1995), pp. 249–61.
67 See, for instance, Derrida, 'Genesis and Structure', p. 162; 'From Restricted to General Economy; an Hegelianism without reserve', in *Writing and Difference*, pp. 251–77; also *Dissemination*, trans. Barbara Johnson (Chicago: University of Chicago Press, 1981), p. 219.
68 See for instance – from a range of philosophical standpoints – Max Black, *Models and Metaphors* (Ithaca, NY: Cornell University Press, 1962); Paul de Man, 'The epistemology of metaphor', *Critical Inquiry*, 5 (1978), pp. 13–30; Eva Feder Kitay, *Metaphor: its cognitive force and linguistic structure* (Oxford: Clarendon Press, 1987); W.H. Leatherdale, *The Role of Analogy, Model and Metaphor in Science* (Amsterdam: North-Holland, 1974); also the works of Bachelard cited in note 42 above.
69 See note 3 above; also Joseph Rouse, *Knowledge and Power: toward a political philosophy of science* (Ithaca, NY: Cornell University Press, 1987).
70 See notes 42–7 above.
71 See Rorty, *Philosophy and the Mirror of Nature.*
72 Descartes, *Meditations on First Philosophy*, in *The Philosophical Works of Descartes*, vol. I, trans. E. Haldane and G. R. T. Ross (Cambridge: Cambridge University Press, 1970), p. 160; cited in Derrida, 'White Mythology', p. 267.
73 See Bachelard, *Le Matérialisme rationnel*; also Roy Bhaskar, *Reclaiming Reality: a critical introduction to contemporary philosophy* (London: Verso, 1989).
74 Cited by Bhaskar, ibid., p. 45.
75 Ibid., p. 45.
76 See entries under note 3 above; also Rouse, *Knowledge and Power* and – from a variety of critical positions – Barry Barnes, *Interests and the Growth of Knowledge*; Lorraine Code, *What Can She Know? feminist theory and the construction of knowledge* (Ithaca, NY: Cornell University Press, 1991); Steve Fuller, *Social Epistemology* (Bloomington: Indiana University Press, 1988); Evelyn Fox Keller, *Reflections on Gender and Science* (New Haven: Yale University Press, 1985); Sandra Harding, *The Science Question in Feminism* (Ithaca, NY: Cornell University Press, 1986); Sandra Harding and Merrill B. Hintikka (eds) *Discovering Reality: feminist perspectives on epistemology, metaphysics, methodology and philosophy of science* (Dordrecht and Boston: D. Reidel, 1983); John Mepham and David-Hillel Ruben (eds), *Epistemology, Science, Ideology* (Atlantic Highlands: Humanities Press, 1979); Hilary Rose and Steven Rose (eds), *The Radicalization of Science: ideology of/in the natural sciences* (London: Macmillan, 1976).
77 See Thomas S. Kuhn, *The Structure of Scientific Revolutions*; also Gary Gutting (ed.), *Paradigms and Revolutions* (Notre Dame, IN: University of Notre Dame Press, 1980); Ian Hacking (ed.), *Scientific Revolutions* (London: Oxford University Press, 1981); John Krige, *Science, Revolution and Discontinuity* (Brighton: Harvester, 1980).
78 See entries under note 42 above; also – by way of contrast with Bachelard's approach – Pierre Duhem, *To Save the Phenomena: an essay on the idea of physical theory from Plato to Galileo*, trans. E. Dolan and C. Maschler (Chicago: University of Chicago Press,

1969); *The Aims and Structure of Physical Theory*, trans. Philip Wiener (Princeton, NJ: Princeton University Press, 1954); Sandra G. Harding (ed.), *Can Theories be Refuted? essays on the Duhem-Quine thesis* (Dordrecht and Boston: D. Reidel, 1976); W. V. Quine, *Word and Object* (Cambridge, MA: Harvard University Press, 1960); *From a Logical Point of View*, 2nd edn (Harvard University Press, 1961); *Theories and Things* (Harvard University Press, 1981).

79 Ian Hacking, *Representing and Intervening: introductory topics in the philosophy of natural science* (Cambridge: Cambridge University Press, 1983); see also Robert Ackerman, *Data, Instruments, and Theory: a dialectical approach to the philosophy of science* (Princeton, NJ: Princeton University Press, 1985); Michael Gardner, 'Realism and instrumentalism in nineteenth-century atomism', *Philosophy of Science*, 46/1 (1979), pp. 1–34; Don Ihde, *Instrumental Realism: the interface between philosophy of science and philosophy of technology* (Bloomington: Indiana University Press, 1991); Josef M. Jauch, *Are Quanta Real? a Galilean dialogue* (Bloomington: Indiana University Press, 1973); J. Perrin, *Atoms*, trans. D. L. Hammick (New York: Van Nostrand, 1923); Wesley C. Salmon, *Scientific Explanation and the Causal Structure of the World* (Princeton, NJ: Princeton University Press, 1984); A. Sudbury, *Quantum Mechanics and the Particles of Nature* (Cambridge: Cambridge University Press, 1986).

80 On the Michelson–Morley experiments, see for instance Rom Harré, *Great Scientific Experiments* (London: Oxford University Press, 1983); also – for a dissenting (anti-realist and strong sociological) view, Harry Collins and Trevor Pinch, *The Golem: what everyone should know about science.*

81 See for instance Rorty, *Consequences of Pragmatism* and Fish, *Is There a Text In This Class?* (cited in note 12 above).

82 Feyerabend, *Science in a Free Society* and *Against Method* (see notes 4 and 10 above).

83 Kuhn, *The Structure of Scientific Revolutions.*

84 See especially Jean-François Lyotard, *The Postmodern Condition: a report on knowledge*, trans. Geoff Bennington and Brian Massumi (Manchester: Manchester University Press, 1984); also – for commentary from various (more or less critical) standpoints – Horace L. Fairlamb, *Critical Conditions: postmodernity and the question of foundations* (Cambridge: Cambridge University Press, 1994); Frank B. Farrell, *Subjectivity, Realism, and Postmodernism: the recovery of the world in recent philosophy* (Cambridge University Press, 1994); Arkady Plotnitsky, *Complementarity: anti-epistemology after Bohr and Derrida* (Durham, NC: Duke University Press, 1994); Thomas, 'Gödel's Thorem and Postmodern Theory'.

85 Lyotard, *The Postmodern Condition.*

86 Quine, 'Two Dogmas of Empiricism', in *From a Logical Point of View*, pp. 20–46; see also Bruce Gregory, *Inventing Reality: physics as language* (New York: Wiley, 1988); Plotnitsky, *Complementarity*; and – for the strong sociological version of this argument – Andrew Pickering, *Constructing Quarks: a sociological history of particle physics* (Edinburgh: Edinburgh University Press, 1984).

87 See note 77 above.

88 See note 79 above; also David Bohm and B. J. Hiley, *The Undivided Universe: an ontological interpretation of quantum theory* (London: Routledge, 1993); J. T. Cushing, C. F. Delaney and G. M. Gutting (eds), *Science and Reality* (Notre Dame, IN: University of Notre Dame Press, 1984); Jarrett Leplin (ed.), *Scientific Realism*

(Berkeley and Los Angeles: University of California Press, 1984); Mary Jo Nye, *Molecular Reality* (London: MacDonald, 1972); Karl Popper, *Quantum Theory and the Schism in Physics* (London: Hutchinson, 1982); Nicholas Rescher, *Scientific Realism: a reappraisal* (Dordrecht: D. Reidel, 1987); Salmon, *Scientific Explanation and the Causal Structure of the World.*

89 See, for instance, J. S. Bell, *Speakable and Unspeakable in Quantum Mechanics: collected papers on quantum philosophy* (Cambridge: Cambridge University Press, 1987); James T. Cushing and Ernan McMullin (eds), *Philosophical Consequences of Quantum Theory: reflections on Bell's theorem* (Notre Dame, IN: University of Notre Dame Press, 1989); Arthur Fine, *The Shaky Game: Einstein, realism, and quantum theory* (Chicago: University of Chicago Press, 1986); John Honner, *The Description of Nature: Niels Bohr and the philosophy of quantum physics* (Oxford: Clarendon Press, 1987); Tim Maudlin, *Quantum Nonlocality and Relativity: metaphysical intimations of modern science* (Oxford: Blackwell, 1993); Michael Redhead, *Incompleteness, Nonlocality and Realism: a prolegomenon to the philosophy of quantum mechanics* (Oxford: Clarendon Press, 1987).

2

Deconstruction and Epistemology: Bachelard, Derrida, de Man

I

In what follows I shall challenge certain widespread misunderstandings in regard to deconstruction and its (supposed) affinity with current anti-realist and cultural-relativist schools of thought. This issue is posed most directly in Jacques Derrida's 'White Mythology: metaphor in the text of philosophy', an essay – I should add – that has very often been construed as a full-scale exercise in the rhetorical deconstruction of such 'logocentric' notions as truth, knowledge, reason and reality.[1] My purpose is therefore to refute the idea of deconstruction as a priori committed to an extreme ('textualist') version of the argument that reality is a purely linguistic construct, that 'all concepts are metaphors', 'all science merely a species of instrumental fiction' and kindred quasi-deconstructive *idées reçues*.[2] It is also to demonstrate that these claims turn out to be strictly unintelligible – or self-subverting – when confronted with the very different kind of sceptical rigour that deconstruction brings to bear in raising such questions.

So can there be – is science capable of delivering or philosophy of conceiving – what Derrida (no doubt 'rhetorically') calls 'a truth of language which would say the thing such as it is in itself, in act, properly'?[3] Not if we construe this claim on the basis of a straightforward correspondence theory, or on the model of a one-to-one guaranteed match between word and object, concept and intuition, or veridical statements and those facts or truths to which they properly correspond. For it is a problem which has long been recognized – and endlessly rehearsed by critics of this argument – that 'facts' are linguistic entities, rather than objects or real-world states of affairs. Thus any talk of matching, correspondence, 'truth-to-the-facts', etc., is sure to generate a vicious circle devoid of explanatory content. However, there is another fallacy – more widespread at present – which treats this criticism as fatal not only to naïve (positivist) conceptions but to *any* philosophy of science which seeks to conserve some intelligible role for notions of truth, reality and scientific fact. 'This truth is not certain,' Derrida continues. 'There can be bad metaphors. Are the latter metaphors? Only an axiology supported by a theory of truth can answer this question, and this axiology belongs to the interior of rhetoric. It cannot be neutral.'[4] But he – like Paul de Man in some essays of comparable subtlety and rigour – is very far from using 'rhetoric' in the commonplace sense of that term which equates (roughly speaking) with language in its suasive or performative aspect, and which thus empties it of all cognitive or epistemological import. This is posssible, de Man writes,

> only because its tropological, figural dimensions are being bypassed. It is as if, to return for a moment to the model of the [classical] *trivium*, rhetoric could be isolated from the generality that grammar and logic have in common and treated as a mere correlative of an illocutionary power.[5]

Hence his double purpose in these essays: on the one hand to assert (as against such reductive uses of the term) that rhetoric has a genuinely critical dimension, an aspect corresponding to what de Man calls the 'epistemology of tropes', and on the other to show how rhetoric thus conceived may turn out to disturb – or call into question – the 'stable cognitive field' that is traditionally thought to extend 'from grammar to logic to a general science of man and of the phenomenal world'. Thus 'difficulties occur only when it is no longer possible to ignore the epistemological thrust of the rhetorical dimension of language' (*The Resistance be Theory*, p. 14).

I have cited these passages from de Man because they offer a concise formulation of issues that Derrida broaches more obliquely in 'White Mythology'. Also, they provide a striking instance of the way that deconstruction calls for both the utmost rigour of linguistic and conceptual analysis *and* – by no means incompatible with this – for a due recognition of the undecidability that inhabits certain of its own crucial terms. Traditionally, the disciplines of the *trivium* – logic, grammar and rhetoric – were supposed to equip students for the study of those other, more scientific branches of knowledge (arithmetic, geometry, astronomy and music) which comprised the classical *quadrivium*. 'In the history of philosophy,' de Man remarks, 'this link is . . . accomplished by way of logic, the area where the rigor of the linguistic discourse about itself matches up with the rigor of the mathematical discourse about the world' (RT, p. 13). And again: '[t]his articulation of the sciences of language with the mathematical sciences represents a particularly compelling version of a continuity between a theory of language, as logic, and the knowledge of the phenomenal world to which mathematics gives access' (ibid.). But to take rhetoric seriously – to give it (so to speak) its epistemological due – is to admit at least some measure of doubt as to whether we can always rely on this homologous relationship between the various orders of analysis concerned. For on a closer reading it may transpire that the model is inadequate to account for those figural (or tropological) dimensions of language which themselves require what de Man – following Nietzsche – pointedly describes as a 'considerable labour of deconstruction'.[6] These functions cannot be reduced to order by treating them – on the standard view – as subsidiary 'adjuncts' of logic and grammar, ornamental features that belong to language in its suasive, poetic or 'merely' rhetorical aspect. For their effect is not only to create a 'disturbance' in some given epistemological field but also – as happens during periods of Kuhnian 'revolutionary' science – to indicate those points at which the current conceptual or explanatory paradigm comes under strain.[7]

It is well to be clear about the extent and the precise nature of de Man's scepticism with regard to scientific and other orders of truth-claim. He is *not* for one moment denying – along with anti-realists, cultural relativists, postmodernists like Lyotard, and proponents of the 'strong programme' in sociology of knowledge[8] – that the continuity thesis *most often* holds good, and that scientists (philosophers of science included) may therefore be justified *up to a point* in assuming the existence of that crucial 'link' between language, logic and reference. Nor is he committed to the strictly preposterous notion – 'preposterous', that is, in the literal sense of putting first what should come after – that reality and truth are wholly constructed in or by the discourse (language game, narrative, tropological paradigm, etc.) which bars any access to that notional world 'outside' the text. It is perhaps understandable, in his case as in Derrida's, that this idea should have taken hold since they both make a point – a philosophical point – of suspending belief in what appears self-evident to commonsense judgement, and thus raising questions that demand something more by way of closely argued critical response. What interests de Man is the way that the correspondence model breaks down in certain highly complex texts, and the fact that those texts none the less provide a *knowledge* – a knowledge of their own linguistic workings, to be sure, but also a knowledge with wider implications for science and epistemology – which can only be had by means of such scrupulous rhetorical exegesis. If these readings always lead up to a moment of somewhat predictable 'undecidability', it is not – or not only – on account of de Man's a priori persuasion in this regard. Still less is it a consequence of his merely refusing, in postmodern-pragmatist fashion, to acknowledge the existence of truths outside or beyond the currency of de facto accredited belief. On the contrary: such readings go against the grain – and generate much resistance – precisely in so far as they remain alert to those obstacles that can always arise to disrupt the smooth passage from logic, via grammar, to 'knowledge of the world', or again (since this process also works in reverse) from observation statements to forms of elaborated scientific theory.

So there is a great difference, I would maintain, between de Man's meticulous analytical procedures in the reading of these problematic passages in various texts and the attitude of *de rigueur* epistemological scepticism evinced in many quarters of postmodern-pragmatist and cultural-relativist debate. This latter view takes it for granted – as standing in no need of proof – that the truth-value of scientific claims is always wholly undecidable or decided only by contingent (e.g., social, political or cultural) factors to which 'science' remains obdurately blind in order to preserve its privileged (truth-telling) status.[9] This is not the place for a full-scale exposition of de Man's approach to these issues. However, there is one essay in particular – 'Pascal's Allegory of Persuasion' – where they figure so centrally as to merit more detailed attention.[10] His aim, simply put, is to deconstruct the terms of that tenacious opposition between scientific and non-scientific (or 'logical' and 'rhetorical') modes of discourse that is often invoked by commentators on Pascal.[11] Of course, this distinction is so deeply ingrained – so much a part of our modern (post-Renaissance) cultural, intellectual and disciplinary framework – that neither they, the commentators, nor indeed Pascal can be expected to renounce it altogether. Moreover, de Man is himself very far from recommending (in postmodern-pragmatist fashion) that we should scrap such obsolete notions of discipline-specific method, knowledge or truth, and thus come around to a view of science – like philosophy – as just another voice in the cultural 'conversation of mankind'.[12] What he *does* call into doubt is the adequacy of any reading of Pascal which takes that distinction at face value. For such a reading will inevitably fail to note the rhetorical (as well as epistemological) problems that arise in connection with Pascal's well-known attempt to vindicate *both* the truth-claims of science *and*, in a realm quite separate from those, the claims of religious truth as a matter of inward, self-authenticating faith.

On this understanding the heart had 'reasons' unknown to the intellect, or to science in its own (properly delimited) field of endeavour. Thus religion might be saved from the otherwise threatening encroachment of a critical drive to establish truth on the basis of unaided human intellect quite apart from all sources of revealed or spiritual wisdom. It is the same move that has often been repeated – most recently by Wittgensteinian philosophers of language – in order to establish a separate domain where religious belief is untouched by those standards of judgement that properly apply in other (e.g., factual, logical and scientific) language games or cultural 'forms of life'.[13] De Man is not rejecting this interpretation in so far as it claims fidelity to the spirit of Pascal's writing, or indeed to its letter in those passages touching on the 'leap of faith' and kindred themes. Rather, he is claiming that matters become more complicated if we attend more closely to the complex interplay of logic, grammar and rhetoric, especially at points of maximal stress where Pascal deploys the full resources of critical reason with a view to exempting religious belief from those same (elsewhere mandatory) standards. In these passages we witness not so much the discomfiture of 'religion' at the hands of 'science' as the way in which this opposition becomes displaced – 'reinscribed', to adopt the deconstructive term – onto a series of radically unstable distinctions which include the key pair 'proof'/'allegory'. 'Persuasion and proof should, in principle, be distinct from each other, and it would not occur to a mathematician to call his proofs allegories' (PAP, p. 2). That is to say, this model cannot allow for the gap that potentially opens up when rhetoric – and 'allegory' as the most self-conscious, most critical and rigorous of rhetorical modes – asserts its claim upon a reading which doesn't take for granted the 'stable cognitive field' supposedly extending from language, through logic, to 'knowledge of the world'.

Only thus can we avoid what de Man calls 'the tendentious and simplistic opposition between knowledge and faith that is often forced upon Pascal' (p. 7). That is to say, we do better to resist the temptation whereby these epistemological problems are 'so often expressed, in Pascal himself and in his commentators, in a tonality of existential pathos' (ibid.). For it then becomes evident that no such clear demarcation can be drawn – in generic or other (e.g., rhetorical or logico-grammatical) terms – between, on the one hand, those texts that apparently belong to Pascal's religious or 'existential' mode of utterance and, on the other, those writings which explicitly subscribe to mathematical or scientific ideals of a priori, demonstrative, analytic or hypothetico-deductive proof. Thus 'for all the sombre felicity of their aphoristic condensation, the *Pensées* are also very systematically schematized texts' (PAP, p. 13). But conversely, Pascal's most rigorous performances in the 'other' (mathematico-scientific) mode are often marked by rhetorical elements – by tropes, figures, analogical swerves from the 'literal' (stipulative) sense of his terms, etc. – which exert a powerful suasive effect, and which cannot be conveniently factored out by treating them as mere dispensable 'adjuncts' to the argument in hand. 'Why is it,' de Man asks, 'that the furthest reaching truths about ourselves and the world have to be stated in such a lopsided, referentially indirect mode?' (p. 2). And again: 'why is it that texts that attempt the articulation of persuasion with epistemology turn out to be inconclusive about their own intelligibility in the same manner and for the same reasons that produce allegory?' (ibid.). Nor are these problems by any means peculiar to a mode of discourse (Pascal's) that seeks to maintain two radically disjunct orders of truth – the mathematico-scientific and the religious – and which thus, one might think, inevitably generates all manner of aporia or paradox. On the contrary, according to de Man: the antinomy of 'reason' and 'faith' in Pascal is a version of that other antinomy – between 'proof' and 'persuasion' – which characterizes all discourse beyond a certain point of rhetorical (or indeed of logical and grammatical) complexity.

This is why – to repeat – de Man rejects any form of the 'tendentious and simplistic opposition between knowledge and faith that is often forced upon Pascal'.[14] His point is not at all to suggest – as Feyerabend does in the case of Galileo – that scientific 'knowledge' is wholly on a par with other modes of belief, or that 'truth' in such matters is itself nothing more than a suasive or rhetorical effect, brought about by a shrewd manipulative grasp of the way that facts, experimental data, theoretical 'proofs' and so forth can always be trimmed to the requirements of this or that (religious or secular) authority.[15] It would take a very long and detailed treatment of de Man's essay on Pascal to bring out the extent of this difference. But readers can gain some idea of what is involved by following the conjoint process of intricate textual analysis and rigorous conceptual critique that de Man brings to bear upon those passages where Pascal attempts to make good his cardinal distinction between 'nominal' and 'real' definitions. The former are – or should be in principle – 'entirely free and never open to contradiction'. Their meaning is purely stipulative, fixed by their role within a system of well-defined terms, postulates, proof procedures, synonymy relations and so forth which ideally admits nothing of the imprecision – the extra-systemic entanglement with contingent matters of fact – that attaches to 'real' definitions. These latter, by contrast, are 'a great deal more coercive and dangerous: they are actually not definitions, but axioms or, even more frequently, propositions that need to be proven' (PAP, pp. 5–6). So it ought to be the case that mathematics – along with the 'pure' sciences – aspires to a condition of truth *more geometrico*, or a state of perfected internal consistency where 'proof' would consist entirely of nominal definitions and the order of necessary (logico-deductive) relations between them. Such is at any rate the standard account of that seventeenth-century rationalist project that is taken to unite thinkers like Descartes, Pascal and Spinoza across and despite their otherwise large philosophical differences of view.

Of course this project finds few defenders nowadays. It is mostly treated – and has been ever since Kant – as a cautionary instance of what goes wrong when reason in its 'pure' (or speculative) mode becomes detached from those anchor-points in phenomenal intuition that provide the only guard against flights of aberrant 'metaphysical' fancy.[16] All the same the issue about 'real' and 'nominal' definitions – whether in Pascalian (rationalist) or Lockean (empiricist) guise – still has enough life in it to connect with a range of continuing philosophical debates. These include (for instance) discussions on the topic of *de re* and *de dicto* necessity; Frege's widely influential distinction between 'reference' and 'sense'; Kripke's revival of a realist (*de re*) theory of naming and necessity, as against descriptivist (Fregean or Russellian) accounts; and Quine's attempt – in his essay 'Two Dogmas of Empiricism' – to show that these debates are wholly pointless since they all presuppose a distinction (that between 'analytic' and 'synthetic' orders of statement) which can muster no defence against the pragmatist case for full-scale meaning holism and ontological relativity.[17] And from Quine's position it is but a short step to those sceptical philosophies (or sociologies) of science that likewise – though in less circumspect fashion – proclaim an end to the old regime of scientific method and truth.

What distinguishes de Man's reading of Pascal is its refusal simply to let go of these issues, or to take the sceptical-relativist line of least resistance, even though that reading shows – with the utmost analytical precision – the difficulties to which such a project inevitably gives rise. For it is, de Man argues, a *demonstrable truth* about Pascal's texts that they are obliged to suspend or elide the difference between 'nominal' and 'real' definitions whenever there is a question as to just how mathematics relates to the realm of phenomenal experience, or how any system of reasoning *more geometrico* can possibly pertain to matters outside its self-enclosed logical domain. Thus 'the nominal definition of primitive terms always turns into a proposition that has to, but cannot, be proven' (PAP, p. 7). And moreover, 'since definition is itself now a primitive term . . . the definition of the nominal definition is itself a real, and not a nominal, definition' (ibid.). Nowhere is this more strikingly the case than in Pascal's writings on the topic of the 'double infinity', the mathematical concepts – if such they may be called – of the infinitely large (or numerous) and the infinitesimally small. Indeed the main question, for Pascal as for present-day philosophers of mathematics, is the issue of just what status such notions possess. Are they 'concepts' capable of adequate grasp by the powers of mathematical reason? 'Intuitions' that would discover some affinity, some analogical ground, in those forms and structures that are already present in the manifold of phenomenal perception? Or are they perhaps 'ideas' in the Kantian sense of that term, notions that surpass the limits of theoretical understanding (where it is the rule that intuitions must be 'brought under' adequate concepts), but which none the less call forth an exercise of reason in its speculative mode?[18]

What de Man brings out in his reading of Pascal is the perpetual 'oscillation' that occurs between these various orders of truth-claim. Thus Pascal has to distinguish between numerical concepts and spatial intuitions – to deny that they can be treated as in any way analogous – since otherwise (as with 'nominal' and 'real' definitions) the latter would contaminate the former, reduce mathematics to the level of mere (phenomenal) cognition, and thereby compromise its status as a rigorous axiomatic-deductive system of proofs. Yet he is obliged to relax (or abandon) this principle by the further – equally pressing – requirement that some link *must* after all be established between mathematics and the natural sciences, or again, between the order of demonstrative (axiomatic) argument and those phenomena that science is called upon to explain. For such, we recall, is the necessary order of relationship that is assumed to hold both within and beyond those disciplines (logic, grammar, and rhetoric) that make up the classical *trivium*. Thus grammar 'stands in the service' of logic, since there exists – or at any rate there ought to exist, so long as rhetoric is kept in its place – a perfect homology between the structure of well-formed logical propositions and the modes of predicative (subject-object) statement that characterize grammatical sentences. And this in turn should lead on, in de Man's formulation, to a 'knowledge of the world' whose chief exemplars are those disciplines of the *quadrivium* (arithmetic, geometry, astronomy and music) which require both an a priori grounding in the categories of logical thought and the extension of those categories – achieved via the truth-preserving forms of grammatical statement – to the realm of phenomenal cognition. Thus Pascal is caught between a twofold (contradictory) set of demands: that mathematics and geometry be conceived analytically, that is to say, as capable of rigorous demonstrative proof quite apart from any appeal to intuition or phenomenal self-evidence, and on the other hand the requirement – in de Man's words – that 'the underlying homology of space and number, the ground of the system, should never be fundamentally in question' (PAP, p. 10).

Not that this problem is in any way peculiar to Pascal, nor again to what is commonly (and loosely) characterized as the discourse of seventeenth-century rationalist thought. Rather, it is a problem in epistemology and philosophy of science whenever there exists a complex, uneven, inconsistent or otherwise strained relationship between the current stage of theoretical advance and the current state of experimental or observational practice. Where responses differ – as between (say) philosophers such as Paul Feyerabend and Gaston Bachelard – is in the interpretation that is placed on such evidence of the failure of theory to match up directly with the findings of empirical research. On the Feyerabend view, such failure is enough to warrant a dismissal of science's truth-telling claims, a switch to social (or ideological) criteria as the sole determinants of 'truth' in any given case, and – following from this – a frankly anarchist recommendation that scientists (or scientific policy-makers) abandon those old methodological fixations and encourage a thousand speculative flowers to bloom.[19] For Bachelard, conversely, there is room for genuine progress in scientific knowledge, despite what very often appears as the underdetermination of theory by evidence, or the lack of adequate (material-technological) resources to vindicate those various advances brought about through the criticism – or conceptual 'rectification' – of existing metaphors and paradigms.[20]

II

I have argued here (and shall develop the case in what follows) that deconstruction should be seen as closely allied to the latter epistemo-critical tradition of thought. And this despite the widespread belief – among literary and cultural theorists especially, but also postmodernist philosophers of science and 'social epistemologists' like Steve Fuller – that deconstruction is just another handy device for cutting science (and philosophy) down to size.[21] In their view it is helpful – strategically valuable – as a clinching demonstration that scientific truth-claims cannot be justified on their own (epistemological or intra-disciplinary) terms. Sociologists can then get on with the more important work of asking how far the different kinds and conceptions of scientific 'truth' have served either to advance or to retard the fulfilment of human hopes and aspirations. Certainly we need not be duped by the seeming self-evidence of science's claim to come up with adequate theories and explanations whose truth is borne out by its manifest achievements in various ('pure' and 'applied') branches of research. For Derrida has shown – has he not? – that all concepts are sublimated metaphors, all truth-talk a species of suasive rhetoric, and, most important for this purpose, all notions of philosophico-scientific rigour or method just a means of securing assent among those predisposed (through *naïveté* or professional self-interest) to credit such claims.[22]

'Not' seems the best short answer to all this, given the evidence of his essay 'White Mythology' and its meticulous working-through of the concept/metaphor distinction as deployed by philosophers of science from Aristotle to Bachelard.[23] But it is worth returning to de Man on Pascal for some further clarification of the difference between a deconstructive approach to these issues and their treatment by thinkers of a postmodern-pragmatist (or a full-scale 'ontological-relativist') persuasion. 'Seventeenth-century epistemology,' he writes,

> at the moment when the relationship between philosophy and mathematics is particularly close, holds up the language of what it calls geometry (*mos geometricus*), and which in fact includes the homogeneous concatenation between space, time and number, as the sole model of coherence and economy. . . . This is a clear instance of the interconnection between a science of the phenomenal world and a science of language conceived as definitional logic, the precondition for a correct axiomatic-deductive, synthetic reasoning. . . . [B]ut this leaves open the question, within the confines of the *trivium* itself, of the relationship between grammar, rhetoric and logic. And this is the point at which rhetoric intervenes as a decisive but unsettling element which, in a variety of modes and aspects, disrupts the inner balance of the model and, consequently, its extension to the outside world as well. (RT, p. 13)

There is no room here for a detailed exegesis of this complex and immensely suggestive passage. I would just make the following points by way of relating it back to my argument so far. First: de Man is by no means recommending that we should henceforth simply give up on the quaint ('seventeenth-century rationalist') idea that truths can be arrived at through a process of relatively abstract reasoning, or – in Bachelardian terms – through a labour of conceptual 'rectification' as applied to images, metaphors and 'commonsense' (phenomenalist) analogies.[24] This is what Spinoza had in mind when he remarked that, given the limits of our perceptual constitution, the Sun must always appear to us as a small reddish-yellow disc-shaped object suspended somewhere in the middle distance and rotating once daily around the earth.[25] Only by abstracting away from such phenomenal appearances – through a critique of naïve sense-certainty allied to more adequate (geometrical, mathematical and astronomical) concepts – can we come to understand both the source of that illusion and its limiting conditions as a product of untheorized (pre-scientific) belief.[26]

Indeed, it is this conception of science as advancing through decisive ruptures – 'epistemological breaks' – with the currency of taken-for-granted (perceptual or commonsense) knowledge that marks out the philosophic line of descent from Spinoza to Bachelard and Althusser. If 'the concept "dog" cannot bark,' as Althusser put it, then no more should we assume that the deliverances of sense-certainty are untouched by those forms of ideological illusion that masquerade as self-evident truth.[27] Nevertheless – and this is the second main point to be gleaned from the above passage of de Man – there is no escaping the requirement that scientific knowledge must at some stage always have reference to the conditions of human perceptual or cognitive grasp. Here we move forward, so to speak, from the discourse of seventeenth-century rationalism to Kant's critical philosophy, a move very frequent in de Man's late texts and clearly signalled, in this case, by his use of the term 'synthetic'.[28] For it was, of course, a cardinal tenet of Kantian epistemology – indeed his chief claim to have achieved a 'Copernican revolution' in philosophic thought – that synthetic a priori knowledge was not only possible but absolutely prerequisite as a means of escaping both the rationalist prison-house of abstract ideas and the sceptical dead-end to which empiricism had led in the case of thinkers like Hume. 'Concepts without intuitions are empty; intuitions without concepts are blind.'[29] In so arguing, Kant was concerned with the conditions of possibility for human knowledge and experience in general, and also – by the same token – with those conditions as applied to the natural sciences. At any rate he would have seen nothing objectionable in de Man's formulation of the necessary link between a 'science of the phenomenal world' and science conceived as 'definitional logic, the precondition for a correct axiomatic-deductive, synthetic reasoning' (RT, p. 13).

However, de Man goes on to raise some questions about the adequacy of this model, questions having to do precisely with its 'extension to the outside world'. He puts this in terms of the logic–grammar–rhetoric triad, with rhetoric – or the 'epistemology of tropes' – intervening as a 'decisive but unsettling element' which thereby 'disrupts the inner balance of the model' (RT, p. 14). This point can also be made with reference to the history of science, its relationship to philosophy and the way that certain problematical concepts – perhaps starting out as speculative metaphors – may likewise exert a disruptive effect upon the 'stable cognitive field'. Thus Kant famously – or notoriously – took the Newtonian laws of space, time and motion as his paradigm case of definitive truths arrived at through the joint application of empirical and theoretical research.[30] This has often been paraded – since the advent of relativity theory – as a cautionary instance of misplaced 'scientism' in philosophy, that is to say, of its chronic proneness to erect a wholesale metaphysics, or system of supposed a priori truths, upon the evidence of some (in its own day) state-of-the-art scientific paradigm which then turned out to be false, unwarranted, or only of limited application. Still there is a sense in which Newton's laws continue to hold good for those middle-range aspects of human knowledge and experience – not only in 'everyday life' but in large areas of the natural sciences – where the concern is neither with micro (sub-atomic) nor with macro (astrophysical or cosmological) phenomena. To this extent at least the space–time coordinates of Newtonian physics – along with Kant's treatment of them as conditions of possibility for experience in general – still have a claim to represent accurately the constraints placed upon our knowledge of the world by our existing as physically embodied creatures with given capacities of cognitive grasp.

Thus Newton's 'laws' are not so much falsified as assigned a henceforth more restricted field of application, one whose validity in certain (well-defined) contexts need not conflict with their supension for other, more advanced – and often highly speculative – scientific purposes. In the same way (returning to the example from Pascal) there is still great difficulty in grasping, from the 'commonsense' standpoint, that infinity is not a single and absolute notion – 'that number or quantity than which nothing greater can be conceived' – but, on the contrary, that there exist *different orders* of infinity, such that mathematicians can deploy them not only as speculative instruments but in forms of calculation susceptible of adequate proof. In both instances one is obliged to accept something like a complementarity principle, that is to say, a principle requiring that one respect the data of intuitive self-evidence within their given (phenomenal) limits while acknowledging that thought may legitimately exceed those limits according to alternative criteria of validity and truth.

What varies considerably – from case to case and from time to time – is the extent to which these two sorts of truth-claim get into conflict. Sometimes, as with Spinoza's example of the Sun, it is a straightforward matter of scientific knowledge *correcting* what was always a false impression brought about by an uninformed reliance on the senses as guarantors of veridical knowledge. Elsewhere – as in the case of Newton versus relativity theory – the 'old' view retains a large measure of validity in so far as it corresponds to what is still, for most practical purposes, our best understanding of the physical world as presented to us under the forms and categories of phenomenal cognition. Rather different are cases like the idea of multiple infinities, or much that has transpired in the more advanced (hypothetical) reaches of present-day particle physics, astronomy, or speculation on the origins of the universe. For here there seems to exist a well nigh unbridgeable gulf between what presents itself as intelligible in Kantian terms – by bringing intuitions under adequate concepts – and what is held out as the best present candidate for scientific truth.[31] Indeed, it is fair to conjecture that Kant would have found no place for such hypotheses within the bounds of science, strictly conceived. This is why he devotes the opening section of the *Critique of Pure Reason* to a discussion of the ways in which thinking goes wrong – falls prey to all manner of insoluble paradoxes and antinomies – when it confuses the realm of pure speculative reason with the realm of theoretical understanding. Hence Kant's well-known image of the dove, thinking to slip the surly bonds of earth and fly to a stratospheric height at which its wings will encounter no resistance, whereas in fact flight would be impossible – and its wings quite useless – lacking such resistance.[32] ('Back to the rough ground!' as Wittgenstein more concisely put it.) This metaphor is appropriate for those periods of Kuhnian 'normal' scientific activity which presuppose a more or less *pari passu* rate of development – or mutually supportive exchange – between advances in the speculative (thought-experimental) and the practical or applied domains. But it fails to account for those other 'revolutionary' periods when theory runs ahead of experiment or experiment turns up results unaccounted-for by any existing theory. At such times there is an obvious problem in maintaining the continuity principle, whether in its Kantian form or in the rationalist version which postulates – on de Man's reading – a structural homology between logic, grammar and their 'extension' to a science of the phenomenal world.

However, the question remains as to why it should be 'rhetoric' which here intervenes as a complicating factor, an obstacle to the otherwise smooth passage from theory to observation, or from the pure (a priori) sciences of number, geometric reasoning, etc., to their supposed manifestation in the realm of natural phenomena. After all, it would appear that Pascal's arguments 'move between spatial and numerical dimensions by means of simple computation (as in the instance of the irrational number for the square root of two), or by experimental representations in space, without the intervention of discursive language' (PAP, p. 8). However, this appearance is belied – on a closer reading – by the fact that Pascal has recourse to postulates or figures of thought which his system absolutely requires in order to preserve its coherence, but for which there exists no adequate justification or proof procedure within that system. Thus, for instance, the idea of infinity is constructed through what de Man terms a process of 'synecdochal totalization', a process whereby, beginning from the 'unit of number, the *one*', it is then possible simply to continue the sequence 'N + 1' to the point of a (notional) infinity. And the infinitesimal is arrived at by applying the same procedure, so to speak, in reverse; that is, by assuming a principle of infinite divisibility which takes the zero as its postulated limit or ideal vanishing-point. Nor are these hypotheses advanced as a matter of purely abstract mathematical convenience, divorced from any possible extension to the realms of phenomenal time and space. For the zero has its correlate in both these realms, translating respectively into the 'instant' as a limit-point of temporal reduction and 'stasis' as its equivalent in terms of space and motion. And, of course, Pascal's idea of the infinite has decisive implications for his thinking about cosmological issues. Hence – as is well known – his proto-existentialist feelings of anguish at the thought of an infinitely extended cosmos wherein human creatures, their hopes and aspirations, occupied so tiny (infinitesimal) a place. Such thoughts run close to the 'absurd' in both usages of that term, the mathematical and the more everyday moral-evaluative sense. There is a kindred ambiguity about the word 'irrational', for instance, as applied to those numerical series – like that corresponding to 'the square root of two' – which are apt to generate a giddying sense of human conceptual limitations.

So one can see why de Man makes a point of rejecting that 'tendentious and simplistic' reading of Pascal which separates issues of 'knowledge' from issues of 'faith', or his writings on mathematics and philosophy of science from his writings on the theme of human finitude in face of these ultimate mysteries. However, we should be clear that de Man, unlike most literary commentators, does not give priority to Pascal's existential dilemmas – that is, to the thematic or 'human' content of his work – as opposed to its wrestling with mathematical, scientific and epistemological issues.[33] On the contrary, he argues: we shall do justice to the rigour of Pascal's thought only if we reverse this conventional order of priority, abandon such readily evocative talk of existential 'pathos', finitude, etc., and focus rather on the conditions of (im)possibility for Pascal's attempt to correlate the rigours of a pure (axiomatic-deductive) science of number with a science of the phenomenal world. This attempt was unsuccessful, de Man concludes, in so far as 'the coherence of the system is seen to be entirely dependent on the introduction of an element – the zero and its equivalences in time and motion – that is itself entirely heterogeneous with regard to the system' (PAP, p. 10). And again: '[t]he continuous universe held together by the double wing of the two infinites is interrupted, disrupted *at all points*, by a principle of radical heterogeneity without which it cannot come into being' (ibid.). But if so then its failure is a matter of rigorous undecidability, of aporias that emerge by virtue of Pascal's thinking them through to the limits of that conceptual paradigm, and not by some unfortunate error or lapse of consequential reasoning.

Historically, the 'invention' of the infinitesimal calculus is credited to subsequent thinkers (Leibniz and Newton) who arrived at it by means of more advanced analytical techniques. And it was likewise much later – in the mid-nineteenth century – that mathematicians refined the calculus of infinite numbers to the point where it became possible to think in terms of multiple (greater or lesser) orders of infinity.[34] However, such advances came about through the same process of conceptual elaboration and critique that characterized Pascal's thinking. Moreover, they often turned out to entail the same tension between, on the one hand, this specialized order of mathematical (axiomatic-deductive) proof, and on the other what appeared self-evident to understanding as it sought – in Kantian mode – to bring intuitions (or phenomenal experience) under the rule of adequate concepts. This is essentially de Man's point about the problems that arise, in Pascal's system, when it is a matter of extending such proofs from the realm of pure mathematics to that of applied science. For it is then no longer possible to maintain any rigorous distinction between 'nominal' and 'real' definitions, such that the former would function purely as analytic terms in a self-contained system of axioms, corollaries or deductive proofs, a system whose rigour could only be compromised by the intrusion of elements – e.g., phenomenal or natural-language elements – beyond its powers of definitional or stipulative grasp. At this stage, de Man writes, 'the sign has become a trope, a substitutive relation that has to posit a meaning whose existence cannot be verified, but that confers upon the sign an unavoidable signifying function' (PAP, p. 7). So it is, in Pascal's text, that nominal definitions give way 'almost imperceptibly' to real definitions, which in turn produce a certain tropological swerve, a necessary movement outside and beyond those conditions laid down within the system for adequate demonstrative proof. Thus the term no longer functions strictly 'as a sign or a name', but rather 'as a vector, a directional motion that is manifest only as a turn, since the target towards which it turns remains unknown' (pp. 6–7).

These statements are characteristic of 'late' de Man in the impression they give of raising fundamental issues in a style of extreme elliptical precision which somehow omits – or disdains to make explicit – the most crucial argumentative moves. Nevertheless, I think that they can best be interpreted with reference to what he has to say elsewhere on the question of scientific knowledge and the problematic 'link' between logic, propositional 'grammar', and its extension to the extra-linguistic domain.[35] There are three main points that need stressing here. One is that de Man sees no possibility of maintaining that classical (whether 'rationalist' or 'empiricist') way of ideas that would pass directly from thoughts in the mind to factual (real-world) states of affairs without going by way of 'natural language'.[36] To this extent he endorses the so-called 'linguistic turn', the argument – common to many schools of present-day philosophical thought – which takes it for granted that facts are in some sense linguistic constructs, and that knowledge cannot 'correspond' to those facts in the ideal absence of all discursive or representational structures. On the other hand, language cannot be conceived as existing in a realm of abstraction altogether cut off from those real-world objects, processes and events which provide its only means of referential purchase. Such a notion is merely the idealist – or the transcendental-solipsist – obverse of a naïve realism which thinks to attain a direct, one-to-one correspondence between facts, propositions and veridical beliefs. Neither theory can provide any workable – scientifically or epistemologically adequate – account of how problems arise to complicate this model whenever some existing paradigm ('conceptual scheme', language game, 'final vocabulary' or whatever) comes up against the limits of its own explanatory powers. For such junctures are defined precisely by the *breakdown* of that normative paradigm or assumed continuity principle. This may occur for various reasons, whether through the emergence of internal (i.e., intra-theoretical) anomalies or through the advent of new observational data that require some more or less radical adjustment to the system of inter-related terms and definitions. But in both cases the system is disturbed by the intrusion of an alien ('heterogeneous') element which cannot be reduced to the previous order of well-defined axiomatic proofs.

Such is de Man's argument about the slippage in Pascal from 'nominal' to 'real' definitions.

> These terms (which include the basic topoi of geometrical discourse, such as motion, number, and extension) represent the natural language element that Descartes scornfully rejected from scientific discourse, but which reappear here as the natural light which guarantees the intelligibility of primitive terms despite their undefinability. (PAP, p. 6)

This is also to claim (my second point) that there is not and cannot be any science – whether of number, motion, extension or language itself – which confines itself entirely to the sphere of 'nominal definition' and which thus excludes any dealing with language in its referential aspect. In Pascal's case, to repeat, 'since definition is itself a primitive term . . . the definition of the nominal definition is itself a real, and not a nominal, definition' (p. 7). Even the most rigorous and self-reflexive of sciences will contain such terms which inescapably *refer* to some entity, concept or other (more 'primitive') term. This is necessary, quite simply, in order for that science to be *about* something, that is, to provide it with a subject-matter beyond the realm of empty or circular (tautologous) definitions. Sometimes this referential moment escapes notice since the terms in question may plausibly be viewed as capable of rigorous definition at the intra-systemic level. Thus, for instance, post-structuralists standardly suppose that the referent has no place – no necessary or functional role – within a dual economy (that of the Saussurean 'signifier' and 'signified') which is presumed adequate to account for the workings of language or semiotic systems in general.[37] But in this they are doubly mistaken. For on the one hand, those definitions refer to certain imputed *properties* of language – i.e., its constitutive structures of sound and sense – which may be conceived, in Saussure's oft-cited phrase, as belonging to an order of contrasts and relationships 'without positive terms', but which none the less claim theoretical and descriptive warrant by virtue of their accurately stating what is the case about language under its structural-synchronic aspect. And on the other hand, that warrant is brought into question by the demonstrable *impossibility* that language should be fully or adequately characterized in terms of a bipolar (signifier/signified) model which on principle takes no account of the referent in its real-world (ontic or extra-linguistic) sense.[38]

One need not doubt the validity of this model in so far as it is adopted – as by Saussure – for the specialized purpose of establishing a certain discipline (namely that of structural linguistics) on a well-defined conceptual and methodological basis. However, it is a different matter when that model is extended to other disciplines – like sociology, historiography, anthropology or philosophy of science – where it is taken as sufficient grounds for rejecting any naïve ('positivist') appeal to the referent, the *hors-texte* or anything beyond the field of discursive representation.[39] These ideas – 'abusive extrapolations' from Saussure, as Perry Anderson has well described them – are so widespread at present as to require no extensive documentation here.[40] What they ignore is the third point brought out with great cogency and precision by de Man: that such all-purpose 'rhetorics' (or textualist remodellings) of this or that discipline are quite devoid of critical purpose or content so long as they merely take for granted this *de jure* 'suspension' of the referent. For there can then be no escaping that latter-day (post-structuralist) variant of transcendental solipsism – the prison-house of language, discourse or representation – which results from imposing such a radical disjunction between word and world. What is required, rather, is a detailed analysis of those 'disturbances of the stable cognitive field' that may arise *in particular cases* when rhetoric is construed not simply as that aspect of language pertaining to its suasive or performative function, but in terms of an 'epistemology of tropes' with far-reaching (if often problematical) consequences for our knowledge of real-world objects and events.

We can now return – for one last time – to de Man's reading of Pascal. For it is a *truth* borne out by that reading, as he claims,

> [that] this rupture of the infinitesimal and the homogeneous does not occur on the transcendental level, but on the level of language, in the inability of a theory of language as sign or as name (nominal definition) to ground this homogeneity without having recourse to the signifying function, the real definition, that makes the zero of signification the necessary condition for grounded knowledge. (PAP, p. 10)

I hope to have shown – through this example of de Man on Pascal – how wide of the mark is that domimant view (dominant at least among Anglo-American 'analytic' philosophers) which treats deconstruction as old sophistry writ large, or as merely an exercise in geared-up 'textualist' mystification.[41] Such criticisms are much better directed towards those other (post-structuralist and postmodernist) variants of the 'linguistic turn' which display nothing like the same degree of philosophical acuity and rigour.

III

Again it is to Bachelard that both lines of thinking point back, albeit a Bachelard whose thinking has been submitted to various degrees of selective emphasis or partial reading. On the one hand, his influence extends, via Foucault, to those varieties of cultural-relativist theory which construe 'truth' as a product of the

epistemic will-to-power as it undergoes successive historical mutations with the shift from one order of discourse to the next.[42] However, this completely fails to explain just why such mutations should ever have occurred or how we could ever be in a position to account for their occurrence as anything more than a process of random cultural drift. What remains operative in de Man – especially his essay on Pascal – is Bachelard's theory of advances in knowledge as involving an order of 'epistemological break', a critique of the commonsense (intuitive or phenomenalist) attitude that perceives no problem in the passage from concepts, via natural language, to an understanding of the world. Thus '[w]hat we call ideology is precisely the confusion of linguistic with natural reality, of reference with phenomenalism' (RT, p. 11). This attitude engenders a false sense of epistemic security, a notion – widely held among disciples of Wittgenstein and proponents of 'ordinary language' philosophy – that such problems may be laid to rest by appealing to the stock of communal wisdom enshrined in our everyday habits of linguistic usage and their associated cultural 'forms of life'.[43] For Bachelard, conversely, science bears witness to a genuine progressive growth in our knowledge of the world which could not have occurred – or whose character would remain wholly unintelligible – if indeed there were no possibility of breaking with the order of commonsense wisdom enshrined in natural language.

His examples range from Galileo and the history of early modern science to relativity theory and quantum mechanics. In each case they involve a detailed study of the way that certain forms of intuitive self-evidence – for instance, those purveyed by everyday discourse, by received traditions of scientific thought, or by naturalized habits of perception – are exposed to alternative (more rigorous or adequate) procedures of scientific concept formation.[44] This is not to say that such advances could ever come about in a realm of purely abstract theoretical conjecture quite apart from phenomenal appearances, on the one hand, or from natural language on the other. Hence Bachelard's particular interest in the process of critical 'rectification' by which metaphors are elaborated into concepts, a process whose stages (whose structural 'genealogy' in the Nietzschean sense of that term) may still be read in the texts that comprise the prehistory of any given science. It is here that his approach differs most sharply from prevailing ideas in the other (i.e., Anglo-American) line of philosophical descent. To this latter way of thinking, the primary task for philosophy of science is to provide an adequately generalized account of the logic of scientific discovery, the structure of scientific revolutions, or the law-like regularities perceived to hold between valid observations and adequate covering theories.[45] That is to say, it is concerned largely or exclusively with issues that arise within the present-day 'context of justification', and very little – if at all – with those *epistemological* questions that marked the emergence of a given problematic at a certain point in the history of science.[46] On this view, there is no need to pursue such 'merely' historical interests since they

have no bearing on the primary issue – for science and philosophy of science alike – as to whether the theory or explanatory paradigm in question is adequate to account for the evidence. Epistemological approaches are misconceived in so far as they commit the error of identifying truth in such matters with what presumably went on in the minds of specific (historically located) seekers after truth. Rather, they should argue to the best possible explanation as given by our current ground-rules or protocols for assessing the validity of scientific truth-claims quite apart from their role within this or that original project of enquiry. Thus philosophy of science should properly have nothing to do with the kind of genealogical questing-back into the origins and emergence of scientific theories which at best provides material for an anecdotal history devoid of substantive (philosophico-scientific) content.

Mary Tiles makes this point in her excellent book on Bachelard when she suggests that both traditions – roughly speaking, the 'Continental' and the 'analytic' – have their origin in Descartes, but that they emphasize very different aspects of Descartes's philosophy.[47] Thus the analytic mode tends to privilege those texts (chiefly the *Regulae* and the *Discourse on Method*) where reason is conceived as an orderly process of logico-discursive thought governed by a set of well-defined procedural rules. This makes it ideally independent of the knowing subject – the Cartesian *cogito* – as the locus of a truth attainable only through reflection on its own epistemological resources. In Descartes's *Meditations*, on the other hand, 'it seems that knowledge is only achieved in clear and distinct perception (the deliverance of natural light) which takes place instantaneously, or at least involves no process of reasoning.'[48] It is not hard to see why philosophers of science in the analytic mode have stressed the one dimension of Descartes's thinking – that which lends itself to covering-law theories, formalized proof procedures, validity conditions, methodological protocols, etc. – while discounting his epistemological appeal to the *cogito* as the ground of all genuine (indubitable) knowledge. For this latter introduces a whole range of problems, among them the well-known difficulties with Descartes's claim to have exorcized the demon of sceptical doubt by recourse to a notion of the thinking subject as source and guarantee of its own existence and, beyond that, of such truths about the external world as were self-evident to reason by virtue of this privileged epistemic access. Hence the very marked anti-foundationalist trend in most recent Anglo-American philosophy of science, a tendency that clearly goes along with the desire to secure that discipline against the encroachments of cultural-relativist thinking. For on this view any appeal to epistemological criteria – to a subject-centred ('Cartesian') paradigm of knowledge as arrived at through forms of internalized representation and critique – is tantamount to declaring science just the product of one particular culture-specific activity of thought.

Such arguments find support, as Tiles points out, in the Fregean distinction – so widely influential across just about every school of analytic philosophy –

between *thoughts* (the proper object of scientific and philosophical analysis) and those other, merely 'psychological' processes of *thinking* whose nature eludes any adequate account in truth-functional terms.[49] 'The former yield logical reconstructions of the conceptual and deductive structures of theories, whereas the latter yield empirically descriptive accounts of historical events. Only the former are the legitimate concern of the philosopher.'[50] Of course it may be said – and with some justice – that this split between 'analytic' and 'Continental' traditions is very largely a fiction, a product of selective hindsight or a failure to perceive the large areas of shared concern which tend to be obscured by differences of idiom and localized professional ethos. But on one point at least there is a real divergence. This is the question – going back to that parting of the ways after Descartes – as to how far philosophy, and especially the philosophy of science, should concern itself with epistemological issues in the history of thought, as distinct from those kinds of 'logical construction' that argue to the best (currently available) criteria of scientific proof, observational warrant, theoretical consistency, and so forth. On Bachelard's account these criteria need not be compromised – and can in fact be best maintained – by enquiring back into the various historically specific contexts of discovery wherein scientific concepts emerged from a previous (quasi- or proto-scientific) matrix of metaphor, analogy, ordinary language, or 'commonsense' (phenomenal) intuition. Only thus can the historian-philosopher of science determine with any accuracy what conditions, theoretical and observational, had at length to be fulfilled in order for some vaguely right-seeming hunch – like the ancient atomists' conception of matter – to be transformed into a working hypothesis capable of adequate conceptualization and proof.[51]

Of course it is not the case that *all* such ideas will present themselves as candidates for scientific truth when subjected to these further, more systematic forms of epistemological enquiry. Bachelard makes the point by way of a contrast between Priestley's ill-fated 'phlogiston' theory of combustion and Black's adoption of the 'caloric' hypothesis which subsequently served as a fruitful (if on its own terms limited) basis for research. In the one case, as Tiles puts it, 'The history of phlogiston theories is no longer the history of legitimate science (and therefore in one sense is no longer legitimate history of science) because such theories are judged to be fundamentally erroneous.'[52] That is to say, legitimate 'history of science' cannot – or should not – be merely anecdotal. This would mean covering the field in a catch-all fashion that allows of no distinction, on philosophico-scientific grounds, between ideas whose validity has been borne out by later advances in thought, and ideas whose erstwhile semblance of truth can now perhaps be better understood in other – e.g., psychological or socio-historical – terms. With these latter (theories of the phlogiston type), 'the only interest which the epistemologist can have . . . is in finding in them evidence of the kind of thought which presents epistemological obstacles.' By contrast,

> even though caloric no longer appears in theories of heat, the notion of specific heat to which it gave rise in Black's work is now a firmly entrenched scientific concept. This means that Black's work will appear as part of the approved history of science, as part of that history which is necessary to an understanding of contemporary concepts.[53]

This example brings out the crucial difference between Bachelard's approach to philosophy of science and those other (broadly 'analytic') conceptions which find little value in historical enquiry except where it bears upon theories or conjectures which are treated as subsumable under present ideas of what counts as adequate scientific method or practice. It is the same sort of 'rational reconstruction' which typifies those histories of philosophy – Bertrand Russell's conspicuous among them – whose habit is to treat previous claimants *either* as contributing (more or less directly) to the current state of the art *or* as salutary instances of what went wrong for lack of such resources.[54] On this view there is no legitimate (philosophically respectable) role for an approach, like Bachelard's, which seeks to understand the modes of thought – the metaphors, analogies, enabling suppositions, partial or approximative modes of conceptualization, etc. – which constitute the prehistory of modern science. What counts is not the process that thinking went through in order to achieve some decisive 'epistemological break', but the extent to which that break – if and when achieved – offers a conceptual point of purchase for present-day understanding.

For Bachelard, such arguments ignore the crucial point: that science is a cumulative process of knowledge formation whose history plays an active and continuing role in shaping our conceptions of adequate ('legitimate') scientific method, theory and practice. It is not enough to apply those conceptions as if they had somehow emerged fully-formed through a process of reasoning ideally exempt from the formative (that is, both enabling and limiting) conditions of scientific enquiry. This covering-law approach – or technique of rational reconstruction – is no more adequate when adopted in philosophy of science than in other fields like social, political or cultural history. To be sure, those disciplines don't involve the same kinds of knowledge-constitutive interest, that is to say, the requirement that a historically oriented philosophy of science should account not only for the temporal sequence of significant episodes to date but also for our knowledge of the growth of knowledge measured by the best available criteria of scientific method and truth. Clearly there is a difference – one that has received much attention from philosophers in the post-Dilthey hermeneutic tradition – between, on the one hand, those modes of reconstructive explanatory argument appropriate to the history and philosophy of the natural sciences and, on the other, those enquiries in the humanistic disciplines that aim towards a deeper understanding of the reasons, motives, meanings or intentions that constitute the 'horizon of intelligibility' for intersubjective dialogue.[55] To collapse this distinction is to hold, like proponents of the 'strong programme' in sociology of knowledge, that scientific truth *just is* whatever it is taken to be according to the interests that happen to prevail within this or that socially-defined 'expert' community.[56] And from here it is no great distance to that modish form of cultural relativism – or all-out epistemological scepticism – which simply ignores the crucial difference between 'context of discovery' and 'context of justification'.

Whence the idea, canvassed by neo-pragmatists like Rorty or 'anarchist' philosophers of science like Feyerabend, that there is simply no distinguishing – in point of validity or truth – between what scientists are pleased to think of as advances in their various disciplines and those occasional switches of cultural paradigm, of metaphor or language game that characterize the arts and humanities.[57] But this is not at all – and the point needs stressing – what Bachelard has in mind when he rejects the covering-law model and argues that philosophy and history of science need to take account of those contingent factors (historical, social, cultural-linguistic) which bear upon the process of scientific knowledge production. Rather, he is insisting that we cannot understand just *how and why* such advances come about – such decisive breaks with the currency of in-place naturalized thought and perception – except by way of a genealogical treatment which attempts so far as possible to reconstruct that process in terms of the various obstacles faced, and the various solutions proposed, by historically situated thinkers.

No doubt it is the case that these obstacles and solutions can be shown to have taken rise within a certain (culture-specific) phase in the history of science. But it is wrong to conclude from this that any resultant hypotheses, theories, observations, truth-claims, etc., are likewise just products of their own time and place, possessing no validity beyond what accrues to them by virtue of the cultural prestige attached to certain ideas at certain historical junctures. For there is still a decisive difference – as with the phlogiston/caloric example – between theories that are no longer a part of the 'legitimate' (philosophically accountable) history of science and those which may have been improved upon or refined in certain respects, but which still play a definite (if limited) role in present-day scientific thought. And we can take stock of that difference, so Bachelard maintains, only if we supplement the covering-law account of what qualifies as an adequate conceptual or explanatory paradigm with a detailed reconstruction of the thought processes – the specific solutions to specific obstacles – that characterized the original context of discovery. For we shall otherwise lack any real understanding of the way that scientific knowledge advances through forms of conceptualization and critique, or by 'rectifying' inadequate ideas (e.g., loosely applied metaphors, animistic residues, elements of naturalized 'commonsense' perception and so forth) which had previously acted as a hindrance to further progress.

IV

This helps to explain the chronic oscillation, in recent philosophy of science, between an approach that counts no previous theory valid – or even of 'legitimate' philosophico-historical interest – if it cannot be made to square with current ideas through some technique of rational reconstruction, and at the opposite extreme those cultural-relativist doctrines which recommend that we junk such obsolete notions ('truth', 'reason', 'method', etc.) and treat all science, present-day science included, as just one voice in the ongoing cultural conversation. Neither position will appear in the least degree plausible if one takes it, like Bachelard, that history and philosophy of science should both be concerned with discovery procedures whose adequacy – now as in the past – can be assessed only by the joint application of procedural (covering-law) and epistemological criteria. For it then becomes clear that the grounds of knowledge – whether in the original context of discovery or in our subsequent attempt to reconstruct that context – must always involve something more than the following of well-defined (rule-governed) protocols of scientific method. These would not be sufficient to account for the activity of science even in its phases of Kuhnian 'normal' (i.e., routine, conventional or consensus-based) activity, let alone during its periods of 'revolutionary' upheaval. What is required also is a detailed knowledge of the various factors – perceptual, observational, theoretical, technological and so forth – which came together under certain historical conditions to produce some particular, more or less adequate mode of scientific understanding. Such is at any rate the claim of Bachelard's critical epistemology, and it is a claim borne out – as I have argued here – by the best evidence to hand.

Mary Tiles raises this issue again with reference to Descartes and his twofold (apparently contradictory) requirement: that thinking proceed both by *rules* – by computation, formalized decision procedures, structures of valid inference, etc., such as reason lays down in advance for its own better guidance – and also by the 'natural light' of a reason whose access to 'clear and distinct ideas' would seem to involve no such process of discursive or methodical self-regulation. However, she suggests, Descartes could better be read

> not as confusedly using two different and incompatible conceptions of reason, but as in fact making the point that it is essential to our nature as finite rational beings that we have to go through discursive reasoning processes in order to acquire knowledge. But going through such processes is not enough; the mere going through the motions, the unreflective observance of rules of procedure (mechanical computation) cannot of itself result in knowledge. The train of thought must be unified, its principle grasped. A precondition of this is that thought not only be ordered, but that there be awareness of the order. It is not enough merely to have arrived at a conclusion (which may be correct); if we are to have any confidence in

> the result we must also have a conception of how we got there. If the emphasis on the reflective awareness of method is excised (as it is in the British empiricists), a quasi-perceptual reason, operating reflectively, can only reflect on and yield reflective awareness of passively received, present mental contents.[58]

This is not to say that Bachelard advocates a return to 'foundationalism' in the strong (nowadays presumptively discredited) sense of that term. His philosophy of science is avowedly 'non-Cartesian' in so far as it rejects any version of the appeal to the *cogito* as a source of self-grounding, indubitable truth. This argument has of course come in for numerous well-aimed criticisms, from Gassendi and the other contemporaries of Descartes who were quick to expose its logical flaws, to the current (linguistic or psychoanalytically oriented) objections which seize upon the rift, the 'split' that opens up within the *cogito* between the 'I' that thinks and speaks (the 'subject of the enunciation') and the 'I' that functions grammatically as the object of that thought (the 'subject of the enounced').[59] Thus Descartes's purported transcendental deduction – along with his derivative argument to the existence of objects, events, other minds, etc. – shows up on this account as a mere linguistic subterfuge, an illicit slide from subjective to objective genitive. Jacques Lacan provides the most succinct (albeit riddling) formulation of this view when he rewrites Descartes's dictum in the form '"*Cogito, ergo sum*" ubi cogito, ibi nonsum' ('Where I think: "I think, therefore I am," that is just where I am not').[60]

For Bachelard, moreover, the history of science provides sufficient counter-instances to refute this idea of the *cogito* as offering an access to truth – a timeless, transcendent, self-validating truth – beyond all the vagaries or limiting conditions of situated human enquiry. To this extent he agrees with sceptics like Quine that there are no a priori guarantees to be had, no ultimate 'laws of thought' (even principles of logic like excluded middle or non-contradiction) so firmly entrenched that their validity might not conceivably be called into doubt by some unforeseen development in scientific theory or practice.[61] Thus Bachelard's thinking is not only non-Cartesian but also (as he repeatedly affirms) non-Euclidean and non-Baconian. That is to say, he rejects any *absolute* constraints upon the past or future development of scientific thought, whether these are taken to consist (after Descartes) in the epistemological grounding of truths self-evident to reason, or (after Euclid) in a system of axiomatic proof procedures conceived *more geometrico*, or yet again (after Bacon) in those rules laid down for conducting experiments, recording observations and arriving at general truths – the very 'laws of nature' – through a process of rational induction. For there are numerous cases, as Bachelard readily concedes, where the historian of science cannot point to any one of these methods (or their joint application) as having brought about some particular advance in knowledge. He is no more inclined than Kuhn – or for that matter Feyerabend – to tidy up the record of scientific progress by viewing it as an orderly succession from stage to stage in a project whose standards of experimental warrant and theoretical consistency are necessarily shared by all competent observers.

On Bachelard's account, it is none the less possible to explain how science makes progress and to hold out against the drift towards forms of cultural-relativist doctrine that would treat all truth-claims as 'socially constructed', or as taking rise within diverse (incommensurable) paradigms. For it remains the case that certain of those claims – such as the phlogiston theory or Aristotle's doctrines of motion and the four elements – have now become a part of the prehistory of scientific thought, rather than continuing to figure (like Newton's account of gravitation) as stages on the path to a larger, more comprehensive theory which still allows of their restricted validity within some specified domain. The fact that we are able to make such distinctions can only be explained, so Bachelard argues, with reference to criteria of trans-paradigm validity and truth which form the basis for all comparisons between past and present modes of conceptualization. Hence his theory of the epistemological break as a real measure of scientific progress rather than a culture-relative shift of Quinean ontologies, Kuhnian paradigms, Wittgensteinian language games, Foucauldian 'discourses', Rortian 'final vocabularies' or whatever. Hence also – as I have argued – the difference of approach that sets deconstruction firmly apart from these and other versions of the current anti-realist or postmodern-pragmatist turn. For there is a closer relationship than might at first appear between Bachelard's work in philosophy of science and the deconstructive (or epistemo-critical) project that de Man undertook in the essay on Pascal and other writings of his last decade.

References

1 Jacques Derrida, 'White Mythology: metaphor in the text of philosophy', in *Margins of Philosophy*, trans Alan Bass (Chicago: University of Chicago Press, 1982), pp. 207–71.

2 For one influential source of such ideas, see Richard Rorty, 'Philosophy as a Kind of Writing: an essay on Derrida', in *Consequences of Pragmatism* (Brighton: Harvester, 1982), pp. 90–109. See also Christopher Norris, 'Philosophy as *Not* Just a "Kind of Writing": Derrida and the claim of reason' and Rorty, 'Two Meanings of "Logocentrism": a reply to Norris', in Reed Way Dasenbrock (ed.), *Re-Drawing the Lines: analytic philosophy, deconstruction, and literary theory* (Minneapolis: University of Minnesota Press, 1989), pp. 189–203 and 204–16. The argument is resumed in Rorty, 'Deconstruction and Circumvention', *Essays on Heidegger and Others* (Cambridge: Cambridge University Press, 1991), pp. 85–106.

3 Derrida, 'White Mythology', p. 241.

4 Ibid., p. 241.

5 Paul de Man, 'The Resistance to Theory', in *The Resistance to Theory* (Minneapolis:

University of Minnesota Press, 1986), pp. 3–20, esp. 18–19. All further references indicated by 'RT' and page number in the text.

6 See also de Man, 'Genesis and Genealogy', 'Rhetoric of Tropes' and 'Rhetoric of Persuasion' (on Nietzsche), in *Allegories of Reading: figural language in Rousseau, Nietzsche, Rilke, and Proust* (New Haven: Yale University Press, 1979), pp. 79–102, 103–18 and 119–31.

7 Thomas S. Kuhn, *The Structure of Scientific Revolutions*, 2nd edn. (Chicago: University of Chicago Press, 1970).

8 See, for instance, Barry Barnes, *About Science* (Oxford: Blackwell, 1985); Harry Collins and Trevor Pinch, *The Golem: what everyone should know about science* (Cambridge: Cambridge University Press, 1993); Paul K. Feyerabend, *Against Method* (London: New Left Books, 1975); Jean-François Lyotard, *The Postmodern Condition: a report on knowledge*, trans. Geoff Bennington and Brian Massumi (Manchester: Manchester University Press, 1984); Richard Rorty, *Objectivity, Relativism, and Truth* (Cambridge: Cambridge University Press, 1991); Steve Woolgar, *Science: the very idea* (London: Tavistock, 1988).

9 See entries under note 8 above; also Paul K. Feyerabend, *Science in a Free Society* (London: New Left Books, 1978); Steve Fuller, *Social Epistemology* (Bloomington, IN: Indiana University Press, 1988) and *Philosophy of Science and its Discontents* (Boulder, CO: Westview Press, 1989); Karin Knorr-Cetina, *The Manufacture of Knowledge: an essay on the constructivist and contextual nature of knowledge* (London: Oxford University Press, 1981); K. Knorr-Cetina and M. Mulkay (eds), *Science Observed* (London: Sage, 1983); Andrew Ross, *Strange Weather: science and technology in the age of limits* (London: Verso, 1991).

10 Paul de Man, 'Pascal's Allegory of Persuasion', in Stephen J. Greenblatt (ed.), *Allegory and Representation* (Baltimore: Johns Hopkins University Press, 1981), pp. 1–25. All further references indicated by 'PAP' and page number in the text.

11 For a fine recent study see Hugh M. Davidson, *Pascal and the Arts of the Mind* (Cambridge: Cambridge University Press, 1993); also A. W. F. Edwards, *Pascal's Arithmetical Triangle* (London: Oxford University Press, 1987); Martin Warner, *Philosophical Finesse: studies in the art of rational persuasion* (Oxford: Clarendon Press, 1989). Most of the texts that de Man discusses – including 'The Mind of the Geometrician' – may be found in *Great Shorter Works of Pascal*, trans. Emilie Caillet (Philadelphia: Westminster Press, 1948). See also the single-volume *Oeuvres complètes*, ed. Louis Lafuma (Paris: Editions de Seuil, 1963).

12 See Richard Rorty, *Consequences of Pragmatism* (Brighton: Harvester, 1982); *Contingency, Irony, and Solidarity* (Cambridge: Cambridge University Press, 1989); *Objectivity, Relativism, and Truth* (Cambridge: Cambridge University Press, 1991).

13 Ludwig Wittgenstein, *Philosophical Investigations*, trans. G. E. M. Anscombe (Oxford: Blackwell, 1953); *Remarks on Frazer's Golden Bough*, ed. Rush Rhees, trans. A. C. Miles (Nottingham: Brynmill Press, 1979); *Culture and Value*, ed. Rhees (Oxford: Blackwell, 1980). See also Peter Winch, *The Idea of a Social Science and its Relation to Philosophy* (London: Routledge and Kegan Paul, 1958); Norman Malcolm, *Wittgenstein: a religious point of view?*, ed. Peter Winch (London: Macmillan, 1993).

14 See Jan Miel, *Pascal and Theology* (Baltimore: Johns Hopkins University Press, 1969).

15 Feyerabend, *Against Method.*

16 Immanuel Kant, *Critique of Pure Reason*, trans. N. Kemp Smith (London: Macmillan, 1964).
17 Gottlob Frege, 'On Sense and Reference', in Max Black and P.T. Geach (eds), *Translations from the Philosophical Writings of Gottlob Frege* (Oxford: Blackwell, 1952), pp. 56–78; Saul Kripke, *Naming and Necessity* (Blackwell, 1980); W. V. Quine, 'Two Dogmas of Empiricism', in *From a Logical Point of View* (Cambridge, MA: Harvard University Press, 1961).
18 Kant, *Critique of Pure Reason.*
19 Feyerabend, *Against Method*; also *Science in a Free Society* (London: New Left Books, 1978).
20 See, for instance, Gaston Bachelard, *La Formation de l'esprit scientifique* (Paris: Corti, 1938); *L'Actualité de l'histoire des sciences* (Paris: Palais de la Découverte, 1951); *L'Activité rationaliste de la physique contemporaine* (Paris: Presses Universitaire de France, 1951); *The Philosophy of No: a philosophy of the new scientific mind* (New York: Orion Press, 1968); *The New Scientific Spirit* (Boston: Beacon Press, 1984).
21 See notes 8 and 9 above.
22 See notes 2 and 12 above.
23 Derrida, 'White Mythology'.
24 See note 21 above.
25 Spinoza, *On the Improvement of Understanding*, in *The Chief Works of Benedict de Spinoza*, ed. and trans. R. H. M. Elwes, vol. II (New York: Dover, 1951), pp. 3–41.
26 For an excellent comparative account of Bachelard's thinking in the wider context of twentieth-century philosophy of science, see Mary Tiles, *Bachelard: science and objectivity* (Cambridge: Cambridge University Press, 1984).
27 See Louis Althusser, *For Marx*, trans. Ben Brewster (London: Allen Lane, 1969); *'Philosophy and the Spontaneous Philosophy of the Scientists' and other essays*, ed. Gregory P. Elliott (London: Verso, 1990).
28 See, for instance de Man, *The Resistance to Theory; The Rhetoric of Romanticism* (New York: Columbia University Press, 1984); 'Phenomenality and Materiality in Kant', in Gary Shapiro and Alan Sica (eds), *Hermeneutics: questions and prospects* (Amherst: University of Massachusetts Press, 1984), pp. 121–44; 'The epistemology of metaphor', *Critical Inquiry*, 5/1 (1978), pp. 13–30; also 'Hegel on the Sublime', in Mark Krupnick (ed.), *Displacement: Derrida and after* (Bloomington: Indiana University Press, 1983), pp. 139–53.
29 See Kant, 'Transcendental Aesthetic', in *Critique of Pure Reason*, pp. 67–91.
30 See Gordon G. Brittan, *Kant's Theory of Science* (Princeton, NJ: Princeton University Press, 1978) and Michael Friedman, *Kant and the Exact Sciences* (Hemel Hempstead: Harvester, 1992).
31 For a detailed historico-philosophical account of this break with intuitively grounded conceptions of mathematics, geometry, logic, epistemology and philosophy of science, see J. Alberto Coffa, *The Semantic Tradition from Kant to Carnap: to the Vienna station* (Cambridge: Cambridge University Press, 1991).
32 Kant, *Critique of Pure Reason.*
33 See, for instance, Robert J. Nelson, *Pascal: adversary and advocate* (Cambridge, MA: Harvard University Press, 1981).
34 William Asprey and Philip Kitcher (eds), *History and Philosophy of Modern Mathemat-*

ics (Minneapolis: University of Minnesota Press, 1988).

35 See especially de Man, *The Resistance to Theory.*

36 See Ian Hacking, *Why Does Language Matter to Philosophy?* (Cambridge: Cambridge University Press, 1975).

37 See Ferdinand de Saussure, *Course in General Linguistics*, trans. Wade Baskin (London: Fontana, 1974); D. Attridge, G. Bennington and R. Young (eds), *Post-Structuralism and the Question of History* (Cambridge: Cambridge University Press, 1987); J. V. Harari (ed.), *Textual Strategies: perspectives in post-structuralist criticism* (London: Methuen, 1980); R. Harland, *Superstructuralism* (London: Methuen, 1987); Robert Young (ed.), *Untying the Text: a post-structuralist reader* (London: Routledge and Kegan Paul, 1981); also – for a shrewd and philosophically informed critique of these ideas – Ora Avni, *The Resistance of Reference: linguistics, philosophy and the literary text* (Baltimore: Johns Hopkins University Press, 1990).

38 See, for instance, Mark Platts (ed.), *Reference, Truth and Reality: essays on the philosophy of language* (London: Routledge and Kegan Paul, 1980).

39 See entries under note 38 above.

40 Perry Anderson, *In the Tracks of Historical Materialism* (London: Verso, 1983).

41 See, for instance, John M. Ellis, *Against Deconstruction* (Princeton, NJ: Princeton University Press, 1989); also Christopher Norris, 'Limited Think: how not to read Derrida', in *What's Wrong with Postmodernism* (Hemel Hempstead: Harvester-Wheatsheaf, 1990), pp. 134–63 and 'Of an Apoplectic Tone Recently Adopted in Philosophy', in *Reclaiming Truth: contribution to a critique of cultural relativism* (London: Lawrence and Wishart, 1996), pp. 222–53.

42 See note 21 above; also Michel Foucault, *The Order of Things: an archaeology of the human sciences* (London: Tavistock, 1970) and *The Archaeology of Knowledge*, trans. A. M. Sheridan Smith (Tavistock, 1972); Dominique Lecourt, *Marxism and Epistemology: Bachelard, Canguilhem and Foucault* (London: New Left Books, 1975).

43 See notes 13 and 14 above.

44 See Tiles, *Bachelard: science and objectivity.*

45 See, for instance, Richard B. Braithwaite, *Scientific Explanation* (Cambridge: Cambridge University Press, 1953); Carl G. Hempel, *Aspects of Scientific Explanation* (New York: Macmillan, 1965); Hans Reichenbach, *Experience and Prediction* (Chicago: University of Chicago Press, 1938).

46 On this 'two contexts' principle, see Reichenbach, *Experience and Prediction*; also C. G. Hempel, *Aspects of Scientific Explanation.*

47 Tiles, *Bachelard.*

48 Ibid., p. 28.

49 See note 18 above.

50 Tiles, *Bachelard*, p. 5.

51 See also J. Perrin, *Atoms*, trans. D. L. Hammick (New York: Van Nostrand, 1923).

52 Tiles, *Bachelard*, p. 14.

53 Ibid., p. 14.

54 Bertrand Russell, *A History of Western Philosophy* (London: George Allen and Unwin, 1945).

55 W. W. Dilthey, *Selected Writings*, ed. H. P. Rickman (Cambridge: Cambridge

University Press, 1976); Dilthey, *Introduction to the Human Sciences*, trans. R. J. Betanzos (Detroit: Wayne State University Press, 1988); H.-G. Gadamer, *Truth and Method*, 2nd ed, rev., trans. J. Weinsheimer and D. G. Marshall (New York: Continuum, 1989); Richard J. Bernstein, *Beyond Objectivism and Relativism: science, hermeneutics, and praxis* (Philadelphia: University of Pennsylvania Press, 1983); R.E. Palmer, *Hermeneutics: interpretation theory in Schleiermacher, Dilthey, Heidegger, and Gadamer* (Evanston, IL: Northwestern University Press, 1969).

56 See entries under note 9 above.

57 See notes 2 and 12 above.

58 Tiles, *Bachelard*, p. 29.

59 See, for instance, Emile Benveniste, *Problems in General Linguistics*, trans. Mary E. Meek (Miami: Miami University Press, 1971); Jacques Lacan, *Ecrits: a selection*, trans. A. Sheridan Smith (London: Tavistock, 1977).

60 Lacan, 'The Agency of the Letter in the Unconscious, or Reason Since Freud', in *Ecrits*, pp. 146–78.

61 Quine, 'Two Dogmas of Empiricism'.

3

Ontological Relativity and Meaning Variance: A Critical-Constructive Review

I

Meaning variance over time, across cultures, and between languages or contexts of enquiry is among the hardest-worked topoi in relativist approaches to philosophy of science. Indeed, it has become the central issue for those who would defend or reject the sorts of claim that are nowadays advanced by an otherwise diverse company of anti-realists, neo-pragmatists, proponents of full-scale ontological relativity and 'strong' sociologists of knowledge.[1] Thus the point is often made that a concept like 'mass' has undergone so many shifts of meaning – from Aristotle to Newton and thence to Einstein – that it is impossible simply to compare or contrast those theories on the basis of 'mass' as a well-defined term with differing descriptive or explanatory roles.[2] In which case we are constrained to acknowledge (1) that observation statements are always couched in such context-relative terms; (2) that these statements are 'theory-laden' in the sense of invoking some presupposed ontology, paradigm, conceptual scheme, etc.; (3) that we must therefore adopt a holistic frame of reference where concepts (or theories) are 'underdetermined' by the evidence; and (4) that a shift in what Quine calls the 'boundary conditions' – the interface between theory and experience – may at length require some revision of our terms, theories, and even (at the limit) our most basic ontological commitments.[3]

These ideas have undoubtedly gained strength from the kinds of conceptual problem thrown up by recent (post-1900) developments in relativity theory and quantum mechanics.[4] However, this does not mean that we are bereft of resources for explaining those anomalies (or limiting conditions) that gave rise to difficulties with the older paradigm, and which thus prompted the quest for more adequate – theoretically sophisticated – models. One alternative is the critical epistemology developed by Gaston Bachelard precisely in order to account for such large-scale paradigm shifts without retreating to a cultural-relativist stance or abandoning the very idea of scientific progress.[5] As Mary Tiles puts it:

> Because theory and experiment are already inextricably linked in the Newtonian framework (theory structures experimental practice), subsequent theoretical development is constrained not merely by having to contain an account of past theory, but also by the body of experiment shaped by that theory. In giving an account of the limitations of Newtonian theory, that experience has to be reinterpreted, but its existence cannot be denied. The development, therefore, to be counted as genuine progress has to be shown (by mathematical methods) to be an advance in the pure theoretical (rational) dimension – an advance in the direction of unification and simplification of basic concepts – and also has to be shown to be empirically more precise, detailed and accurate in its predictions. The Newtonian framework has to be shown to be adequate up to a certain level of experimental accuracy and under a limited range of conditions (which must include those normal for terrestrial experience).[6]

I have cited this passage at length because it offers an admirably clear-headed statement of the case against those forms of wholesale ontological relativism (and epistemological scepticism) that currently enjoy wide acceptance. Such doctrines start out from the valid observation that terms like 'mass', 'gravity', 'substance', 'matter', 'element' and so forth have played different roles – and thus acquired different meanings – down through the history of scientific thought. But from this they derive what Bachelard shows to be a threefold set of erroneous conclusions. First, they assume that the 'theory-laden' character of scientific discourse is such as to preclude all meaningful comparison between theories in point of their scientific rigour, consistency or experimental warrant. Second, they suppose that any talk of 'progress', 'rationality', 'precision', 'experimental accuracy', etc., must itself be interpreted with reference to some particular culture-specific paradigm, one that disposes of its own immanent criteria for meaning and truth, and which therefore cannot be applied to other (presumptively less adequate) paradigms. Hence (third) the proposal that philosophy of science give up its attachment to modes of reasoning – covering-law theories, rational reconstructions, hypothetico-deductive arguments, inferences to the best explanation and so forth – all of which manifest this same myopic tendency to equate 'truth' and 'progress' with whatever counts as such by their own present-day cultural lights.[7]

Up to a point Bachelard is willing to concede the force of these arguments. Certainly he rejects any version of the claim that scientific theories can be analysed out into observation statements directly based on the data of empirical research and those forms of inferential (inductive or hypothetical-deductive) reasoning that bring such statements under some suitably generalized covering-law principle.[8] In this respect at least he agrees with Quine: that logical empiricism fails to explain how observation could ever take place – or achieve articulate form – in the absence of some given 'ontological scheme', some pre-existing theory of what should count as an adequate (scientifically admissible) observation statement. Hence Quine's celebrated attack, in his essay 'Two Dogmas of Empiricism', on what he sees as the various residual forms of Kant's distinction between synthetic and analytic statements, or those that involve some element of empirical knowledge and those whose predicate is 'contained in' their subject, and which thus have the character of purely logical (tautologous and empirically vacuous) truths.[9] Where Bachelard nevertheless parts company with Quine is on the issue of 'ontological relativity', construed very often as a flat denial that we can ever talk of 'progress' – of advances in scientific theory or practice – except by ignoring this ineluctable fact of the non-commensurability between rival paradigms, conceptual schemes or whatever. For even if it is the case that observations (and observation statements) are always in this sense 'theory-laden', still there are instances – like the concept of 'mass' in its various usages from Aristotle to the present – where progress consists precisely in our being able to explain *both* where the older theories went wrong *and* where the current state of knowledge offers a more exact, more rigorous and adequate means of conceptualization. That is to say, the term 'mass' may indeed be assigned different meanings in different theoretical contexts. Yet it still remains the focus of an ongoing process of enquiry which provides sufficient criteria to individuate 'mass' as a shared (trans-paradigm) topic of investigation, and also to show – no doubt with benefit of hindsight – where problems arose with earlier theories.[10] The same applies to Aristotle's theory of the elements. On this account there were four such basic substances (earth, air, fire and water), none of which was ever encountered in its pure, i.e., truly 'elemental' form, but all of which entered – in varying proportions – into the physical constitution of different objects in the world. The trouble with this theory, as Bachelard points out, is that it offers no means of verification (or falsification) since its claim is perfectly compatible with *any* result that might be turned up through experiment or observation. In short, it is simply not a candidate for scientific truth, if by 'science' is meant the testing of well-defined hypotheses against the best available evidence, and by 'truth' the property of surviving such tests and (moreover) of accounting for that evidence in adequate causal-explanatory terms.

This is not to deny that subsequent work in the field – from Boyle to Newton, from Priestley and Lavoisier to Mendeleyev's periodic table – has often involved a large measure of divergence as to what should count as an 'adequate' theoretical framework. Nor is it blind to the fact, most often adduced with regard to modern (post-Mendeleyev) developments, that science may sometimes proceed through forms of purely hypothetical conjecture – like the positing of as yet undetected or undiscovered elements – which answer more to theoretical needs (of symmetry, elegance, taxonomic completeness, etc.) than to anything currently possible in the way of straightforward empirical observation. Thus, as Tiles points out,

> the full list of elements is not dictated by naturally occurring substances whose initial identification is empirical. Elements such as Rhenium, Francium and

> Technitium were discovered, or created, because according to the theory they should have existed. Their empirical discovery/creation marks the realisation of theoretical concepts. This means that from the outset the criteria for their identification were theoretical; the empirical criteria used were derived from the theory.[11]

Modern science offers many such examples of theory combining with new-found technological resources to produce some advance – some 'discovery/creation' – that could not have come about through any process of empirical observation or of inference from given (i.e., naturally occurring) objects, substances or events. One case often cited is that of recombinant DNA techniques whose advent raises not only some acute ethical problems but also the question, for philosophers of science, as to where (if anywhere) the line can be drawn between discovery and creation. Another is that of high-energy particle physics in which theory postulates the existence of entities – to begin with, purely hypothetical or virtual entities – which then become the basis for further research and the development of new technologies (accelerators, electron-positron colliders, etc.) whereby to track their passage and effects under controlled laboratory conditions. Clearly there is a sense in which such cases pose a sizeable challenge to any realist ontology which seeks to uphold the distinction between observed (verifiable) phenomena and the theories that supposedly account for them. For here the phenomena can only be conceived as themselves products of theory, brought into being by a joint process of speculative thought and technological advances in the means of knowledge production.

On one view, the only philosophy of science consistent with cases like this is a thoroughgoing instrumentalist conception which holds that truth *just is* whatever science is enabled to discover (or create) by deploying such techniques as and when they become available. This view was espoused by Pierre Duhem and its influence can be seen not only in Bachelard's thinking but also in Quine's arguments against logical empiricism and other such variants on the scheme/content, theory/observation, or analytic/synthetic dualism.[12] For there is, according to Quine, no end to the possible ways in which a theory can be adjusted – or its predicates redistributed – so as to take account of some problem confronted in the course of scientific research. And conversely, there is always the possibility of construing the anomalous data (or observation statements) in a way that resolves any perceived conflict with established theoretical truths. From which it follows – on Quine's holistic account – that no single statement (no single item of theory or evidence) can ever be conclusively verified or falsified. For if the meaning of a statement is given by its truth-conditions, and if these are interwoven with the entire 'fabric' of currently accepted beliefs, then it becomes impossible to fix any limit to the range of possible truth-preserving adjustments. Or again, 'to change the metaphor', as Quine famously puts it,

> total science is like a field of force whose boundary conditions are experience. A conflict with experience at the periphery occasions readjustment in the interior of the field. Truth values have to be redistributed over some of our statements . . . [But] no particular experiences are linked with any particular statements in the interior of the field, except indirectly through considerations of equilibrium affecting the field as a whole.[13]

There is probably no passage in recent philosophy of science that has been so often quoted – or tacitly invoked – as an authorizing source for the widespread turn towards various forms of relativist, pragmatist or anti-foundationalist thinking. What they have made of it has depended very largely on prior convictions as to how far philosophy can claim to adjudicate in matters of scientific method and truth, or indeed how far it continues to exercise any such distinctive role *vis-à-vis* the natural or human sciences.

Thus, for instance, Richard Rorty cites Quine in support of his case that epistemology – including its analytic variants – has now been shown up as a misconceived endeavour, characterized to begin with by a handful of delusive metaphors (like the mind as a 'mirror of nature'), and latterly by its lingering attachment to notions of truth as correspondence, privileged epistemic access, explanatory warrant and so forth.[14] On his view, we should simply abandon such ideas and acknowledge – with Quine among others – that this entire tradition, culmimating in the programme of logical empiricism, has at last run up against insoluble problems in the nature of its own undertaking. That is to say, there exist as many ways of relating observation to theory – of making internal adjustments to the 'web' or the 'fabric' of relations between them – as there exist language games, 'final vocabularies' or cultural 'forms of life'. From which he and others derive the lesson that science (and a fortiori philosophy of science) can do no better than come up with various, more or less productive or adventurous modes of self-description in order to play a continuing if modest role in the 'cultural conversation of mankind'. For if 'truth' drops out – along with notions of 'reality', 'progress' and constructive (problem-solving) endeavour – then we can well afford to be rid of those tedious constraints upon the freedom of science and philosophy to invent new projects in answer to changing social and cultural needs.

Rorty's is of course just one response – an extreme example of the kind – to those issues raised by Quine's attack on the analytic/synthetic distinction, and by kindred arguments (like Wilfrid Sellars's 'myth of the given') which also recommend a holistic approach, one that renounces any appeal to the logical-empiricist idea of matching observational data with theories, concepts or covering-law generalizations.[15] Others are more circumspect, avoiding the upshot of Rorty's claim that truth just is whatever it is taken to be within this or that local community of belief. Mary Hesse puts the case as follows in a passage that effectively paraphrases Quine:

> [N]o feature in the total landscape or functioning of a descriptive predicate is exempt from any modification under pressure from its surroundings. That any empirical law may be abandoned in the face of counterexamples is trite, but it becomes less trite when the functioning of every predicate is found to depend essentially on some laws or other and when it is also the case that any 'correct' situation of application – even that in terms of which the term was originally introduced – may become incorrect in order to preserve a system of laws and other applications.[16]

Hesse is careful, here and elsewhere, to avoid suggesting (in company with out-and-out sceptics like Feyerabend) that truth is pretty much up for grabs, or – what Rorty takes to distinguish a pragmatist from a wholesale relativist position – that the range of admissible truth-claims at any given time is constrained only by consensual norms within some given 'interpretive community'.[17] Still, she follows Quine in viewing ontological relativity as a simply inescapable conclusion, given the three main tenets (incommensurability, holism, and the radical 'underdetermination' of theory by evidence) which she takes to characterize the current situation in philosophy of science. And if this much is conceded then the way is wide open for those other, less guarded or carefully formulated versions of the argument.

Some of them, Rorty's included, even go so far as to enlist Heideggerian depth hermeneutics as a pointer towards the kind of post-epistemological, post-analytic, non-truth-fixated culture that would welcome science – once shorn of its grandiose pretensions – into the mainstream of present-day cultural debate.[18] And this despite all the evidence, in Heidegger's writing, of a fixed antipathy to science that derived in equal part from his viewing it as the upshot of a deep-laid 'metaphysical' will-to-power over man and nature, and from his claim that 'science does not think,' since it is concerned with those merely 'ontic' problems whose solution is always given in advance by the procedures of rational-calculative technique.[19] Or, perhaps more to the point, one may well suspect that the source of Heidegger's appeal is precisely the scope it offers for doctrines that would relativize science (and epistemology) to current ideas of what is good in the way of belief. Hence no doubt the fashion for revisionist, mostly neo-pragmatist readings of Heidegger that play down his depth-ontological talk of Being, *aletheia*, the 'essence' of technology, etc., and emphasize rather the notion of science – along with all our other ways of being-in-the-world – as a practice whose meaning (and hence whose 'truth') entails nothing more than its playing a role in some particular cultural 'form of life'.[20] This also has the advantage that Wittgenstein can be enlisted as yet another salutary influence, one who shows philosophy the way back down from its delusive 'metaphysical' heights.[21] For it then becomes possible to push right through with the relativist argument and to assert that *every* such life-form – including (for instance) the Azande ritual practices reported by Evans-Pritchard and taken up by Wittgensteinian social philosophers like Peter Winch – must be allowed to possess as good a claim to truth ('what works for them') as our own equally contingent or culture-bound ideas of scientific knowledge.[22] Thus, whatever their differences of 'final vocabulary', Heidegger and Wittgenstein can both be made to line up on the side of a pragmatist outlook whose claims – despite Rorty's protestations to the contrary – are open to all the criticisms standardly advanced against relativism in its wholesale (ontological or cultural) varieties.[23]

II

In Quine's case, these statements sit oddly alongside others which profess what amounts to a robust, indeed a commonsense-realist position. On this more sanguine view – as likewise for Hume in his hours of unreflective social leisure – scepticism tends to retreat before the sheer self-evidence that scientific knowledge exists; that it displays a well-attested capacity for growth through various procedures of experiment, criticism and theoretical elaboration; and moreover – 'ontological relativity' notwithstanding – that we can and do dispose of adequate criteria for judging when such progress has occurred. What cannot be maintained is any version of the logical-empiricist claim that we could ever have grounds for asserting the correspondence – or lack of it – between *individual* sentences, statements, predictive hypotheses, etc., and some discrete (observable) state of affairs that would assign them a determinate truth-value. For, according to Quine's holistic conception, it is the entire range of such candidate sentences – of beliefs held true at a given time – which has to confront the sum total of experience at those peripheral points where theory is tested against the evidence.[24] More precisely: since theories are always 'underdetermined' by the evidence, and since the evidence (as captured in observation statements) is always to some extent 'theory-laden', therefore it is impossible to pair off neatly individuated facts and statements in the manner required by logical empiricism.[25] 'Taken collectively, science has its double dependence upon language and experience; but this duality is not significantly traceable into the statements of science taken one by one.' ('Two Dogmas of Empiricism', p. 42) Furthermore, '[t]hat there is such a distinction to be drawn at all is an unempirical dogma of empiricism, a metaphysical article of faith' (p. 37). But Quine sees this as no great problem, either for science – which can still carry on making trade-offs between theory and observation as and when required – or indeed for philosophy of science, once relieved of its pointless ('metaphysical') craving for some theory that would fix their boundary conditions.

Still there is a difficulty, as I argued above, in squaring the two sorts of statement to be found – often in close proximity – throughout Quine's essay. Thus he often adopts an attitude of sturdy pragmatist indifference to such bother-headed scruples, an outlook that leads him to declare, quite simply, that 'science is a continuation of common sense,' since where successful it 'continues the common-sense expedient of swelling ontology to simplify theory' (TDE, p. 45). That is to say, 'ontology' can take up the slack – be adjusted or expanded pretty much to order – just so long as we adopt a sensible (pragmatist) line and don't suppose that some *particular* ontological scheme is capable of cutting nature at the joints, or rendering our statements incorrigibly 'true to the facts'. After all, '[a]ny statement can be held true, come what may, if we make drastic enough adjustments elsewhere in the system' (TDE, p. 43). The only reason why philosophers tend to fret at such claims is that they still hang on to some version of the scheme/content (or logical-empiricist) dualism, some means of preserving the putative 'laws of thought' – at the very least those of non-contradiction and excluded middle – from any possible revision brought about under pressure of empirical counter-evidence. But this effort is in vain if – as Quine avers – that evidence is always already 'theory-laden', and if the logical ground rules are themselves nothing more than relatively fixed procedural conventions that happen to occupy a niche somewhere near the centre of our (currently prevailing) web of beliefs. If we just let this assumption go then it will no longer seem problematic that science should get along, now as in the past, by once in a while making some adjustment to our sense of what counts as an adequate observation statement or a legitimate logical inference. Thus: 'Revision even of the logical law of the excluded middle has been proposed as a means of simplifying quantum mechanics.' Moreover, 'what difference is there between such a shift and the shift whereby Kepler superseded Ptolemy, or Einstein Newton, or Darwin Aristotle?' (TDE, p. 43). In each case, according to Quine, the 'shift' came about through a process of large-scale ontological revision which in effect reconfigured the existing epistemic field, from its observational 'periphery' right down to those supposed a priori constraints that were (and in some quarters still are) held to constitute the simply unrevisable core of valid scientific reasoning.

In his sanguine moments Quine is quite content to take all this as a propaedeutic exercise meant to rid science of its excess 'metaphysical' baggage, thus clearing the way to a straightforward reassertion of the commonsense pragmatist view. However, as I have said, there are statements in his essay – statements more in keeping with its overall drift – which would problematize any such easy 'adjustment' to the conditions of what science (whether normally or exceptionally) claims to achieve. The following passage is a fair enough sample. 'As an empiricist,' he writes,

> I continue to think of the conceptual scheme of science as a tool, ultimately, for predicting future experience in the light of past experience. Physical objects are conceptually imported into the situation as convenient intermediaries – not by definition in terms of experience, but simply as irreducible posits comparable,

> epistemologically, to the gods of Homer. For my part I do, qua lay physicist, believe in physical objects and not in Homer's gods; and I consider it a scientific error to believe otherwise. But in point of epistemological footing the physical objects and the gods differ only in degree and not in kind. Both sorts of entities enter our conception only as cultural posits. (TDE, p. 44)

No doubt this self-denying ordinance – this refusal to prejudge the ontological issue – follows consistently enough from Quine's arguments as outlined above. That is to say, he has to make room for Homer's gods (along with every other past or present 'mythical' entity, from centaurs to phlogiston, or from the Ptolemaic cycles to Newton's conceptions of gravity and space) as so many diverse 'cultural posits' completely on a par, ontologically speaking, with anything that contemporary science has to offer.

Still, it is an odd position to adopt for one, like Quine, who otherwise exhibits a sturdy confidence in science's capacity to get things right. All the more so in that Quine's main efforts as a logician and philosopher of science have focused on precisely those regions of enquiry – at the putative 'centre' and 'periphery' of the web of belief – which he would here have us treat as products of convenience, quite devoid of any grounding in the 'laws of thought' or the nature of physical reality. For 'reality' can show up in as many shapes as there exist ontological schemes – or alternative ways of distributing predicates – whereby to preserve the validity of certain statements *vis-à-vis* the entire range of currently accepted beliefs, theories, observational posits, etc. And logic (mathematical logic included) is in principle no more exempt from revision under pressure of 'recalcitrant' evidence. Thus:

> The abstract entities which are the substance of mathematics – ultimately classes and classes of classes and so on up – are another posit in the same spirit. Epistemologically these are myths on the same footing with physical objects and gods, neither better nor worse except for differences in the degree to which they expedite our dealings with sense experience. (TDE, p. 45)

From which Quine concludes – as conclude he must, given this radically holistic conception of meaning and truth – that '[t]otal science, mathematical and natural and human, is likewise but more extremely underdetermined by experience' (p. 45).

However, there is a real question whether Quine's own theory can accommodate such internal adjustments (or revisions of meaning on a sentence-to-sentence basis) as would enable its various component parts to hang together in a coherent fashion. This strain shows clearly at numerous points in the passages cited above. Thus it is hard to make sense of those repeated appeals to the witness of 'experience', given his thesis that experiential data (as captured in observation statements) are always to some extent theory-laden, and hence on a footing with the whole range of 'cultural posits', from Homer's gods to mathematical classes or brick houses on Elm Street. Nor is the matter much clarified by Quine's assertion that for his part he, 'qua lay physicist', prefers to believe in the existence of classes and houses rather than in Homer's gods; moreover, that he thinks it a 'scientific error' to espouse an ontology where the gods – and not the physical objects – take pride of place. For then it would appear that the choice between them comes down to mere local prejudice or to the fact of his professional interest in positing the one and not the other kind of entity. And the way is thus open for 'strong' sociologists of knowledge to press home their argument that the truth-claims of science amount to *nothing more* than a product of the interests and motives at work within this or that society, communal life-form, or scientific research project.[26]

Not that Quine himself comes anywhere near accepting this grossly reductive view. For there are, as I have remarked, a good many passages in 'Two Dogmas of Empiricism' that take for granted both the existence of logical constraints upon the conduct of scientific thought and the availability of evidential grounds to which those standards properly apply. As he puts it in the essay's closing sentence: 'Each man is given a scientific heritage plus a continuing barrage of sensory stimulation; and the considerations which guide him in warping his scientific heritage to fit his continuing sensory promptings are, where rational, pragmatic' (TDE, p. 46). However, this is desperate talk indeed if one construes it in accordance with Quine's express doctrines of ontological relativity, the underdetermination of theory by evidence, and the impossibility of 'radical translation' between rival (incommensurable) paradigms. For it is hard to understand, on his own account, how those 'sensory promptings' could register as such except under some interpretation or according to a sense of their relative priority – their epistemological salience – when set against the background of beliefs accepted at present. Yet Quine goes so far towards holistically levelling the difference between experiential data and pre-existent theories (or ontological schemes) that there is simply no room for conflicts to develop, or for the emergence of new data that require something more than a process of adjustment at this or that stress-point located somewhere between 'centre' and 'periphery'. In which case there is a problem in explaining how a scientific heritage might be 'warped' (or its predicates redistributed under pressure from 'recalcitrant' items of experience) if these latter can be expressed only in the form of observation statements whose meanings – or whose operative truth-conditions – are determined by their role in preserving equilibrium within the system as a whole. 'Conservatism figures in such choices,' Quine writes, 'and so does the quest for simplicity' (TDE, p. 46). But, if anything, this makes it all the more difficult to conceive how science could ever make progress – let alone undergo periods of Kuhnian revolutionary change – whether prompted by 'experience' (i.e., by anomalous or discrepant experimental data) or by problems encountered in the quest for theoretical consistency, rigour and precision. For on Quine's avowedly 'conservative' account, such problems would always be smoothed away by some process of adjustment aimed at preserving the overall coherence of the system. Any 'fact' can in the end be squared with any 'theory' just so long as we acknowledge their reciprocal underdetermination as required by the thesis of ontological relativity.

That Quine is not altogether happy with this result is evident at several points in his essay where he offers what appears a grudging concession to the opposite (realist or anti-conventionalist) view. 'Some issues do,' he grants, 'seem more a matter of convenient conceptual scheme and others more a question of brute fact.' Thus, for instance,

> the issue over there being [mathematical] classes seems more a question of convenient conceptual scheme; the issue over there being centaurs, or brick houses on Elm Street, seems more a question of fact. But I have been arguing that this difference is only one of degree, and that it turns upon our vaguely pragmatic inclinations to adjust one strand of the fabric of science rather than another in accommodating some particular recalcitrant experience. (TDE, p. 46)

But again one has to ask what grounds there could possibly be for counting any experience 'recalcitrant' – or even 'particular' – given Quine's doctrine of across-the-board ontological relativity. That is to say, such a datum could be particularized only in relation to some ontological scheme whose fitness for the purpose would in turn be a function of its accommodating just that datum with least revision to the structure of in-place (scientific or commonsense) belief. The choice of scheme would come down to a matter of construing the experience so far as possible in line with previous habits of response. The aim would be always to minimize conflict through a redistribution of predicates over the system, whether these fell close to the observational periphery or near the heart of its (presumed) a priori grounding in the logical 'laws of thought'. In which case, such adjustments to the system cannot be explained or justified beyond an appeal to 'vaguely pragmatic inclinations', those that induce us – on the holistic principle of economy – to save appearances, reweave the 'fabric' where needed, and otherwise leave well enough alone. But there is still that question as to what Quine could mean by 'recalcitrant' items of experience if – as he claims – we can always restore equilibrium by making some ontological trade-off betwen truths of fact and truths of reason, or observation statements and statements held true by virtue of their logically necessary (analytic) character. For any resistance at the periphery would then have to come from raw – as yet uninterpreted – sense data at odds with the entire existing set of beliefs, statements, logical axioms, etc. And there is clearly no room, on Quine's holistic account, for such a straightforward appeal to the data as supplied through 'sensory prompting' quite apart from any given ontological or conceptual scheme.

On Quine's view, meaning holism is not so much an option as a necessary outcome once the way has been cleared by abandoning the analytic/synthetic distinction, along with all its attendant 'metaphysical' baggage. This process began with the Frege/Russell argument that terms possessed meaning only in so far as they functioned in the context of a well-formed (truth-bearing) proposition. But difficulties arose, so he argues, when these thinkers treated propositions as the fundamental units of significant discourse, rather than taking the obvious next step, i.e., pursuing this contextualist principle to the point where the meaning (or truth-value) of individual statements could itself be viewed as a function of their role within the entirety of sentences held true at some given time. Thus:

> The idea of defining a symbol in use was, as remarked, an advance over the impossible term-by-term empiricism of Locke and Hume. The statement, rather than the term, came with Bentham to be recognized as the unit accountable to an empiricist critique. But what I am now urging is that even in taking the statement as unit we have drawn our grid too finely. The unit of empirical significance is the whole of science. (TDE, p. 42)

But this doctrine can be seen to lead to some awkward conclusions. It leaves Quine totally unable to explain (1) how science makes progress in certain specific regions of enquiry; (2) how we could ever have knowledge of the growth of knowledge; (3) how any item of 'recalcitrant' data – or anomalous observation statement – could conflict with the body of existing beliefs so as to bring about a definite change in those beliefs; (4) what could count as an adequate reason for revising some well-entrenched theory in response to such evidence; (5), conversely, how it may sometimes happen that theories attain the elegance, the explanatory yield or conceptual power to override the counter-evidence of anomalous experimental findings; and (6) – at the limit – what grounds we can have for rejecting a principle of ontological parity between (say) centaurs, Homer's gods, set-theoretical classes, irrational numbers and brick houses on Elm Street. Not that Quine *really* believes in this principle, disposed as he is – *qua* lay physicist, logician, and sometime advocate of commonsense realism – to range these entities on a scale extending from physical objects via logico-mathematical functions to purely fictitious or imaginary constructs. However, there is no place for such distinctions when he speaks as a proponent of full-scale ontological relativity, or – what amounts to the same thing – of meaning holism pushed to the point where the truth of our statements is simply a matter of their hanging together with the entire 'fabric' of currently accepted beliefs. For in that case they are all 'cultural posits' whose position on the scale can be assigned only in accordance with this or that culture-specific range of beliefs and priorities.

III

This seems to me a striking example of the problems that arise with 'ontological relativity', even when that doctrine is propounded by a thinker (like Quine) who otherwise exhibits a high respect for the achievements of the physical sciences. What results is a chronic divorce between his strong realist commitments – as manifest at various points throughout the essay – and the self-denying ordinance that treats such claims as products of descriptive convention, adopted out of preference for some given 'ontology' that happens to suit our current purposes. The following passage is especially revealing in this regard.

> The salient differences between the positing of physical objects and the positing of irrational numbers are, I think, just two. First, the factor of simplicity is more overwhelming in the case of physical objects than in the numerical case. Second, the positing of physical objects is far more archaic, being indeed coeval, I suspect, with language itself. For language is social and so depends for its development upon intersubjective reference.[27]

All three arguments – from simplicity, archaic origins and the communality of language – are liable to look rather weak if one should ask what possible grounds they can offer for distinguishing the orders of mythic, everyday-commonsense, and scientific knowledge. Simplicity is hardly an adequate criterion to account for those developments – like non-Euclidean geometry, relativity theory and quantum mechanics – which have come about precisely through the advent of complicating factors in a given (hitherto well-ordered or adequately axiomatized) field of thought. Nor is there much philosophical comfort to be had from the argument that our positing of physical objects is more 'archaic' (i.e., reaches further back into the formative prehistory of commonsense wisdom) than the other candidates listed above. After all, it may be said – and in a spirit quite in keeping with Quine's ontological-relativist principle – that centaurs and Homer's gods have at least as strong a claim to exist on those grounds.

The case is yet weaker as regards his appeal to the sociality of language and the fact that communities of speakers *just do* have this habit of referring to physical objects by way of establishing a shared ('intersubjective') realm of experience. No doubt this is true in a Kantian transcendental (or 'conditions-of-possibility') sense, i.e., in so far as such objects provide all the spatio-temporal coordinates by which we are assured of inhabiting a world of intelligible waking reality.[28] But of course Quine has no use for such arguments, involving as they do a dimension of supposed a priori synthetic judgement which cannot be reconciled with his doctrine of ontological relativity, that is, his belief that even the most firmly entrenched 'dogmas' – like Kant's transcendental deduction of the categories of space and time or his requirement that intuitions be 'brought under' adequate concepts – may always be open to revision in the face of 'recalcitrant' experience. All that is left, once again, is the 'vaguely pragmatic inclination' to save appearances, along with the equally vague appeal to language as a social medium whose tendency is to conserve existing (communally sanctioned) habits of usage and reference. However, this ignores the most striking fact about the history of science, namely its capacity – as noted by Bachelard – to achieve 'epistemological breaks' with the currency of in-place (commonsense) belief, and to do so, moreover, precisely by questioning those assumptions that characterize pre-scientific (e.g., mythical, metaphorical or naturalized) habits of thought.[29]

Undoubtedly Quine is well aware of all this as a matter of plain self-evidence, at least when he allows such evidence to count despite (and against) his doctrinal persuasion on the issues of meaning holism and ontological relativity. However, these concessions have an air of unguarded liberality, and are promptly revoked whenever Quine sees fit to reassert his relativist credentials. For on this view, such talk of 'epistemological breaks' would appear just a sign of our subscribing to the myth that science gets things right – or achieves progress – through its ability to offer a more adequate, accurate or truthful descriptive-explanatory account of objects and events in the physical domain. And we are then driven back (so it seems) to some version of the correspondence theory, the idea that scientific knowledge accrues through the process of somehow matching propositions, hypotheses, observation statements, etc., with a realm of objective, real-world 'facts' taken to exist independently of all such discursive representations. At which point the doctrine of meaning holism re-enters Quine's argument, along with the denial that we can ever have grounds – aside from 'vaguely pragmatic inclinations' – for reweaving the total fabric of belief so as to admit some items (and exclude others) as plausible candidates for belief.

So a great deal hangs on the validity of Quine's argument that statements can have meaning (or be assigned some determinate truth-value) only in the widest possible context of beliefs held true at some particular time. In their recent critical survey of the field (*Holism: a shopper's guide*) Jerry Fodor and Ernest LePore conclude that the case is still very much open since no really decisive or convincing argument has yet emerged to settle the issue either way.[30] Nevertheless they maintain – with a good show of evidence – that on balance the doctrine of meaning holism generates more problems than it offers constructive or workable solutions. It seems to me that this verdict is fully borne out by the difficulties encountered with Quine's attempt, in 'Two Dogmas', to save the appearances of scientific method while espousing a form of radical empiricism linked to a doctrine of wholesale ontological relativity. Still, one can see why certain passages from that essay – in particular his metaphor of the 'web of belief' – have so often been cited (directly or obliquely) by cultural theorists, sociologists of knowledge, 'post-analytic' philosophers like Rorty and others with an interest in demoting science, along with philosophy of science, from its erstwhile privileged status.

This not to say that Quine would be at all happy in their company, nor again to suggest that 'Two Dogmas' offers *carte blanche* for the kinds of ultra-relativist or 'strong' sociological approach that acknowledge no difference – in pragmatist terms, no difference that makes any difference – between truth and what is currently 'good in the way of belief'. Nor is it to deny those residual elements of robust commonsense realism that are likely to bring such readers up short if they don't pass rapidly over them in search of more congenial passages. All the same, these latter give a strong handle for anyone wishing to recruit Quine on the side of a thoroughgoing postmodern-pragmatist or cultural-relativist outlook.

Thus Rorty, for one, sees nothing in the least incongruous about placing him – as a 'post-analytic' philosopher – in the same camp as Dewey, Heidegger, the later Wittgenstein and, of course, Rorty himself.[31] What these thinkers have in common, despite their manifest differences, is a desire to talk philosophy down from its delusions of epistemological grandeur, and to point the way forward to its new-found role as just one (strictly non-privileged) voice in the ongoing 'cultural conversation of mankind'. That this involves a highly selective reading of Heidegger – not to mention Quine – is a minor problem since, after all, the main interest is not to get these thinkers 'right' in some old-fashioned (truth-fixated) sense but to see whether they have something useful to say in the present context of debate. So one can cheerfully ignore all that high-flown Heideggerian talk of *Dasein*, Being, 'fundamental ontology' and the like, and coax him back into the pragmatist fold as one who has an interesting story to tell about the various wrong turns in 'Western metaphysics' from Plato and Aristotle to Descartes, Kant, Husserl and beyond.[32] In Quine's case likewise, the best line to take is a strong revisionist approach that pays no heed to those occasional lapses into a posture of 'commonsense' realism, but which pushes all the way with his holistic (i.e., pragmatist) conception of truth as purely and simply what is 'good in the way of belief'.

As I say, there is good reason to suspect that Quine would have no wish to be coopted into Rorty's broad-church alliance. Still less could he relish the coupling with Heidegger, given the latter's view that science 'does not think' (*think*, that is, in the authentic, depth-ontological sense of that word), and his blanket diagnosis of modern technology as a product of the will-to-power within 'Western metaphysics' pursued to its furthest, most inhuman and destructive extreme.[33] As for Wittgenstein, there is little enough in common – except on Rorty's promiscuously levelling view – between Quine's appeal to the sheer self-evidence of scientific knowledge once delivered from those false dualisms of synthetic/analytic, truths of fact *versus* truths of reason, etc., and Wittgenstein's idea of the language game of science as just one cultural 'form of life' among others, quite devoid of any privileged epistemic or explanatory warrant.[34] The same goes for Rorty's other main exemplars of the present-day linguistic, hermeneutic or post-analytic turn. Among them are Dewey as elective precursor of this whole neo-pragmatist movement of thought, and Hans-Georg Gadamer as proponent of an 'edifying' discourse that relativizes truth to the various interpretive 'horizons' whose momentary fusion in the act of understanding is the sole means of access to a knowledge conceived in radically historicist, culture-relative and depth-hermeneutical terms.[35] On the face of it nothing could be more at odds with Quine's express (if problematical) commitment to a physics-led, science-based ontology and an extensionalist semantics with no room for such – as he would regard them – fuzzy subjectivist notions. But the fact that Rorty's recruitment strategy has at least a prima facie plausibility is a clear indication of the way that Quine's holistic or ontological-relativist talk undermines his otherwise strong commitment to science as the paradigm instance of our dealings with a sometimes 'recalcitrant' reality.

The main reason for this, I suggest, is the widely shared view among contemporary thinkers in various disciplines – philosophy of science included – that language (or 'discourse') constitutes the absolute horizon of intelligibility for everything we can possibly claim to know about ourselves and the physical world.[36] Of course this is true in the trivial (self-evident) sense that scientists and others are dependent upon language for propounding hypotheses, writing up their research proposals, describing their experimental results and communicating any knowledge thus gained in a form comprehensible to themselves and other members of the relevant 'interpretive community'. Moreover, there is a certain justification for the stronger claim – whether couched in Wittgensteinian, Foucauldian or Quinean-holistic terms – that the limits of our language are (in some sense) the limits of our world.[37] For, clearly, language is the precondition for extending our knowledge of that world, whatever place such knowledge may be thought to occupy in the 'fabric' extending from empirical observation to the a priori 'laws of thought'. Without language, that is to say, we should possess no means of articulating these or any other distinctions, let alone of pursuing science (and philosophy of science) as an enterprise aimed towards better understanding of those knowledge-constitutive interests. But there is a large – and highly questionable – leap of thought from this point about the ubiquity of language in our dealings with the world to the further (cultural-relativist) argument which views that world, or our knowledge of it, as *nothing more* than a construction out of the various language games, discourses or scientific 'paradigms' that happen to prevail at some particular time and place. For one may grant, in the former (unobjectionable) sense, that without language there could be no possibility of advancing truth-claims, observation statements, explanatory theories and so forth, while *not* going on to draw the extravagant conclusion – with proponents of cultural relativism – that those claims, statements and theories can only make sense according to certain localized (language- and culture-specific) criteria. It is this further extension of the argument that marks out Quine's and other versions of the case for a full-fledged holistic theory of meaning and truth.

IV

Thomas Kuhn's *The Structure of Scientific Revolutions* is probably the text which, after Quine's 'Two Dogmas', has done most to propagate this idea of knowledge as 'relative to' (or 'constructed by') some particular linguistic or cultural framework of belief. Such is at any rate the notion of a Kuhnian 'paradigm' that has entered the wider currency of debate, most often – I shall argue – with untoward consequences for the various disciplines (or pseudo-disciplines) concerned.[38] For one major problem with this term is the ease with it translates, whatever Kuhn's original intentions, into a range of roughly equivalent ideas which exploit its more extreme cultural-relativist implications while ignoring (or dismissing) those aspects of his theory that resist such assimilative treatment. Thus talk of 'paradigms' can readily stand in for Wittgensteinian talk of 'language games', cultural 'forms of life', etc.; for Rorty's idea of 'final vocabularies' as the furthest we can get in justifying scientific or other sorts of truth-claim; and again, for the all-purpose notion of a 'discourse' as it figures in those various archaeologies (or genealogies) of knowledge pursued under the joint aegis of Nietzsche and Foucault.[39] In each case the analogy with Kuhn enters by way of claiming scientific/philosophical warrant for the view that all facts, theories, observation sentences, etc., relate to some language, discourse or cultural paradigm whose criteria determine what shall count as a valid statement at any given time, and whose limits remain necessarily invisible to those who think within them. What these doctrines all have in common is the holistic tendency to relativize truth-values – analytic, a priori, synthetic, observational or those that (supposedly) hold between these various orders of judgement – to an overall 'web' or 'fabric' of belief where no such distinctions can any longer be drawn. And so it comes about (in Quine's words) that 'our statements about the external world face the tribunal of sense experience not individually but only as a corporate body' (TDE, p. 41).

I have remarked already on the curious disjunction between this aspect of Quine's thinking and the approach that he adopts in his text-book treatment of issues in the philosophy of logic.[40] There he insists very firmly – perhaps for pedagogical purposes – on the need to 'regiment' natural language according to the principle of non-contradiction and the basic requirements of first-order quantified predicate logic. Thus, for instance, the interpreter confronted with a native informant who appears to respond in the affirmative to two contradictory statements will be better advised to try out some alternative translation than to suppose that that principle does not hold for the language or belief-system concerned. This argument is stated most clearly in his essay 'Carnap and Logical Truth'.

> Let us suppose it claimed that . . . natives accept as true certain sentences of the form '*p* and not *p*'. Or – not to oversimplify too much – that they accept as true a

> certain heathen sentence of the form '*q* ka bu *q*', the English translation of which has the form '*p* and not *p*'. But now just how good a translation is this, and what may the lexicographer's method have been? If any evidence can count against a lexicographer's adoption of 'and' and 'not' as translations of 'ka' and 'bu', certainly the natives' acceptance of '*q* ka bu *q*' counts overwhelmingly.[41]

From which Quine concludes that 'prelogicality is a myth invented by bad translators,' since on this view there could never be adequate reason for adopting so unlikely a hypothesis. In the case of quantum mechanics likewise, Quine rejects any premature resort to non-standard, deviant or many-valued logics in order to accommodate such anomalous data as quantum nonlocality or superposition.[42] This puts him at odds with those proponents of quantum logic – Reichenbach and Putnam among them – who regard it as a worthwhile or necessary trade-off in order to keep the physics relatively simple.[43] However, it also conflicts with the view that Quine himself expresses in 'Two Dogmas of Empiricism': namely, that 'no statement is immune to revision', since 'revision even of the law of the excluded middle has been proposed as a means of simplifiying quantum mechanics' (TDE, p. 43). Of course it could be said – and is indeed held by anti-realist thinkers such as Dummett – that non-contradiction and excluded middle are principles of different range and scope, so that giving up the former is wholly unacceptable (for reasons well-known since Aristotle) while giving up the latter may well be justified on intuitive, theoretical or observational grounds.[44] But it is clear enough from the context of Quine's statement cited above that he means to place no such stipulative limit on what might be required by way of revision to the ground-rules of logic or the 'laws of thought' as hitherto conceived.

Such is at any rate Quine's view in 'Two Dogmas' and also the position most often identified with Kuhn's account of scientific paradigms, despite those passages in his 1969 postscript that suggest a more qualified (less wholesale-relativist) approach to the issue of paradigm change.[45] For even here Kuhn takes it – very largely on Quine's authority – that observation, experiment, theory and ontology comprise a single field whose 'boundary conditions' are always underdetermined by the evidence, and which thus gives scope for any amount of internal readjustment relative to this or that chosen ontological scheme. On Kuhn's account, this follows from the 'theory-laden' character of observation statements, or the interpretive moment that occurs with the passage from 'stimulus' to 'sensation'. That is to say, his argument presupposes a crucial difference of epistemic status between those incoming sensory data (or stimuli) and the construction that observers are disposed to place upon them through scientific training, cultural background or commitment to some given paradigm. However, the difficulty here – as we saw previously with Quine – is that it leaves him unable to explain how observers (or theorists) could ever have *reasons* (as distinct from mere 'pragmatic inclinations') to shift their allegiance from one to another (more adequate) paradigm under pressure of anomalous experimental data or perceived shortcomings in the theory currently to hand. Moreover, there is the problem, resulting from Kuhn's incommensurability thesis, that nobody could be in a position to judge what this shift really amounted to, since each and every component of the earlier paradigm – from its posited realia, through its observation sentences, to its logical 'laws of thought' – might possibly have undergone some radical change of status or meaning in the passage to a new ontological scheme.

Kuhn addresses the problem in his 1969 postscript by invoking Quine on the topic of 'radical translation' and suggesting that this provides a possible way out of the dilemma imputed by his critics. But there is, as I have suggested, reason to doubt whether any 'solution' along these lines can offer much help once the thesis of ontological relativity has been pushed so far as to deny any recourse to 'standards of enquiry' independent of prevailing habits of belief within this or that interpretive community. 'Since translation, if pursued, allows the participants in a communication breakdown to experience vicariously something of the merits and defects of each other's point of view, it is a potent tool both for persuasion and for conversion' (*The Structure of Scientific Revolutions*, p. 202). But this amounts to no more than a vague faith in the hermeneutic circle of shared understanding, the idea that some saving process of adjustment will occur whereby discrepant beliefs can always be reinterpreted so as to smooth away the differences between them.

Elsewhere – again taking a lead from Quine – Kuhn distinguishes 'stimuli' from 'sensations', the former presumed to obtain for all possible contexts of human observation, while the latter exhibit some degree of cultural, linguistic or paradigm-inflected construal. But this distinction breaks down (as it does in Quine) as soon as the argument gets under way. 'Notice,' Kuhn requests us,

> that two groups, the members of which have systematically different sensations on receipt of the same stimuli, do *in some sense* live in different worlds. We posit the existence of stimuli to explain our perceptions of the world, and we posit their immutability to avoid both individual and social solipsism. About neither posit have I the least reservation. But our world is populated in the first instance not by stimuli but by the objects of our sensations, and these need not be the same, individual to individual or group to group. To the extent, of course, that individuals belong to the same group and thus share education, language, experience, and culture, we have good reason to suppose that their sensations are the same. How else are we to understand the fulness of their communication and the communality of their behavioral responses to their environment? They must see things, process stimuli, in much the same ways. But where the differentiation and specialization of groups begins, we have no similar evidence for the immutability of sensations.

> Mere parochialism, I suspect, makes us suppose that the route from stimuli to sensation is the same for the members of all groups. (*SSR*, p. 193)

I have quoted this passage at length because it brings out very clearly all the problems that arise when a physicalist (stimulus-response) model of what occurs in the act of observation is joined directly to a holist construal of the paradigm – or overall interpretive scheme – within which such acts are taken to acquire meaning, salience, and (so it appears) assignable truth-values. In fact it is very hard to make sense of Kuhn's argument, as can be seen if one presses his point about different observers living 'in different worlds'. Kuhn is well aware of the difficulties with this, since one main purpose of the postscript is to answer his critics on the point about 'individual and social solipsism', i.e., the vexed issue of shared understanding across and between paradigms. His answer is to claim that observers may be in receipt of 'the same stimuli' even though their 'sensations' – that is, the objects or the posited realia which they take as causes of those stimuli – may vary from one observational context to the next. But he then promptly inverts this order of priority by arguing that 'our world is populated *in the first instance* not by stimuli but by the objects of our sensations'; moreover, that these latter 'need not be the same, individual to individual or group to group'.

In which case the incommensurability thesis returns with full force to undermine his commonsense realist claim that we *can* in fact have grounds – pragmatically adequate grounds – for qualifying that thesis to the point where it answers the objections of his critics. For even on Kuhn's concessionary account, those stimuli are construed as cultural 'posits', that is to say, as convenient operative fictions adopted on the one hand to explain how we could ever (as individual observers) have knowledge of the world, and on the other to explain – through their supposed 'immutability' – how such knowledge could be open to testing and revision as applied (nominally) to 'the same' stimuli by participants in an ongoing scientific enterprise. 'About neither posit', Kuhn flatly declares, 'have I the least reservation' (*SSR*, p. 193). The trouble is that his argument leaves no room for this saving assurance, requiring as it does that both sensations *and* stimuli are always already 'theory-laden' (or relativized to the framework of existing beliefs), with the consequence that nothing survives of the appeal to stimuli as a source of invariance conditions taken to hold across otherwise unbridgeable differences of paradigm, language, ontology, world-view, cultural 'form of life', or whatever. All that is left to Kuhn is the weak communitarian thesis that '[since] individuals belong to the same group and thus share education, language, experience, and culture, we have good reason to suppose that their sensations [and, a fortiori, their stimuli] are the same' (p. 193). For on his view, how else should we account for the 'fulness of their communication and the communality of their behavioral responses to their environment?' (ibid.).

However, the problems crowd in once again with this reintroduction of the Quinean appeal to a behaviorist (stimulus-response) model for handling the problem of 'radical translation' and a holistic (cultural-relativist) account of what occurs in the passage from stimuli to sensations, from sensations to observation sentences, and thence to the tribunal of agreed-upon criteria for what qualifies as scientific knowledge. For Kuhn, as indeed for Quine, this process really works in reverse, with the whole of science – the entire inter-articulated 'web of belief' – deciding what internal adjustments are needed in order for those sentences, sensations and stimuli to produce least conflict with other (more entrenched) components of the system. This is still wide open to the objection raised by Kuhn's first-round critics, namely that it offers no adequate account of how 'communication' takes place across paradigm shifts or between research communities, apart from his insistent (but under-argued) claim that '[t]hey *must* see things, process stimuli, in much the same way.' In any case, this could apply only to members of the same community, defined as such by their subscribing to the tenets of some given research paradigm during a period of 'normal' (i.e., consensus-based and predominantly 'puzzle-solving') activity. Beyond that – where conflicts of theory exist or during periods of Kuhnian 'revolutionary' science – there is absolutely no guarantee of shared understanding between adherents to rival ('incommensurable') paradigms. For, of course, this follows from his Quinean premises with regard to 'radical translation', ontological relativity and meaning holism. That is to say, the passage from one to another paradigm might always have involved such deep-laid shifts in the entire configuration of knowledge as to render impossible any term-by-term comparison or any attempt to sustain invariance conditions at the level of sensory data, observation sentences, or even the most 'fundamental' rules of logic.

Kuhn still thinks that his critics can be answered – and the charge of full-fledged relativism averted – by holding the line between 'stimuli' and 'sensations'. On this account, the former would be given directly through perceptual acquaintance, while the latter must be taken as already involving some degree of ontological commitment, some interpretive schema that divides the perceptual field into objects, events, processes and other such (albeit primitive) constructs. Thus, according to Kuhn, it is 'mere parochialism' – or our attachment to some localized 'commonsense' ontology – which makes us suppose that 'the route from stimuli to sensation is the same for all members of all groups' (*SSR*, p. 193). What is needed to secure at least the minimum necessary conditions of trans-paradigm grasp is the fact that different observers can in principle have access to the same kinds of sensory imput, no matter how great their subsequent divergence of views. However, this argument breaks down, on Kuhn's own submission, since those 'stimuli' are themselves items whose existence we 'posit' – as a matter of convenience – in order to 'explain our perceptions of the world', and whose 'immutable' character is likewise a product of our need to avoid 'both individual and social solipsism' (p. 193). In other words, the thesis of ontological relativity cannot be contained within those limits prescribed by the need to make sense of science as a communal truth-seeking enterprise. No doubt Kuhn is right to acknowledge this point – under pressure from the book's early critics – and to seek some way of saving his thesis from its own ultra-relativist upshot. But it is far from clear that his stated beliefs, however emphatically expressed ('about neither posit have I the least reservation'), are adequate to meet those criticisms.

V

This also leads to problems with the Kuhnian distinction between 'normal' and 'revolutionary' phases of scientific thought. For that distinction makes sense only on the premise that one can tell such periods apart by the degree of divergence to be found at some particular time between various conflicting (at the limit 'incommensurable') paradigms. Within 'normal' science these disputes would involve nothing more than the kind of disagreement that could always be resolved by adjusting observation to theory (or vice versa) so as to remove some puzzling anomaly or localized source of trouble. For 'revolutionary' science, conversely, such anomalies would accrue to the point of communicative breakdown where rival paradigms disposed of no common criteria for comparing observations, assessing evidence, or even – as Quine and Kuhn both maintain – for adjudicating issues of logical consistency and truth. Needless to say, it is this aspect of his thought which has captured most attention, whether from anarchist philosophers of science like Feyerabend or from postmodern thinkers (Lyotard among them) who effectively invert Kuhn's order of priority by taking 'revolutionary' science as their norm and asserting the radical heterogeneity of language games, discourses, phrase genres, etc.[46]

It should hardly need saying – at least to those who have read (as distinct from read about) his book – that one cannot blame Kuhn for some of the wilder doctrines propounded in his name. Still, there is a sense in which the book lies open to just this kind of misreading, a prospect that accounts for his marked shift of emphasis in the 1969 postscript. And even there – as I have argued – it is difficult to see how Kuhn can maintain anything like his original thesis while building in refinements, concessions or qualifications which would meet his critics on the main point at issue, i.e., that of ontological relativism and its self-disabling paradoxes. For this would require *first* that he make good his claim for the prevalence of 'normal' (constructive or problem-solving) science; *second*, that this claim be grounded in something more than the vague (Quinean-pragmatist) assurance that 'they [members of 'the same group'] must see things, process stimuli, in much the same way'; and *third* – prerequisite for each of the above – that he hold that line between invariant 'stimuli' and paradigm-specific (or theory-laden) 'sensations'. But as we have seen already, this distinction is upheld only by stipulative warrant, and not by anything in the nature of scientific truth, knowledge or the logic of enquiry as Kuhn standardly conceives them.

Thus, on Kuhn's ontological-relativist account, it might always be 'mere parochialism' which makes us suppose not only that 'the route from stimuli to sensation is the same for the members of all groups,' but also – what he here takes as self-evident – that the reverse route (from sensation to stimulus) provides an adequate hedge against charges of downright anarchism. This problem can be seen most clearly in his moving straight on (in consecutive sentences) from the flat declaration about scientists in a 'normal' paradigm processing stimuli in much the same way to his statement that, 'where the differentiation and specialization of groups begins, we have no evidence for the immutability of sensations' (*SSR*, p. 193). For this begs the obvious question as to how far back that process begins, or whether – even during 'normal' periods – there might not be deep-laid differences of view (of paradigm, ontological scheme or whatever) which affected the interpretation of stimuli and hence the possibility of communicative grasp between different observers. Here, as with Quine, there simply comes a point where ontological relativity is deemed to have an end and where the argument falls back on a crudely behaviourist (stimulus-response) psychology that cannot bear anything like the required weight of theoretical justification.

Kuhn appears sometimes willing to accept this consequence – that is, the underdetermined character of 'stimuli' or 'data' – in the case of 'revolutionary' periods in science or of paradigm switches that involve some far-reaching change in the meaning of even the most basic observational terms. His favourite examples are the switch from Aristotelian to Galilean physics and the issue between Priestley and Lavoisier as regards the chemical process of combustion. In such cases, he writes,

> what occurs during a scientific revolution is not fully reducible to a reinterpretation of individual and stable data. In the first place, the data are not unequivocally stable. A pendulum is not a falling stone, nor is oxygen dephlogistated air. Consequently, the data that scientists collect from these diverse objects are . . . themselves different. More important, the process by which either the individual or the community makes the transition from constrained fall to the pendulum or from dephlogistated air to oxygen is not one that resembles interpretation. How could it do so in the absence of fixed data for the scientist to interpret? Rather than being an interpreter, the scientist who embraces a new paradigm is like the man wearing inverted lenses. Confronting the same constellation of objects as before and knowing that he does so, he nevertheless finds them transformed through and through in many of their details. (*SSR*, pp. 121–2)

This poses the problem that I noted above as to how Kuhn can state (as if the fact were self-evident) that such a scientist 'confronts the same constellation of objects as before', and moreover 'know[s] that he does so'. For as Kuhn here presents it the paradigm shift in each case – from Aristotelian 'constrained fall' to Galilean pendular motion, or from Priestley's 'dephlogistated air' to Lavoisier's account of the role of oxygen in the process of combustion – was such as to induce a wholesale change in the meaning of observational terms, in the theory (or entire ontological scheme) that assigned a descriptive or explanatory role to those terms, and also (inescapably) in the very 'data' that were adduced by way of evidence.

Of course it may be said – and this would seem to be the line that Kuhn adopts in his postcript – that 'data' are more like 'sensations' than 'stimuli', i.e., that one can admit a high degree of relativity (of theory-ladenness or paradigm inflection) in the data while maintaining the invariance of stimulus-response as a hedge against full-scale epistemic relativism. But again, such an argument fails to stand up if taken along with Kuhn's other statements in this context. Thus, '[w]hen Aristotle and Galileo looked at swinging stones, the first saw constrained fall, the second a pendulum.' From which Kuhn derives the (somewhat reassuring) conclusion that 'though the world does not change with a change of paradigm, the scientist afterwards works in a different world' (p. 121). But it seems a somewhat arbitrary line of demarcation that conserves 'swinging stones' as invariant objects of perception while evoking paradigm change (or ontological relativity) in order to explain how an account of those stones as instances of 'constrained fall' gave way to an account of them as instances of pendular motion. For it would surely follow from Kuhn's holistic premises that the verb 'to swing' underwent a decisive semantic shift – along with manifold other related terms – in the passage from one paradigm to the other. Moreover, it is conceivable (or not ruled out by the strong relativist thesis) that even object-words like 'stone' could acquire a quite different meaning – or a different functional-explanatory role – as the result of such a wholesale paradigm change.[47] In which case there is little (if anything) left of the datum – or the sensory stimulus – that supposedly serves as a minimal anchor-point for the comparison and assessment of rival paradigms.

What is at issue here is the question whether Kuhn can be justified, by his own (ontological-relativist) lights, in maintaining on the one hand, sensibly enough, that 'the world does not change with a change of paradigm' and on the other that 'the scientist afterwards works in a different world.' One should note the crucial ambiguity of meaning that allows this latter sentence to slip across from a weaker to a stronger version of the claim as required by various contexts of argument. Thus Kuhn is very often taken to hold – and there are many passages which bear this out – that adherents to rival paradigms must indeed work 'in different worlds' (on the strong interpretation) since those paradigms in each case determine what shall count as a 'world', that is, an ontological domain of objects, processes and events whose identifying features – together with the logic of enquiry best fitted to explain them – allow for no possible term-for-term translation across such differences of view. Even so, he might respond, 'the world does not change,' only our particular conceptions of it or (in Quinean terms) the values we assign to this or that variable as decided by its place in this or that scheme of ontological priorities. Such would be the weak (observer-relativized) version of the thesis and one that falls pretty much square with the commonsense realist view. On this account, as Kuhn puts it, 'Priestley and Lavoisier both saw oxygen, but they interpreted their observations differently; Aristotle and Galileo both saw pendulums, but they differed in their interpretations of what they had both seen' (*SSR*, pp. 120–1). However, this is *not* Kuhn's view of the matter, but a viewpoint he ascribes to the resisting reader whose commonsense convictions may well be affronted by the relativist thesis in its strong (ontological) form. Such a reader 'will surely want to say' – and will surely have intuitive grounds for believing – that 'what changes with a paradigm is only the scientist's interpretation of observations that themselves are fixed once and for all by the nature of the environment and of the perceptual apparatus' (p. 120). But they will none the less be wrong, Kuhn argues, since until the appropriate paradigm was invented 'there were no pendulums, but only swinging stones, for the scientist to see.' And again: 'Pendulums were brought into existence by something very like a paradigm-induced gestalt switch' (ibid.)

It seems to me that Kuhn skews the terms of this debate by casting his fictive interlocutor in the role of an unreconstructed positivist, one who believes that observations are fixed 'once and for all' both by 'the nature of the environment' and by the 'perceptual apparatus' that supposedly affords direct knowledge of that environment. It is the same argumentative move that relativists have often exploited by way of scoring easy points off a typecast naïve opposition. But in this case the move is more subtle, since it allows Kuhn to maintain his position (that such knowledge is paradigm-dependent or relative to the observer's ontological scheme) by avoiding any talk of stimulus-response – which would raise all the difficulties mentioned above – and instead bringing in that notion of a 'perceptual apparatus' naïvely presumed to match up with reality by virtue of its innate constitution. For he can then invoke all manner of evidence – chiefly from Gestalt psychology – to show that such perceptions are always already theory-laden, and hence that Priestley and Lavoisier (or Aristotle and Galileo) were not so much interpreting their data differently as seeing different objects in accordance with divergent (incommensurable) paradigms of knowledge. The great advantage of this, from Kuhn's point of view, is that it doesn't close off but merely brackets the appeal to a yet more basic (stimulus-response) level of perception which the relativist can hold in reserve as a last line of defence against the charge of 'individual or social solipsism'.

Granted, there is an obvious (historically well-attested) sense in which the 'laws of nature' have indeed shown up very differently from time to time and thereby given rise to correspondent changes in what scientists (and others) have standardly viewed as the reliance to be placed upon – or the limits imposed by –

our shared 'perceptual apparatus'.[48] Thus Kuhn can treat it as merely self-evident that, for instance, 'after discovering oxygen, Lavoisier worked in a different world.' For by very definition a change had come about in the 'nature' of his working environment, a change that involved, among other things, the disappearance of phlogiston as a (pseudo)-substance with various supposedly observable properties and effects, and the arrival of oxygen as a 'new' substance with transformative implications for science as a whole. Even so – and despite his ontological-sounding talk of 'different worlds' – Kuhn still wants to think of the scientist after such a radical paradigm shift as 'confronting the same constellation of objects as before' and as 'knowing that he does so' despite finding them 'transformed through and through in many of their details' (*SSR*, p. 122). This trick is brought off by the Quinean expedient of positing a yet more primitive order of stimulus-response – or of 'stimulus-synonymous' observation statements across otherwise divergent schemes – which can then provide the needful (if minimal) measure of trans-paradigm intelligibility.

Hence Kuhn's suggestion that, '[r]ather than being an interpreter, the scientist who embraces a new paradigm is like the man wearing inverted lenses' (p. 122). To call him an *interpreter* would err too much on the side of supposing that there existed a world of invariant data – 'laws of nature' or whatever – which could ultimately serve to fix what should count as truth at the end of enquiry. The inverted-lens metaphor is presumably meant to block such a reading by its suggestion that the scientist's entire perceptual and conceptual 'world' – including his or her sense of that knowledge-constitutive horizon – is altered out of all recognition with the passage to a new paradigm. But it does leave room for a different understanding, one that avoids any commitment to the thesis of wholesale (ontological) relativity. For no matter how drastic its effect upon the contents of her visual field, it would still be the case that the wearer of such a lens inhabited the same world and indeed – as Kuhn says – 'confront[ed] the same constellation of objects as before'. In short, the metaphor works only at an epistemological (and not an ontological) level. Thus it fails altogether if it is meant to make the point that thinkers like Aristotle and Galileo, or Priestley and Lavoisier, were so far apart with respect to their most basic working beliefs that they *really* (in the strong sense) inhabited 'different worlds'. What the metaphor suggests, rather, is that scientists – as well as historians or philosophers of science – may often have to cope with sizeable differences of cultural or paradigm perspective if they wish to discover some common basis for comparing and assessing rival theories.

So it is not just a lapse of consistency or a local imprecision in his argument that leads Kuhn to adopt this ambivalent metaphor. On the contrary: it is part of a larger pattern of likewise ambivalent (sometimes contradictory) statements which allow him to maintain *both* the major thesis of his book, i.e., that adherents to rival paradigms inhabit 'different worlds', *and* the saving clause that one can none the less make sense of their differences by appealing to certain agreed standards of jointly scientific and historical enquiry. Take the following passage – one of many – where the strong relativist thesis is in effect undermined by Kuhn's use of terms which scarcely permit such a reading. His topic, again, is the wholesale 'transformation of vision' that purportedly occurs when some particular finding – in this case Lavoisier's discovery of oxygen – requires far-reaching adjustments elsewhere in the field of scientific knowledge. Lavoisier, he writes,

> saw oxygen where Priestley had seen dephlogistated air and where others had seen nothing at all. In learning to see oxygen however, Lavoisier also had to change his view of many other more familiar substances. He had, for example, to see a compound ore where Priestley and his contemporaries had seen an elementary earth, and there were other such changes besides. At the very least, as a result of discovering oxygen, Lavoisier saw nature differently. And in the absence of some recourse to that hypothetical fixed nature that he 'saw differently', the principle of economy will urge us to say that after discovering oxygen Lavoisier worked in a different world. (*SSR*, p. 118)

The rhetoric of this passage would bear a good deal of analysis, especially as concerns its own (albeit muted) ontological commitments and the tension they generate when drawn into the ambit of Kuhn's relativist argument. What sense can we attach to his talk of Lavoisier's 'discovery', which after all implies that there must have been something *there* to be discovered, not only the 'new' element oxygen but also other substances now found to be compounds and not elements as previously thought? How far is Kuhn willing to grant ontological parity – as his Quinean principles must surely dictate – to Lavoisier's oxygen hypothesis on the one hand and Priestley's 'dephlogistated air' theory on the other? Or again, to the former's view of earth as a 'compound ore' and the latter's idea of it as an elementary (uncompounded) substance? On the face of it the answer to both questions is 'all the way', to the point (that is) where 'discovery' becomes just a matter of switching perspectives, of 'learning to see' things differently or taking up provisional residence 'in a different world'.

Such would at any rate seem to be the upshot when Kuhn invokes the language of Gestalt psychology – along with the Wittgensteinian notion of 'seeing-as' – by way of undermining any last appeal to that naïve hypothesis of a 'fixed nature' that would somehow subsist through all these changes of interpretive viewpoint. In which case the 'principle of economy' requires that we give up thinking like this and acknowledge the impossibility of comparing paradigms in point of their observational accuracy, their explanatory power, theoretical consistency or whatever. For if Priestley and Lavoisier were indeed working in 'different worlds' then there is simply no deciding the issue between them except from our own present-day perspective. And of course this perspective is so deeply marked by the history of modern (post-Lavoisier) chemistry and physics that it cannot provide any comparable understanding of what it meant to inhabit that previous world. The best we can do – like Quine's radical translator – is relativize truth-claims or observation statements to 'the whole of science' at any given time, and assume that those other-world denizens (like us) must have achieved at least a workable trade-off between stimuli, statements, and ontological scheme.

However, this would entail a degree of cognitive scepticism on Kuhn's part which the passage quoted above fails to sustain, and which he certainly abandons when answering his critics in the 1969 postscript. For there could then be no question of Lavoisier's having 'discovered' anything, much less of his having set enquiry on the path towards a better (more adequate) theory of combustion, one that resolved certain problems with the old theory and produced a genuine advance in our scientific knowledge. That Kuhn cannot quite bring himself to let go of such claims is evident not only from his talk of 'discovery' but also in the strong implication – carried by verb phrases of achievement like 'learning to see' – that the oxygen hypothesis did indeed mark a decisive stage of advance, rather than merely a Gestalt switch induced by whatever obscure combination of cultural or circumstantial factors. It is also quite clear that his own ontological commitments incline him more towards Lavoisier's view than towards a belief in 'dephlogistated air'. What is not so clear – here or indeed in the postscript – is whether his theory in fact leaves room for any such saving retreat.

VI

If I have pressed rather hard on these problems with Kuhnian paradigm talk, it is because they point the way to a broader diagnosis of what is wrong with much present-day 'advanced' thinking in philosophy of science and other disciplines. Kuhn himself seems almost to endorse this view in a revealing passage of his book where the methodological question takes on a certain edge of despair and self-doubt.

> Is sensory experience fixed and neutral? Are theories simply man-made interpretations of given data? The epistemological viewpoint that has most often guided Western philosophy for three centuries dictates an immediate and unequivocal, Yes! In the absence of a developed alternative, I find it impossible to relinquish entirely that viewpoint. Yet it no longer functions effectively, and the attempts to make it do so through the introduction of a neutral language of observations now seem to me hopeless. (*SSR*, p. 126)

What I think this passage reveals is not some predicament confronting philosophy of science in general, still less any 'crisis' in the theory or practice of science itself. Rather, it points to particular problems that Kuhn has inherited from a certain (fairly recent) tradition of thought, problems that are rendered more acute by the turn towards language – holistically conceived – as a wished-for means of resolution. They arise from the conjunction of a radical empiricist outlook – one that rejects all appeals to a priori knowledge, truth-claims, validity conditions, etc. – with a thoroughgoing holistic or contextualist approach that distributes truth-values over the entire 'fabric' or 'web' of beliefs held true at any given time. But there is just no way of squaring these two commitments, as emerges in the above (rather plaintive) passage of Kuhn. For it follows from the second line of argument – if fully and consistently applied – that *nothing* is exempt from contextual redefinition, even that (supposedly) bedrock register of sensory stimulus input to which Quine and Kuhn both have occasional resort.

Hence Kuhn's rhetorical questions ('Is sensory experience fixed and neutral? Are theories simply man-made interpretations of given data?'), clearly inviting a negative response yet drawing from Kuhn the frank admission that they represent a viewpoint he is unable entirely to relinquish. Elsewhere, as we have seen, he does attempt to hold the line between presumptively invariant 'stimuli' and paradigm-specific (or ontology-relative) 'sensations'. But this distinction proves untenable on the holist view, since it finds no place for nominal entities like 'sense data' except in so far as they figure in observation sentences which themselves possess meaning – or truth conditions – only by virtue of their playing some role in the overall pattern of beliefs. So one can quite understand Kuhn's rueful verdict that this model 'no longer functions effectively', and that attempts to salvage it by introducing a 'neutral language of observations' now seem to him simply 'hopeless'. Indeed, his whole approach in *The Structure of Scientific Revolutions* seems predestined to arrive at just such a point of well-nigh terminal aporia.

But there is still room for doubt as to whether Kuhn is justified in finding this conclusion forced upon him 'in the absence of any developed alternative'. Such may perhaps have appeared to be the case when *The Structure of Scientific Revolutions* was first published in 1962. For at that time philosophers in the broadly 'analytic' camp, Quine among them, were engaged upon just the kind of large-scale revisionist exercise with regard to prevailing methodological assumptions which Kuhn treats as symptomatic of science in its crisis-prone or 'pre-revolutionary' periods. In part this had to do with the growing sense that developments in recent scientific thought – especially quantum mechanics – had brought about a situation where none of the existing ground rules any longer applied, and where even the most 'elementary' principles of logic – together with notions of an observer-independent 'reality' – might be subject to drastic revision.[49] But it was also the result of problems encountered in other, more familiar regions of philosophy of science, chiefly those concerning the scope and limits of inductive inference, covering-law (deductive-nomological) theories, and the status of causal explanations *vis-à-vis* the logic of scientific enquiry.

These problems have received a good deal of attention from recent commentators, notably Wesley C. Salmon.[50] They include all the well-known puzzles about induction first stated by Hume and revived, in more sharply paradoxical form, by Nelson Goodman and others;[51] the difficulties that arise with any version of the covering-law approach when it comes to establishing the evidential grounds that would count decisively for (or against) some given candidate law;[52] the similar problem with compatibilist arguments, since every item of scientific fact or observation is 'compatible with' a limitless range of causally unrelated (hence explanatorily vacuous) theories and hypotheses;[53] and the kindred objection to Popperian (fallibilist) approaches, namely that in order to falsify a hypothesis one needs to prove it invalid, at which point one encounters all the same problems with regard to evidence, inductive warrant, 'truth to the facts' and so forth.[54] As I say, these problems have been well documented in recent (post-1960) writings on philosophy of science, and they help to explain why Kuhn should have found it so difficult to envisage an alternative way forward.

Nevertheless such alternatives did exist, both in 1962 when his book was first published and – even more – when the postscript was appended just seven years later. In fact, one could argue that they have existed since a time very near the emergence of science and philosophy of science as disciplines recognizably akin to what is nowadays practised in their name. That is to say, there is strong Aristotelian precedent for the view that any adequate account of scientific enquiry must entail both a causal and a logical component. On the one hand, it will seek to explain natural phenomena with reference to their intrinsic properties, their constitutive natures, elemental composition, characteristic behaviour under given conditions, etc. On the other, it will provide a detailed analysis of the justificatory arguments – the various interlinked orders of inductive or deductive reasoning – whose validity is a matter of their inferential warrant and causal-explanatory power.[55] Such is I think the 'developed alternative' which Kuhn rejects out of hand and whose 'absence' he views – not without regret – as the simply inevitable upshot of recent developments in science and thinking about science.

From thirty years on that judgement must appear, to say the least, somewhat premature. It can now be seen as the outcome of a short-term 'crisis' in philosophy of science, a crisis brought about by the convergence of various doctrines – ontological relativity, meaning holism, the linguistic or hermeneutic 'turn' – whose advent was not so much a matter of logical (or science-led historical) necessity but more the result of problems engendered within a certain tradition of thought. One alternative is the causal-realist approach developed by Salmon and – albeit with some differences of emphasis – by British philosophers such as Rom Harré and Roy Bhaskar.[56] Another, as I have suggested already, was the mode of epistemo-critical philosophy and history of science practised by Gaston Bachelard.[57] Above all, Bachelard sought to demonstrate how changes in scientific theory come about through a process of conceptual 'rectification' – of progressive refinement, elaboration and critique – starting out (very often) from intuitive metaphors or forms of naïve sense certainty.[58] His work thus provides a needful corrective to the various epistemic confusions (as between 'data', 'stimuli', 'sensations', etc.) that bedevil Kuhn's thinking on the topic of ontological relativity and his attempt to make sense of the extravagant claim that scientists quite literally 'live in diferent worlds' before and after some crucial paradigm change.

Reference

1 See, for instance – from a range of philosophical positions – Paul K. Feyerabend, 'On the "Meaning" of Scientific Terms', *Journal of Philosophy*, 62 (1965), pp. 266–74; Thomas S. Kuhn, *The Structure of Scientific Revolutions*, 2nd edn (Chicago: University of Chicago Press, 1970); Michael E. Levin, 'On theory-change and meaning-change', *Philosophy of Science*, 46 (1979), pp. 407–24; David Papineau, *Theory and Meaning* (London: Oxford University Press, 1979); W. V. Quine, 'Two Dogmas of Empiricism', in *From a Logical Point of View*, rev. edn (Cambridge, MA: Harvard University Press, 1961), pp. 20–46; Crispin Wright, *Realism, Meaning and Truth* (Oxford: Blackwell, 1987).

2 See especially Kuhn, *The Structure of Scientific Revolutions*.

3 Quine, 'Two Dogmas of Empiricism'; also *Word and Object* (Cambridge, MA: MIT Press, 1960) and *Ontological Relativity and Other Essays* (New York: Columbia University Press, 1969).

4 See, for instance, Arthur Fine, *The Shaky Game: Einstein, realism, and quantum theory* (Chicago: University of Chicago Press, 1986); Peter Gibbins, *Particles and Paradoxes: the limits of quantum logic* (Cambridge: Cambridge University Press, 1987); J. Polkinghorne, *The Quantum World* (Harmondsworth: Penguin, 1986).

5 See Gaston Bachelard, *La Formation de l'esprit scientifique* (Paris: Corti, 1938); *The Philosophy of No: a philosophy of the new scientific mind* (New York: Orion Press, 1968); *The New Scientific Spirit* (Boston: Beacon Press, 1984).

6 Mary Tiles, *Bachelard: science and objectivity* (Cambridge: Cambridge University Press, 1984), pp. 177–8.

7 See, for instance, Barry Barnes, *About Science* (Oxford: Blackwell, 1985); Harry Collins and Trevor Pinch, *The Golem: what everyone should know about science* (Cambridge: Cambridge University Press, 1993); Paul K. Feyerabend, *Against Method* (London: New Left Books, 1975); Steve Fuller, *Philosophy of Science and its Discontents* (Boulder, Co: Westview Press, 1989); Andrew Pickering (ed.), *Science as Practice and Culture* (Chicago: University of Chicago Press, 1992); Andrew Ross, *Strange Weather: culture, science and technology in the age of limits* (London: Verso, 1991); Steve Woolgar, *Science: the very idea* (London: Tavistock, 1988).

8 See Richard B. Braithwaite, *Scientific Explanation* (Cambridge: Cambridge University Press, 1953); Carl G. Hempel, *Fundamentals of Concept Formation in Empirical Science* (Chicago: University of Chicago Press, 1972); Hans Reichenbach, *Experience and Prediction* (University of Chicago Press, 1938).

9 Quine, 'Two Dogmas of Empiricism'.

10 See Hartry Field, 'Theory change and the indeterminacy of reference', *Journal of*

Philosophy, 70 (1973), pp. 462–81; also Field, 'Quine and the correspondence theory', *Philosophical Review*. 83 (1974), pp. 200–28; 'Conventionalism and instrumentalism in semantics', *Nous*, 9 (1975), pp. 375–405; 'Logic, meaning, and conceptual role', *Journal of Philosophy*, 74 (1977), pp. 379–409; also Arthur Fine, 'How to compare theories: reference and change', *Nous*, 9 (1975), pp. 17–32. On the radical meaning variance argument, see C. R. Kordig, *The Justification of Scientific Change* (Dordrecht: D. Reidel, 1971).

11 Tiles, *Bachelard*, p. 173.

12 See Pierre Duhem, *The Aims and Structure of Physical Theory*, trans. Philip Wiener (Princeton, NJ: Princeton University Press, 1954); *To Save the Phenomena: an essay on the idea of physical theory from Plato to Galileo*, trans. E. Dolan and C. Maschler (Chicago: University of Chicago Press, 1960); Sandra G. Harding (ed.), *Can Theories be Refuted? essays on the Duhem-Quine thesis* (Dordrecht and Boston: D. Reidel, 1976); also Quine, 'Two Dogmas of Empiricism'.

13 Quine, 'Two Dogmas of Empiricism', pp. 42–3.

14 Richard Rorty, *Philosophy and the Mirror of Nature* (Oxford: Blackwell, 1980); also *Consequences of Pragmatism* (Brighton: Harvester, 1982) and *Objectivity, Relativism, and Truth* (Cambridge: Cambridge University Press, 1991).

15 See Wilfrid Sellars, *Science, Perception and Reality* (London: Routledge and Kegan Paul, 1963).

16 Mary Hesse, *The Structure of Scientific Inference* (Berkeley and Los Angeles: University of California Press, 1974), p. 16. See also Hesse, *Revolutions and Reconstructions in the Philosophy of Science* (Brighton: Harvester, 1980).

17 See entries under note 14 above.

18 See Rorty, *Essays on Heidegger and Others* (Cambridge: Cambridge University Press, 1991); also Hubert L. Dreyfus, *Being-in-the-World: a commentary on Heidegger's Being and Time, Division One* (Cambridge, MA: MIT Press, 1991); Mark Okrent, *Heidegger's Pragmatism: understanding, being, and the critique of metaphysics* (Ithaca, NY: Cornell University Press, 1988); Joseph Rouse, *Knowledge and Power: towards a political philosophy of science* (Cornell University Press, 1987).

19 See Martin Heidegger, *The Question Concerning Technology and Other Essays*, trans. William Lovitt (New York: Harper and Row, 1977).

20 See entries under note 18 above.

21 See, for instance, Stephen Mulhall, *On Being in the World: Wittgenstein and Heidegger on seeing aspects* (London: Routledge, 1990).

22 See principally Peter Winch, *The Idea of a Social Science and its Relation to Philosophy* (London: Routledge and Kegan Paul, 1958); also David Bloor, *Wittgenstein: a social theory of knowledge* (New York: Columbia University Press, 1983) and Derek L. Phillips, *Wittgenstein and Scientific Knowledge: a sociological perspective* (London: Macmillan, 1977).

23 See, for instance, Martin Hollis and Steven Lukes (eds), *Rationality and Relativism* (Oxford: Blackwell, 1982); Larry Laudan, *Science and Relativism: some key issues in the philosophy of science* (Chicago: University of Chicago Press, 1990).

24 Quine, 'Two Dogmas of Empiricism'. All further references indicated by 'TDE' and page number in the text.

25 For further discussion see Harold Brown, *Perception, Theory, and Commitment* (Chicago:

University of Chicago Press, 1977); Richard E. Grandy (ed.), *Theories and Observation in Science* (Englewood Cliffs, NJ: Prentice-Hall, 1973); Harding (ed.), *Can Theories be Refuted?*; E. Nagel, S. Bromberger and A. Grunbaum, *Observation and Theory in Science* (Baltimore: Johns Hopkins University Press, 1971); David Papineau, *Theory and Meaning*.

26 See entries under note 7 above.

27 This passage was cut from the revised (1961) version of Quine's 'Two Dogmas of Empiricism'. It appears in the earlier text as published in T. Olshewski (ed.), *Problems in the Philosophy of Language* (New York: Holt, Rinehart and Winston, 1969), pp. 398–417; p. 415.

28 Immanuel Kant, *Critique of Pure Reason*, trans. N. Kemp Smith (London: Macmillan, 1974). See also Gordon G. Brittan, *Kant's Theory of Science* (Princeton, NJ: Princeton University Press, 1978) and Michael Friedman, *Kant and the Exact Sciences* (London: Harvester, 1992).

29 See entries under note 5 above.

30 Jerry Fodor and Ernest LePore, *Holism: a shopper's guide* (Oxford: Blackwell, 1991).

31 See Rorty, *Consequences of Pragmatism*.

32 Martin Heidegger, *Being and Time*, trans. John MacQuarrie and Edward Robinson (New York: Harper and Row, 1962).

33 See Heidegger, *The Question Concerning Technology*.

34 Ludwig Wittgenstein, *Philosophical Investigations*, trans. G. E. M. Anscombe (Oxford: Blackwell, 1958).

35 See Rorty, *Philosophy and the Mirror of Nature*.

36 See, for instance, entries under notes 3, 7, 14, 18 and 22 above; also George W. Grace, *The Linguistic Construction of Reality* (London: Croom Helm, 1987); Bruce Gregory, *Inventing Reality: physics as language* (New York: Wiley, 1988); Alan G. Gross, *The Rhetoric of Science* (Cambridge, MA: Harvard University Press, 1990). For counter-arguments from a range of philosophical perspectives, see Michael Devitt, *Realism and Truth* (Oxford: Blackwell, 1984); Jarrett Leplin (ed.), *Scientific Realism* (Berkeley and Los Angeles: University of California Press, 1984); David Papineau, *Reality and Representation* (Blackwell, 1987); Edward Pols, *Radical Realism: direct knowing in science and philosophy* (Ithaca, NY: Cornell University Press, 1992 – on 'seven dogmas of the linguistic consensus'); Roger Trigg, *Reality at Risk: a defence of realism in philosophy and the sciences* (Brighton: Harvester, 1980); Peter J. Smith, *Realism and the Progress of Science* (Cambridge: Cambridge University Press, 1981); Gerald Vision, *Modern Anti-Realism and Manufactured Truth* (London: Routledge, 1988).

37 See entries under Note 22 above; also Michel Foucault, *The Order of Things: an archaeology of the human sciences* (London: Tavistock, 1972); Wittgenstein, *Philosophical Investigations*; Quine, *Ontological Relativity and Other Essays* (New York: Columbia University Press, 1969).

38 Thomas S. Kuhn, *The Structure of Scientific Revolutions*. See also Gary Gutting, *Paradigms and Revolutions* (Notre Dame, IN: University of Notre Dame Press, 1980); Ian Hacking (ed.), *Scientific Revolutions* (London: Oxford University Press, 1981); Paul Horwich (ed.), *The World Changes: Thomas Kuhn and the nature of science* (Cambridge, MA: MIT Press, 1993); John Krige, *Science, Revolution and Discontinuity* (Brighton:

Harvester, 1980); Kuhn, *The Essential Tension: studies in scientific tradition and change* (Chicago: University of Chicago Press, 1977).

39 See, for instance, Richard Rorty, *Contingency, Irony, and Solidarity* (Cambridge: Cambridge University Press, 1989); Michel Foucault, *Language, Counter-Memory, Practice*, trans. D. F. Bouchard and S. Weber (Oxford: Blackwell, 1977); also – from a different though related standpoint – John Dupré, *The Disorder of Things: metaphysical foundations of the disunity of science* (Cambridge, MA: Harvard University Press, 1993).

40 See, for instance, W. V. Quine, *Philosophy of Logic* (Englewood Cliffs, NJ: Prentice-Hall, 1970).

41 Quine, 'Carnap and Logical Truth', in *The Philosophy of Rudolf Carnap*, ed. Paul A. Schilpp (La Salle, IL.: Open Court, 1963), p. 387. Reprinted in Quine, *The Ways of Paradox* (New York: Random House, 1966).

42 See Quine, *The Philosophy of Logic*; also Susan Haack, 'Quantum Mechanics', in *Deviant Logic* (Cambridge: Cambridge University Press, 1974), pp. 148–67.

43 See especially Hilary Putnam, 'How to think quantum-logically', *Synthèse*, 29 (1974), pp. 55–61 and *Mathematics, Matter and Method* (Cambridge: Cambridge University Press, 1979); John Gibbins, *Particles and Paradoxes: the limits of quantum logic* (Cambridge: Cambridge University Press, 1987); M. Gardner, 'Is quantum logic really logic?', *Philosophy of Science*, 38 (1971), pp. 508–29; Peter Mittelstaedt, *Quantum Logic* (Princeton, NJ: Princeton University Press, 1994).

44 See Dummett, *Truth and Other Enigmas;* also entries under notes 42 and 43 above. For further discussion see Michael Luntley, *Language, Logic and Experience: the case for anti-realism* (London: Duckworth, 1988); N. Tennant, *Anti-Realism and Logic* (Oxford: Clarendon Press, 1987); Alan Weir, 'Dummett on meaning and classical logic', *Mind*, 45 (1986), pp. 465–77; Timothy Williamson, 'Knowability and constructivism: the logic of anti-realism', *Philosophical Quarterly*, 38 (1988), pp. 422–32.

45 Kuhn, *The Structure of Scientific Revolutions*, 2nd edn. All further references indicated by '*SSR*' and page number in the text.

46 See Paul K. Feyerabend, *Against Method* (London: New Left Books, 1975) and *Science in a Free Society* (New Left Books, 1978); Jean-François Lyotard, *The Postmodern Condition: a report on knowledge*, trans. Geoff Bennington and Brian Massumi (Manchester: Manchester University Press, 1984) and *The Differend: phrases in dispute*, trans. Georges van den Abbeele (Manchester University Press, 1988).

47 For further discussion see entries under note 10 above; also Horwich, *The World Changes*.

48 For a variety of arguments from widely differing (realist and anti-realist) standpoints, see Peter Achinstein, *Law and Explanation: an essay in the philosophy of science* (Oxford: Clarendon Press, 1971); D. M. Armstrong, *What Is a Law of Nature?* (Cambridge: Cambridge University Press, 1983) and *Universals and Scientific Realism* (Cambridge University Press, 1978); John W. Carroll, *Laws of Nature* (Cambridge University Press, 1994); Nancy Cartwright, *How the Laws of Physics Lie* (London: Oxford University Press, 1983) and *Nature's Capacities and their Measurement* (Clarendon Press, 1989); Bas C. van Fraassen, *Laws and Symmetry* (Clarendon Press, 1989); Rom Harré and E. H. Madden, *Causal Powers* (Oxford: Blackwell, 1975); J. L. Mackie, *The Cement of the Universe* (Clarendon Press, 1974); Wesley C. Salmon, *Scientific Explana-*

tion and the Causal Structure of the World (Princeton, NJ: Princeton University Press, 1984); E. Sosa (ed.), *Causation and Conditionals* (Oxford University Press, 1975); J. J. C. Smart, *Philosophy and Scientific Realism* (London: Routledge & Kegan Paul, 1963).

49 See notes 43 and 44 above.

50 See, for instance, Adolf Grunbaum and Wesley C. Salmon (eds), *The Limitations of Deductivism* (Berkeley and Los Angeles: Unversity of California Press, 1988); Salmon, *Four Decades of Scientific Explanation* (Minneapolis: University of Minnesota Press, 1989); Salmon (ed.), *Hans Reichenbach: logical empiricist* (Dordrecht: D. Reidel, 1979); also Quine, 'Two Dogmas of Empiricism'.

51 See especially Nelson Goodman, *Fact, Fiction and Forecast* (Cambridge, MA: Harvard University Press, 1955).

52 See note 50 above.

53 For a causal-realist approach to these issues of inductive warrant, see Armstrong, *What Is a Law of Nature?* and other entries under note 48 above.

54 See Karl Popper, *The Logic of Scientific Discovery* (New York: Harper and Row, 1959) and *Conjectures and Refutations* (Harper and Row, 1963).

55 See, for instance, Baruch Brody, 'Towards an Aristotelian theory of explanation', *Philosophy of Science*, 39 (1972), pp. 20–31; David-Hillel Ruben, *Explaining Explanation* (London: Routledge, 1992).

56 See Salmon, *Scientific Explanation and the Causal Structure of the World*; also Roy Bhaskar, *A Realist Theory of Science* (Leeds: Leeds Books, 1975), *Scientific Realism and Human Emancipation* (London: Verso, 1986) and *Reclaiming Reality: a critical introduction to contemporary philosophy* (Verso, 1989); Rom Harré, *The Principles of Scientific Thinking* (Chicago: University of Chicago Press, 1970) and *The Philosophies of Science* (London: Oxford University Press, 1972); J. Aronson, R. Harré and E. Way, *Realism Rescued: how scientific progress is possible* (London: Duckworth, 1994); Harré and Madden, *Causal Powers*.

57 See note 5 above; also Tiles, *Bachelard: science and objectivity*.

58 See also Georges Canguilhem, *Etudes d'histoire et de philosophie des sciences* (Paris: Vrin, 1968) and *La Connaissance de la vie* (Vrin, 1969); Dominque Lecourt, *Marxism and Epoistemology: Bachelard, Canguilhem and Foucault* (London: New Left Books, 1975).

4

Quantum Mechanics: A Case for Deconstruction?

I

In this chapter I shall be looking at some recent (post-1970) developments in philosophy of science which offer a hopeful way forward from the various well-known problems (of meaning variance, paradigm change, ontological relativity and the like) bequeathed by logical empiricism. Most promising among them – in my view – is the theory of critical realism which draws inspiration from the work of Rom Harré and whose chief exponent during the past two decades has been Roy Bhaskar.[1] In North America the emphasis has fallen rather differently, with no such overt or programmatic link between the two major arguments that find expression in the title of Bhaskar's best-known book, *Scientific Realism and Human Emancipation*.[2] Nevertheless there is a clearly marked ethical and sociopolitical dimension to the current widespread renewal of interest in causal realism among US philosophers of science. This aspect is important, I shall argue, partly on account of the challenge it offers to the kinds of Kuhnian relativist thinking that have attained near-orthodox status elsewhere in the social and human sciences.

Its most prominent spokesman is Wesley C. Salmon, whose book *Scientific Explanation and the Causal Structure of the World* (1984) lays out a 'new' agenda for philosophy of science which in fact belongs a tradition stretching back to Aristotle. Salmon puts the case most succinctly as follows:

> To explain a fact is to furnish an explanans. We may think of the explanans as a set of explanatory facts; these general and particular facts account for the explanandum. Among the explanatory facts are particular events that constitute causes of the explanandum, causal processes that connect the causes to their effects, and causal regularities that govern the causal mechanisms involved in the explanans. In presenting the explanans, we use statements that detail the explanatory facts – we may refer to them as explanans-statements.[3]

The most crucial point here – and what sets this argument decisively apart from all versions of the covering-law (deductive-nomological) approach or its sceptical (conventionalist) progeny – is the necessary link that Salmon asserts between the *causal structure* of those objects or events to be explained and the *logical structure* of the theory that purports to explain them.[4] For otherwise there is no reason why even the most 'adequate' theory – one that 'accounts for' or is 'compatible with' the best evidence to hand – should be thought of as actually *explaining* those particular items of evidence, rather than merely bringing them under a certain law-like description. In short, what Salmon is here proposing is a strong counter-argument to Kuhn's thesis that our knowledge of the world is always relative to some given (communally sanctioned) paradigm, ontological scheme, interpretive framework or whatever.[5]

Such arguments have achieved wide currency not only among post-structuralists, postmodernists, Foucauldians and others in the cultural-theory camp but also among many philosophers of science who have arrived at them by the alternative (Quinean–Kuhnian) route of a radical empiricism joined to a holistic conception of meaning and truth.[6] The result, in each case, is to fix what appears to be an insuperable gulf between word and world, theory and reality, or – as Salmon more precisely puts it – between the order of causal-explanatory thought and those various causally explicable objects, processes and events which constitute its proper domain. For post-structuralists this follows from Saussure's doctrine of the 'arbitrary' nature of the sign, usually (and wrongly) taken as warrant for renouncing any concern with language in its referential aspect.[7] In Foucault it amounts to a form of thoroughgoing nominalism, an 'archaeology' (or Nietzschean 'genealogy') of knowledge that views all truth-claims, whether in the natural or the human sciences, as products of an epistemic will-to-power configured through various discursive formations that undergo successive paradigm shifts beyond reach of rational explanation.[8] So it is not hard to see why commentators on Foucault often make reference to Kuhn (as well as to Quine's thesis of ontological relativity) by way of recruiting further philosophical support.[9]

One could offer various examples of the current *rapprochement* between philosophers of science in the post-analytic camp and cultural theorists with an eye to capturing the high ground of interdisciplinary debate. Those who have read Jean-François Lyotard's book *The Postmodern Condition* will know what becomes of 'science' when its governing values (truth, reason, observational warrant, theoretical consistency, etc.) are viewed as belonging to just one 'phrase genre' (the cognitive) whose privileged status has now gone the way of all such delusory and authoritarian 'meta-narrative' ploys.[10] On this account, scientific truth is no longer – if ever it was – a matter of attaining more accurate knowledge or providing better (more adequate) explanations of objects and events in the physical domain. On the contrary, Lyotard argues: we have now moved on into a postmodern epoch where 'truth' is defined solely according to performative (as opposed to constative) criteria, and where the measure of success for a scientific discourse, paradigm or research programme is simply the degree to which it wins assent – or attracts funding – from members of the relevant 'expert' community. Moreover, this transformation extends to the very nature of (so-called) physical 'reality', rather than pertaining – as more orthodox thinkers would believe – to the limits of our current state of knowledge. After all, don't we have it on the 'authority' of contemporary science that in some cases there is simply no deciding between rival hypotheses (e.g., wave and particle theories of light or matter); that in sub-atomic physics the act of observation may crucially affect the phenomenon observed, thus giving rise to an uncertainty principle which undermines the very notion of 'objective' reality; that quantum mechanics may force a wholesale revision of the classical 'laws of thought', those that require a strictly distributive (bivalent) logic of truth and falsehood; or again, following Quine, that there exist as many possible worlds or ontologies as there exist ways of distributing predicates over the range of beliefs (or cultural 'posits') in place at any given time?[11] And if further corroboration is needed, then we need only look to more recent ideas at the forefront of speculative thinking in mathematics and the physical or life sciences. For these offer numerous examples of the way that research, 'by concerning itself with such things as undecidables, the limits of precise control, conflicts characterized by incomplete information, "*fracta*", catastrophes, and pragmatic paradoxes . . . is theorizing its own evolution as discontinuous, catastrophic, non-rectifiable and paradoxical'.[12]

For Lyotard these signs all point to the advent of a postmodern science for which 'performativity' (not truth) is the chief criterion, and whose parameters can no longer be set by the aim to achieve rational consensus as regards both its objects and its proper methods of enquiry. Rather, such enquiry should seek to maximize the degree of *dissensus* – or the variety of competing (incommensurable) paradigms – which keeps science open to new ideas and prevents the imposition of dogmatic truth-claims or orthodox methodological norms. Hence Lyotard's demotion of the cognitive phrase genre to the role of a strictly non-privileged participant voice in the open multiplicity of discourses, language games or phrase genres that characterize the postmodern condition. In which case, he urges, we had best give up thinking in terms of those rigid (yet purely conventional) lines of demarcation that have hitherto been drawn between the natural and the human sciences, or again, between 'science' as a mode of knowledge accountable to rigorous (cognitive) standards of enquiry and art as a realm of creative or imaginative freedom subject to no such restrictive conditions. Thus for Lyotard contemporary science – or his favoured branches of it, those concerned with 'undecidables, conflicts, . . . "fracta", catastrophes, pragmatic paradoxes', etc. – may be seen as embarked upon the same kind of enterprise that is currently pursued by avant-garde figures in the literary, visual and musical arts. That is to say, both ventures partake of the 'sublime' in what Lyotard construes as its authentic (sharply paradoxical) Kantian sense. This involves the idea of 'presenting the unpresentable', of the sublime as somehow giving access to a realm of 'suprasensible' experience where intuitions cannot be brought under adequate concepts (as is the rule for judgements in the cognitive mode), and where thinking thus encounters the maximal challenge to its powers of recuperative grasp.[13] So it is that the sublime bears ultimate witness to the 'heterogeneity' of phrase genres, or – in Lyotard's parlance – to the narrative *differend* that always and inevitably opens up between various 'phrases in dispute'.[14] Such rival claims cannot be subject to any higher-level arbitrating order of judgement since they invoke wholly different (incommensurable) criteria. And this is just as much the case for science, he argues, as in those other domains – ethics, politics, aesthetics – where the only rule is to ensure that no one set of phrases be allowed to monopolize the discourse of truth or justice.

I have written elsewhere about the confusions engendered by this currently widespread postmodernist cult of the Kantian sublime.[15] What it amounts to, in short, is a pre-emptive move against the philosophic discourse of modernity, a reading of Kant – inspired very largely by Heidegger's example – which fastens on those no doubt problematic passages which lend themselves to treatment in this strong revisionist mode.[16] One can see the same bias at work in Foucault's long sequence of tactical engagements with Kant, starting out with some brilliant (if tendentious) pages in *The Order of Things* and ending up with the remarkable essay 'What Is Enlightenment?', where Foucault reads Kant – alongside Baudelaire! – as pointing the way towards his own project for an ethics based on the values of aesthetic self-fashioning.[17] There is no room here for a detailed account of this latest chapter in the fortunes of the Kantian sublime at the hands of its postmodern exegetes. Sufficient to say that it serves their purpose by enabling a large-scale shift of emphasis away from Kant's cognitive or epistemological concerns, as addressed mainly in the first *Critique*, and towards that more obscure ('suprasensible') region of thought where certain of his more oblique or cryptic formulations can then be read back – so to speak – as symptomatic of the problems that supposedly arise with the cognitive requirement that intuitions be brought under adequate concepts.[18]

Such are the passages that Heidegger singles out, among them Kant's notoriously vague appeal (in the 'Transcendental Aesthetic' of the first *Critique*) to an 'art concealed in the depths of the soul', a synthesizing power that alone can accomplish the required link between intuitions and concepts, yet whose nature inherently eludes conceptual definition.[19] For Heidegger – as also for Foucault and Lyotard – those passages stand as promissory notes whose value can be redeemed only through Kant's treatment of the aesthetic (in his third *Critique*) as a *sui generis* modality of judgement. Above all, they are taken up in his reflections on the sublime as a figure of maximal 'heterogeneity' which in effect, so far from making good that promise, turns out to disarticulate the entire structure of Kantian epistemological priorities and truth-claims.[20] So it is that Foucault can annex Kant to his own strong revisionist project, and that Lyotard can likewise claim Kantian warrant for demoting the cognitive phrase-genre, along with that of 'science', as a discourse presumptively aimed towards truth at the end of enquiry. In each case the sublime does service as a kind of all-purpose deconstructive lever for prising apart the various modes of judgement (cognitive and evaluative, constative and performative, epistemological, ethical and aesthetic) and then systematically disrupting or inverting their supposed orders of relationship. Thus science can be viewed as a 'performative' genre, an open multiplicity of language games whose criteria no longer have anything to do with truth conceived – in constative terms – as a matter of bringing empirical observations under adequate theoretical or explanatory concepts.

All of which helps to explain why some commentators, Rorty among them, have sought to construct a broad alliance of interests between 'post-analytic' philosophers like Quine, relativist historians of science such as Kuhn, proponents of a depth-hermeneutical approach after Heidegger, and postmodern thinkers – Lyotard especially – who denounce the 'philosophic discourse of modernity', together with science's privileged role within that discourse as a paradigmatically truth-seeking mode of enquiry.[21] This alliance can be made to look fairly plausible if one accepts Rorty's debunking view of the Western philosophical tradition to date as just a series of wrong turns brought about by its attachment to delusory (epistemological) ideas of reason, knowledge and truth. For the 'linguistic turn' can then be presented not only as a cultural *fait accompli* but as a wished-for release from all our self-induced philosophical afflictions, or – in Wittgenstein's famous metaphor – as a means of showing the fly the way out of the fly-bottle. However, there is a growing awareness that this idea of language (or discourse) as the bottom line of enquiry has created more problems – especially for philosophy of science – than it provides genuine or workable solutions. Hence, as I have argued, the need to take account of those alternative resources offered by more recent (causal-explanatory) theories.

'As I construe the term "fact",' Salmon writes, 'facts are not linguistic entities – unless, of course, we are talking about language, and not just using language to talk about other sorts of facts. Facts are features of the physical world that exist objectively whether or not we try to explain them.'[22] And again, lest this crucial point be ignored or misunderstood:

> The term 'explanation' may be construed in either of two ways. It may be taken as the combination of the explanans and the explanandum – the explanans having its place in the causal pattern or the causal nexus in which it is objectively embedded. The term may also be used to refer to the combination of explanans-statement and explanandum-statement – the linguistic entity that is used to present the objective facts. [The] term 'explanatory text' is an entirely suitable expression to employ when we want to talk about the statements used to formulate an explanation.[23]

Salmon offers numerous – to my mind convincing – examples of such causal explanation at work. They include the established (epidemiological) account of how the Great Plague was caused by a known type of bacterium, observable under a microscope, whose transmission occurred through its first taking hold among a migrant rodent population and then being carried by fleas who bit both rats and human beings. This account satisfies all the criteria for an adequate causal-explanatory theory. Thus 'it does not invoke submicroscopic entities', nor any notions of (for example) mysterious action at a distance or disease as a punishment for human sins and transgressions. To be sure, it establishes no *absolute and necessary* causal link from stage to stage along *every* such line of transmission. For, as Salmon notes, '[n]ot all residents of London contracted the disease; some people seem to have greater resistance than others.' From which we may infer that '[n]ot every flea that bites an infected organism actually transmits the disease to another organism, and not every organism that comes into contact with an infected flea becomes infected.'[24] But this is just to say that causal explanations may very well involve – and can very well cope with – a measure of statistical probability weighting. Its effect is not to impair their explanatory force but to introduce further requirements in the way of specific data related to specific times and places of occurrence.

Thus, in Salmon's words,

> We would want to know what ship or ships from what particular locale brought the infected rats to London. Whether we are treating the explanation of a particular epidemic or dealing with such epidemics in general, these antecedent conditions will be etiological factors. . . . The explanation of statistical regularities consists in exhibiting them as the statistical result of an aggregate of causal processes and interactions. Such internal analysis of the explanandum – the epidemic – contributes to the constitutive aspects of the explanation.[25]

In short, there is no reason – doctrinaire scepticism apart – to doubt that this offers an adequate explanation of the various causal factors involved. And it does so precisely by rejecting that major premise of the sceptic's argument which holds (on Humean or other grounds) that causal explanations cannot be justified by any appeal to criteria of logical consistency or truth. In Salmon's view this is a false dilemma, shown up as such by the numerous cases where science has demonstrably managed to account for some natural phenomenon in terms that possess both explanatory force and an adequate degree of inferential warrant. That is to say, such problems only arise – at least in so sharp and intransigent a form – if one takes it that there is nothing *in the nature* of the given explanandum which justifies certain procedures of reasoning to the best, most adequate explanation.

One could cite many instances to similar effect, since this form of jointly causal and logical reasoning is paradigmatic for the natural sciences and for every discipline that seeks to go beyond observed regularities to the deeper (causal) structures and propensities that explain them. Thus, for instance, it is the properties of nuclear radiation, together with their known physiological effects, which account for the far above-average incidence of leukaemia among people exposed to high levels of radioactivity following the nuclear explosions over Hiroshima and Nagasaki. As Salmon remarks, these are not 'chance clusterings' but statistical abnormalities that demand something more than a weak 'explanation' in terms of random sampling and numerical distribution. For enough is now known about the various causal factors involved – about nuclear reactions, particles, radiation fields, fall-out patterns and the impact of these upon the biochemistry of living organisms – to make it possible for scientists to explain very precisely how and why those clusters should have occurred as they did. Thus '[t]he causal processes probabilistically linking the nuclear explosions to the physiological damage that eventuates in leukaemia are the photons and other particles travelling from the nuclear reactions to the bodies of the victims.'[26]

Of course this gives a handle for the sceptic to argue that such purported explanations amount to no more than a matter of assigned probability, or again – in Humean terms – that those putative causal 'forces', 'propensities', 'dispositions', etc., are just so many occult essences which we (the observers) attribute to phenomena that otherwise exhibit only 'constant conjunction' or patterns of repeated co-occurrence. In which case Salmon could only be mistaken when he argues for a necessary linkage between, on the one hand, the structure of scientific theories (i.e., the logic of explanation) and, on the other, those deep-lying structures of causal efficacy which are thought of as intrinsic to the various objects, processes or events that constitute the domain of scientific enquiry. On this account there is simply no bridging the gulf, no means of arguing from observed regularities in nature to a theory that would establish adequate criteria for assessing the validity of causal explanations. The best that science can do, therefore, is to offer covering-law (deductive-nomological) generalizations which take in as many as possible of the observed data but which reject any talk of hypothetical 'causes' as merely an illicit metaphysical appeal to unexplained forces, essences, 'deep further facts' or whatever.[27]

As I have said, this was, until recently, the dominant view within mainstream Anglo-American philosophy of science. It is the principle adopted by Carnap and other advocates of the logical-empiricist approach when they introduced the notion of semantic 'ascent' from the material to the formal mode, or from matters of first-order empirical observation (as given in natural language) to a higher-level discourse that disposed of its own meta-linguistic criteria of logic, consistency and truth.[28] Otherwise put, it is a version of the old scholastic distinction between matters of *de re* and *de dicto* necessity, or properties ascribed to physical realia by virtue of their inbuilt nature or structure and analytic truths of definition strictly devoid of empirical content. From here it is but a short step to Quine's (and Kuhn's) dissolution of the logical-empiricist paradigm into a doctrine of wholesale ontological relativity that regards all truth-claims – from empirical observations to the so-called logical 'laws of thought' – as open to revision should need or opportunity arise.[29] Thus, according to Quine, it is a matter of redistributing predicates over the existing 'fabric' of beliefs, or adopting some different ontological scheme in response to some emergent theoretical anomaly or item of 'recalcitrant' evidence.

This is perhaps the most striking irony of recent (post-1930) developments in philosophy of science. It is the irony whereby a programme that started out with a strong (albeit disjunctive) commitment to empirical truths of observation on the one hand and analytic truths of definition on the other should then have given rise to a relativist doctrine that treats both notions as wholly contingent upon the 'language games', 'paradigms' or culture-specific habits of belief that happen to prevail within some given community of knowledge. But in a sense this outcome was inevitable once logical empiricism had driven its Humean wedge between the order of discrete observation sentences and the order of covering-law theories. For its programme quickly ran up against the threefold objection: (1) that all 'facts' (or empirical observations) are to some degree theory-laden; (2) that such facts are linguistic (or discursive) constructs, rather than somehow existing 'out there' as real-world objects or entities awaiting discovery; and (3) – following from this – that scientific 'progress' is a hopelessly deluded idea since it always rests on some version of the correspondence theory, or some jointly fact-based and regulative notion of truth at the end of enquiry.

Thus the way was open to all those sceptical doctrines that soon came to occupy the high ground in philosophy of science. They include Quine's theses of ontological relativity and meaning holism; Kuhn's idea of paradigm shifts as involving a radical 'incommensurability' between rival scientific theories; Feyerabend's all-out 'anarchist' assault on truth-claims or methodological constraints in whatever – always authoritarian – guise; and again, where philosophy of science meets cultural theory, Lyotard's talk of multiple heterogeneous 'phrase genres', along with his view of scientific truth as a matter of purely 'performative' efficacy. For some commentators – Rorty among them – this is quite simply where we are now, at the end of all those old-fashioned and fruitless epistemological endeavours. However, such arguments can seem plausible only from an internalist perspective on recent (post-logical-empiricist) philosophy of science which mistakes the dilemmas of that particular programme for the predicament of science – or philosophy of science – in general. For otherwise, as Salmon rightly argues, it will look much more like a bad case of tunnel vision, a parochial view brought about by its fixation on just those self-induced problems and its refusal to envisage alternative resources.

II

Such was Bertrand Russell's early position, in his 1929 book *Mysticism and Logic*, where he espoused the logical-empiricist line against any talk of 'causes' or kindred metaphysical abstractions. Thus:

> All philosophers, of whatever school, imagine that causation is one of the fundamental axioms or postulates of science, yet, oddly enough, in advanced sciences such as gravitational astronomy, the word 'cause' never occurs. . . . To me it seems that the reason why physics has ceased to look for causes is that, in fact, there are no such things. The law of causality, I believe, like much that passes muster among philosophers, is the relic of a byegone age, surviving, like the monarchy, only because it is erroneously supposed to do no harm.[30]

And of course this argument had long been applied not only to talk of causality but also to other terms – like 'force' in dynamics or in Newton's theory of gravitational attraction – which supposedly served no real explanatory purpose and should therefore be reckoned out of the equation. But, as Salmon points out, Russell's scepticism in this regard was fairly short-lived. For by 1948, in *Human Knowledge, its Scope and Limits* he can be found defending a strong thesis with regard to the necessity of causal explanations in science. On this later account, what justifies such reasoning is the existence of 'causal lines' which connect various phenomena (or temporal stages of the same phenomenon), and which thus provide the otherwise missing link between observed co-occurrence – or Humean 'constant conjunction' – and the order of conceptual-explanatory grasp.

Thus, in Russell's words, a causal line may be taken to consist in 'a temporal series of events so related that, given some of them, something can be inferred about the others whatever may be happening elsewhere'. And again: 'A causal line may be regarded as the persistence of something – a person, a table, a photon, or what not. Throughout a given causal line, there may be constancy of quality, constancy of structure, or a gradual change of either, but not sudden changes of any considerable magnitude.'[31] This latter requirement becomes problematic in the context of quantum mechanics where – at least on some interpretations – it may be necessary to readmit concepts like that of non-local simultaneous causality, ideas which had long been expelled (so far as possible) from the discourse of reputable scientific thought.[32] But even here, as in the famous series of debates between Einstein and Bohr, what is at issue is not so much the question of whether science can or should attempt to explain such phenomena, but the question whether they fall under certain classical schemes of descriptive or causal-explanatory grasp.[33] Clearly Russell was aware of this when he included photons – along with persons and tables – in his short list of items persisting throughout some given 'causal line'. For the main point about the photon, in this context of debate, is that it figures as a highly paradoxical entity, an instance of the wave/particle dualism whose behaviour under different conditions (very often in speculative thought-experiments of the kind conducted by Einstein and Bohr) varies according to the observational framework or the sorts of theoretical question posed. Hence all the well-known puzzles – like that of Schrödinger's perhaps ill-fated cat – which result from trying to assign determinate outcomes (or even probability weightings) to the so-called 'collapse of the wave-packet' or other such quantum-mechanical phenomena.[34]

These ideas have exerted widespread appeal through their diffusion in a range of popularizing narrative accounts. Thus it is often assumed – not least by theologians with an eye to the main chance – that theoretical physics has at last given up on the quest for adequate causal explanations and come around to an anti-determinist stance with room for a speculative 'God of the gaps' or other such genial hypotheses.[35] Nor is this tendency by any means confined to thinkers of a religious persuasion. As we have seen, the 'new physics' also does service for those postmodern thinkers – like Lyotard – who invoke uncertainty, complementarity, quantum phenomena, fractals, chaos theory and so forth as proof that contemporary science is moving in just the direction marked out by their own reading of the cultural signs. All of which gives a certain pointed relevance to Russell's inclusion of photons – that is to say, of a quantized wave/particle 'packet' – on his list of phenomena that satisfy the conditions for counting as participants in a causal line. For by this time (1948) Russell was already witnessing the early symptoms of an anti-positivist backlash that would soon emerge, across various disciplines, under the guise of a 'linguistic turn' wedded to doctrines of ontological relativity, semantic holism and language-game incommensurability. His attitude towards these developments was uniformly critical, whether in the case of American pragmatism, Wittgensteinian (or Oxford-style) 'ordinary-language' philosophy, or those currents in philosophy of science – inspired by the new physics – which responded to the perceived 'crisis' in old ways of thought by counting reality a world well lost for the sake of more adventurous hermeneutic paradigms. Thus Russell came down very firmly on the side of an outlook that treated such puzzles and paradoxes as artefacts of our own presently limited state of knowledge, rather than as mysteries somehow inhering in the nature of things. 'That there are such more or less self-determined causal processes is [he wrote] . . . one of the fundamental postulates of science.' Moreover, '[i]t is in virtue of the truth of this postulate – if it is true – that we are able to acquire partial knowledge in spite of our enormous ignorance.'[36]

In short, Russell had moved a long way from his early scepticism with regard to the status – or the very possibility – of causal explanations. For in default of such resources one can make no sense of the scientific enterprise as aimed towards a better, more adequate grasp of the underlying mechanisms that constitute the order of physical reality, nor indeed of our own place in that order as creatures equipped to perceive and comprehend them. I shall not here attempt to adjudicate on the still highly controversial issue of realist *versus* anti-realist philosophies of quantum mechanics.[37] All the same, there would appear to be several good arguments for the truth of Russell's causal postulate and for the realist interpretation that follows from it. One is the fact, as noted above, that some of the more crucial debates on this topic – like those between Einstein and Bohr – took the form of elaborate thought-experiments presumed to model what would actually occur under given (closely specified) laboratory conditions.[38] These experiments were of course devised *faute de mieux*, that is to say, in the absence of empirical test-procedures or for want of equipment (e.g., particle accelerators or electron microscopes) sufficiently powerful or fine-tuned to provide the needed observational data. But they were none the less taken as genuine evidence for or against rival hypotheses, for instance, in various, progressively more elaborate versions of the famous 'two-slit' experiment set up to determine how the photon would behave (i.e., as a wave-like or a particle-like phenomenon) should technology advance – as indeed it did – to the point where such experiments might actually be carried out.

Nor was this in any sense a novel departure in the history of scientific thought. For that history offers numerous examples of the way in which theory can progress beyond the limits of current observational technique, or come up with conjectures – most often on the basis of mathematical or geometrical reasoning – which only later become subject to empirical confirmation. Thus Galileo and Kepler each went far beyond the evidence to hand in framing hypotheses (like Kepler's second law of planetary motion, that the radius vector of the Sun to a planet sweeps over equal areas in equal times) whose discovery preceded – but also made possible – the kinds of observation that would subsequently bear them out. This in turn enabled Newton, starting from his theories of force, mass, acceleration and gravitational attraction, to derive the laws of planetary motion and hence a whole series of strong predictive hypotheses, including the existence and exact location of as yet unobserved celestial bodies. So it was that the discovery of Neptune came about through observed deviations in the orbit of Uranus, perceived as anomalous – and therefore as standing in need of explanation – by applying Newton's laws. A similar case was Einstein's prediction that the path of light would be bent by the presence of a neighbouring heavy body. This result was first arrived at on purely hypothetical grounds, that is, as one of the consequences entailed by Einstein's theory, itself – at this stage – a theory unconfirmed by any kind of experimental evidence or data. But the situation changed when, in 1919, the opportune occurrence of a total eclipse of the sun made it possible to test and vindicate Einstein's prediction.[39]

In each of these cases the sought-after evidence was such as to provide *post-hoc* empirical warrant for a hypothesis whose truth had hitherto been matter for conjecture, albeit conjecture of a highly disciplined (hypothetico-deductive) kind. In Russell's terms, it established a 'causal line' that linked the phenomenon in question to a large body of existing theoretical and observational knowledge. Of course, there remain large areas of dispute with respect to the status of quantum-mechanical explanations and the extent to which they can – or cannot – be brought under any 'classical' logic of enquiry. On one interpretation these phenomena are such as to require the development of a new kind of quantum logic. This would be a mode of thinking able to cope with paradoxes like the wave/particle dualism, the momentary superposition of discrete physical states, and the non-distributive – undecidable – relation that holds between values of truth or falsehood in any such case.[40] Then again other physicists, David Bohm among them, have postulated the existence of some 'hidden variable', some as yet unknown property or characteristic that would show these puzzles to have resulted from the present limits of our knowledge or powers of observation.[41] In which case one could still maintain, like Russell, that quantum mechanics is compatible with – at any rate does not preclude – a commitment to ontological realism and to the existence of determinate 'causal lines' as a necessary premise for all scientific enquiry.

However, this argument runs up against further problems if one asks how quantum phenomena could ever be *describable* (or lend themselves to adequate theoretical representation) given their presumed lack of fit with the concepts and categories – the logical grammar – of any available observation language. Niels Bohr was thus driven to the melancholy conclusion that we simply lack the conceptual and linguistic resources to characterize those phenomena in other than fuzzy or highly metaphorical terms.[42] And this for the reason, as Peter Gibbins puts it, that 'all the data by means of which we test our theories must be expressed in the language of classical physics.' To the question 'Why so?' Gibbins responds, again following Bohr:

> Because the language of classical physics is ideally suited to describing the macrophysical world, the world of medium-sized objects, in which we learn and constantly test natural language. The species has evolved classical concepts (of position, speed and so on) for so many generations that they are now an intrinsic part of our worldview. However far we travel from the world of everyday reality we are stuck with classical concepts.[43]

So there is a sense in which Bohr's 'realism' about quantum phenomena – his refusal to accept that they might be just artefacts of our limited cognitive or explanatory powers – leads him around to an anti-realist position as regards the possibility of our *ever* deciding between these rival interpretations. This is why Gibbins describes him, borrowing the phrase from Heisenberg, as 'the philosopher who operated in the mode of theoretical physicist' and as one possessed of a 'darkly metaphysical mind'.[44] For on Bohr's account of it, quantum mechanics seems to point towards regions of speculative thought which offer no hold for conceptual understanding or for any explanation that would not give rise to intractable paradoxes and aporias. In the end he is obliged, like Kant, to posit the existence of a noumenal realm, one to which we can never have access by virtue of our dependence on just those 'classical' concepts – of space, time, position, velocity, causal relations, distributive (bivalent) truth-values, and so forth – which constitute the limits of 'natural language' and of any science which at some point is constrained to operate within those limits.

Thus even Salmon finds it necessary to enter certain caveats with regard to quantum mechanics despite his otherwise vigorous defence of a causal-explanatory approach. On balance, he concludes, 'I accept the *moderate* view that "anyone who isn't worried about this problem has rocks in their head."'[45] Nevertheless, Salmon gives plenty of reasons elsewhere in his book for rejecting the argument that quantum phenomena are so radically indeterminate – or so alien to all our existing notions of logic, consistency and truth – as to place them forever outside and beyond the realm of causal explanation. After all, as he remarks, 'quantum mechanics has had more explanatory success than any other theory in the history of science.' And again, more specifically:

> Classical physics could not explain the distribution of energy in the spectrum of blackbody radiation; quantum mechanics provided the explanation. Classical physics could not explain the photo-electric effect; quantum mechanics could. Classical physics could not explain the stability of atoms; quantum mechanics could. Classical physics could not explain the discrete spectrum of hydrogen; quantum mechanics could. Unless one wants to retreat to the old position that science never explains anything, it seems implausible in the extreme to deny that quantum mechanics has enormous explanatory power.[46]

As we have seen, it is not only unreconstructed positivists or adherents to some version of the 'old', deductive-nomological (covering-law) paradigm who might wish to deny that claim. For the idea has gained ground among postmodernists also – as witness Lyotard's remarks cited above – that quantum mechanics can always be relied upon (along with talk of undecidability, complementarity, fractals, chaos theory and the like) to demolish all ideas of scientific enquiry as aimed towards establishing causal explanations of well-defined physical phenomena.[47] It is also a part of Feyerabend's case for an anarchistic theory of paradigm change that would radicalize Kuhn's (by comparison) fairly moderate account to the point of rejecting every last standard of truth, method, logic, observational warrant, etc.[48] And there are others in the broadly deconstructionist camp – Arkady Plotnitsky among them – who have sought to enlist Derrida along with Bohr (plus a range of associated 'post-metaphysical' thinkers on the borders of speculative science and cultural or literary theory) in order to press this case for the passage to an epoch of radical undecidability across all the present-day 'advanced' disciplines of thought.[49] What these thinkers share is an understanding of quantum mechanics and related phenomena as leaving no room either for causal explanations or for anything that even remotely resembles the 'classical' logic of scientific enquiry.

Plotnitsky is undoubtedly right – up to a point – in his claim that Derrida's thinking has developed in close proximity to various topoi in the recent philosophy of mathematics and theoretical physics. Indeed, he can muster some persuasive textual evidence for this claim, including Derrida's allusions to Gödel's theorem by way of providing a formalized statement of the aporias – or the logics of undecidability – that emerge through a deconstructive reading of texts in the Western philosophical tradition.[50] Plotnitsky also has some good points to make about Derrida's and Bohr's strikingly cognate reflections on the limits imposed by 'classical concepts'; the problematic relation between grounding (a priori) intuitions and constructivist proof procedures, especially in post-Euclidean geometry;[51] and the extent to which quantum-mechanical 'explanations' require a departure from established norms of logico-deductive or causal-explanatory thought. But I should still maintain that his reading – both of Derrida and of these issues in the philosophy of physics and mathematics – is more postmodernist than deconstructive in its general line of approach. That is, Plotnitsky tends to play down both the logical rigour and the textual specificity of Derrida's readings, taking it always as a presupposition (almost an article of faith) that the 'restricted economy' of determinate truth-values is everywhere caught up in a 'general economy' of *différance*, supplementarity, undecidability, complementarity, dissemination and so forth.

Plotnitsky's main point of reference for the use of these terms is Derrida's early essay on Hegel, 'From Restricted to General Economy: a Hegelianism without reserve'.[52] Here his intertextual approach is via Bataille and the notion of an infinitized expenditure, an 'economy' – so far as that term still applies – of the pure, uncalculating gift without thought of recompense or return, such as would open Hegel's text to a reading beyond the various forms of premature dialectical closure.[53] However, we should also bear in mind – as always when Derrida invokes such notions as 'unlimited semiosis', dissemination, the 'free play' of the signifier, etc. – that he does so with a view to questioning established (*de jure*) protocols of reading and method, and not through some postmodern libertarian idea that reading should *in practice* be unconstrained by standards of logic, consistency or truth. Of course Plotnitsky is aware of this, with regard not only to deconstruction but also to the kinds of loose analogical talk that are often substituted for serious discussion in popularizing accounts of the 'new physics'. Thus he cannot be accused – like many 'literary' deconstructionists – of completely ignoring the conceptual rigour of Derrida's arguments and exploiting his texts for whatever they are worth in the way of new-found hermeneutic licence. But there are some passages in his book – especially those where Bataille's influence looms large – that could easily be taken as lending credence to just such a view of the dizzying prospects henceforth opened up by a passage beyond the naïve certitudes of truth, logic, 'Western metaphysics', or the history of 'logocentric' thought.

At this point it is worth recalling what Derrida has to say, in his essay 'The Double Session', about the pertinence of Gödel's Theorem and the precise specification of that 'undecidability' that a deconstructive reading has here brought to light in the texts of Plato and Mallarmé. 'An undecidable proposition,' he writes, 'is [one] which, given a system of axioms governing a multiplicity, is neither an analytical nor a deductive consequence of those axioms, nor in contradiction with them, neither true nor false with respect to those axioms. *Tertium datur*, without synthesis.'[54] One can see why Plotnitsky cites this passage as an indicator of the manifold associative links between (1) 'undecidability' in its Gödelian (mathematical or set-theoretic) sense, (2) Derrida's usage of the term in various (also highly specific) contexts of argument, and (3) the whole range of scientific, literary and cultural phenomena which can also be described – albeit with some loss of analytical precision – as partaking of a generalized 'undecidability'. In that final sentence, moreover, Plotnitsky might plausibly claim to detect a confirmation of his own symptomatic reading of Hegel as the last thinker of Absolute Reason and also, following Bataille, as the first thinker of radical difference, heterogeneity or a 'general economy' without limit or conceptual closure. Thus there seems to be at least some warrant for his view that Derrida – along with Bohr, Gödel and other prophets of postmodernity – is engaged upon a wholesale undoing of those values (like the 'classical' truth/falsehood distinction) that once held sway in the discourse of the Western natural and human sciences.

To be sure, this answers to one prominent aspect of Derrida's thinking, that is, his refusal to impose a priori restrictions upon the range of meanings or interpretive possibilities that exist for any given text. Hence his insistence that we cannot (or should not) assume too readily that we know what should count as an adequate – properly 'Hegelian' – reading of Hegel, one that would respect the priorities vested in Hegel's dialectical conception of reason as aimed towards some ultimate truth at the end of enquiry. To this extent deconstruction may rightly be seen as rejecting the appeal to established – philosophical or scholarly – protocols of meaning, authorial intention, logical entailment, contextual specificity, etc. Such is Derrida's purpose in adopting a rhetoric of 'dissemination', 'unlimited semiosis', or (in his occasional more generalized usage of that term) textual 'undecidability'. But it will soon become clear to any attentive reader that his exegetic practice is far removed from the kinds of all-licensing rhetorical 'free play' that this is often taken to imply. Indeed, it is worth pointing out that the metaphor of *free play*, as Derrida deploys it, derives from the terminology of precision engineering. In that context it signifies the exact degree of tolerance (or 'play') necessary for a bearing – or other such mechanical device – to function with maximum efficiency, or with least energy loss through frictional resistance. This usage is about as remote as possible from the *au courant* (pseudo-Derridean) idea of deconstructive reading as wholly unconstrained by the requirements of logic, consistency or truth. The confusion here can best be brought out by comparing the equivalent argument as applied (say) to Gödel's Theorem or Heisenberg's uncertainty principle. Thus it will scarcely be claimed – unless (perhaps) by postmodernists like Lyotard or by popularizing adepts of the 'new physics' – that these theories are themselves exempt from criteria of logical or demonstrative proof since, after all, they give reason to doubt the possibility of proof or accurate measurement beyond a certain point.

No doubt such arguments do give rise to certain problems of 'self-reflexivity', problems that have lately received much attention not only from logicians but also – with different ends in view – from cultural theorists keen to demolish the distinction between first-order 'natural' language and the second-order discourse of 'meta-linguistic' analysis or critique.[55] But it is equally clear that the very possibility of raising such questions depends upon a prior (prerequisite) commitment to probative standards of validity and truth, in the absence of which any talk of 'undecidability', 'complementarity', the 'uncertainty principle' and so forth amounts to no more than a species of rhetorical imposture. The point can best be made with regard to deconstruction by citing the passage that immediately follows Derrida's summary of Gödel's Theorem, cited above. 'Undecidability,' he writes,

> is not caused here by some enigmatic equivocality, some inexhaustible ambivalence of a word in a 'natural' language. . . . It is not a matter of repeating what Hegel undertook to do with German words like *Aufhebung*, *Urteil*, *Meinen*, *Beispiel*, etc., marveling over that lucky accident that instals a natural language within the element of speculative dialectics. What counts here is not the lexical richness, the semantic infiniteness of a word or concept, its depth or breadth, the sedimentation that has produced inside it two contradictory layers of signification [continuity and discontinuity, inside and outside, identity and difference, etc.]. What counts here is the formal or syntactical *praxis* that composes and decomposes it.[56]

If one picks up a distant rumbling here it is most likely the sound produced as whole schools of Derrida commentary collapse under each of those repeated negative characterizations. Thus the passage leaves as little room for an orthodox Hegelian (dialectical) reading as for a Heideggerian harking-back to the etymopoeic sources and resonance of those German words, or again, for a facile postmodernist rhetoric of radical 'difference', 'discontinuity', infinitized 'free play' and the like.

Of course, such declarations would count for nothing were they not borne out by Derrida's practice in the reading of particular texts. However – as I have argued here and elsewhere – the above passage describes very exactly what is entailed by a deconstructive reading in the rigorous ('formal or syntactical') mode.[57] It would apply, for instance, to the role of those various deconstructive key terms – *pharmakon* in Plato, *supplément* in Rousseau, *parergon* in Kant, *différance* (or differing-deferral) in Husserl – whose undecidability is *not* a matter of 'lexical richness' or 'semantic infiniteness', nor yet of their gesturing towards some purely notional realm of open-ended textual 'free play'.[58] Rather it is the logical grammar of these words – more precisely, their role in complex (often contradictory) chains of logico-syntactic entailment – that enables them to function as internal points of leverage for a deconstructive reading. And it is here, I would argue, that we can best look for signs of a genuine convergence – rather than a loose set of ad hoc analogies – between Derrida's project and developments in recent philosophy of science.

III

The most promising text for this purpose would be his early book-length introduction to Husserl's essay 'The Origin of Geometry'.[59] Briefly summarized, Derrida's argument has to do with the question (first raised by Pythagoras and Plato) of whether there exists a realm of a priori geometrical truths – of 'essences', 'forms', 'primordial intuitions' or 'ideal objectivities' – that could serve to guarantee their accurate and faithful transmission from mind to mind down through the history of mathematics, science and kindred disciplines of thought. On the one hand, this process would necessarily involve the positing of an 'original' – e.g., Euclidean – moment of discovery or access to truth, a moment which thereafter could always in principle be repeated (or 'reactivated') by any thinker in possession of those same a priori indubitable grounds of knowledge. Such had been Plato's doctrine of *anamnesis*, of knowledge as a kind of 'unforgetting' or thinking-back to primordial truths that are somehow latent in the mind (or soul) of every rational person, and which a wise teacher – such as Socrates – can elicit through a process of 'education' in the strict (etymological) sense of that word. Hence the set-piece scene of instruction in the *Meno* where Socrates claims to demonstrate this point by showing how a slave-boy – hitherto 'ignorant' of such matters – is none the less able to intuit certain basic principles of geometry with minimal prompting from his tutor. But on the other hand, as Derrida shows in his reading of Husserl, it is strictly impossible to conceive such truths as taking rise from repeated acts of a priori intuition that would somehow be revealed to the individual mind in a state of receptive grace. For even in the case of reasoning *more geometrico* there is always a certain context of acquired knowledge – of rules, guidelines or criteria for what should count as an adequate demonstrative proof – which necessarily precedes the moment of achieved understanding and which constitutes its very condition of possibility.

Thus Socrates can bring off his neat pedagogical trick with the slave-boy only by contriving – as he does so often in Plato's dialogues – to leave his interlocutor with no real choice but to follow in the steps of his (Socrates') subtly concealed instruction. And with Husserl likewise there is a constant slippage from the idea of geometrical truths as arrived at through a sequence of repeated primordial intuitions to the idea of such truths as necessarily preserved and transmitted through a history of thought which depends upon writing – or some means of reiterable graphic inscription – for its very existence. Of course this is 'writing' in Derrida's greatly extended sense of that term, a sense that is developed – refined and elaborated – in his texts on Plato, Rousseau, Husserl, Saussure, Austin, Lévi-Strauss and others.[60] In brief, it refers to that order of a priori structural necessity which enables language to signify – or communication to take place – in the absence of any direct appeal to the utterer's intention, intuitive self-evidence, or truth as existing in a realm of Platonic ideas above or beyond the realm of material inscription.

This is precisely Derrida's point about the problems encountered by Husserl in his essay on 'The Origin of Geometry'. That essay has two main objectives. It seeks to provide apodictic justification for certain of the grounding intuitions – and hence the axiomatic deductive procedures – that characterize the domain of geometrical thought. But it also sets out, most importantly for Husserl, to show that any *ultimate* justification must pass beyond this level of (presumed) objective self-evidence to elucidate the various structures of consciousness that make such thinking possible. That is to say, there must always be a further (phenomenological) appeal to modes of self-conscious reflective understanding which explain how it is that reason can achieve a comprehension of those 'absolute ideal objectivities' that would otherwise exist in a realm quite apart from the order of humanly accessible truth. Such is the project of transcendental phenomenology, as Husserl conceives it: a discipline of thought which 'suspends' or 'brackets' the data of naïve sense-certainty, along with any concepts that cannot be explicated by means of a primordial intuition which provides them with a grounding in the a priori structures of knowledge and experience.[61]

However, it is precisely in the tension between 'genesis' and 'structure' that Derrida locates the chief problem with Husserl's approach. What emerges from a deconstructive reading of this and other Husserlian texts is 'the principled, essential, and structural impossibility of closing a structural phenomenology'.[62] That is to say, there always comes a point of aporia where Husserl is obliged to navigate a course between 'the Scylla and Charybdis of a logicizing structuralism and a psychologistic geneticism'.[63] The first has to do with the requirement that thinking have access to those permanent truths – 'absolute ideal objectivities' – which constitute the domain of a priori knowledge in geometry, mathematics and the sciences in general. The second (equally imperative for Husserl) points back towards the psychologistic 'genesis' of such truths in the forms and modalities of lived experience which render them both humanly intelligible and subject to the process – the investigative rigours – of phenomenological reflection and critique.

Thus, as Derrida puts it, '[Husserl's] is a philosophy of essences always considered in their objectivity, their intangibility, their apriority.' But by the same token it is 'a philosophy of experience, of the temporal flux of what is lived, which is the ultimate reference'.[64] This applies not only to Husserl's arguments in 'The Origin of Geometry' but also – as Derrida shows elsewhere – to issues raised by his philosophy of language and his phenomenological enquiries into the forms and modalities of internal time-consciousness.[65] In each case there is a deep and inescapable conflict – an *aporia* in the strictest sense of that term – between the claims of 'absolute ideal objectivity' and the claims of lived (expressive or temporal) experience. And in each case, moreover, there is a covert appeal to *writing* or some analogue thereof as a means of explaining how ideas given in the mode of intuitive self-evidence can yet attain that level of 'ideal objectivity' required in order for such ideas to be communicated from one mind to another.

Hence Derrida's claim, cited above, that writing (or the very *possibility* of writing, conceived as an actual or virtual resource within thought, language and culture) is that which assures the 'traditionalization' of knowledge or the objects of knowledge. This it does – to repeat – by 'emancipating sense from its actually present evidence for a real subject and from its present circulation within a determined community'.[66] As concerns Husserl's philosophy of language, this argument takes the form of a rigorous demonstration – on transcendental or 'conditions of possibility' grounds – that Husserl cannot make good his claim for the absolute priority of 'expressive' over (merely) 'indicative' signs, since the latter are prerequisite to any conceivable sign-system, language or discourse of articulate thought. Thus: 'Whenever the immediate and full presence of the signified is concealed, the signifier will be of an indicative nature.'[67] And again: 'although discourse would not be possible without an expressive core, one could almost say that the entirety of speech is caught up in an indicative web.'[68] There is a similar case to be made with regard to the grounding suppositions of Husserl's book *The Phenomenology of Internal Time-Consciousness*.[69] Here also it is a matter of showing – again through a form of negative transcendental deduction – that such consciousness cannot be defined or represented by appealing to a moment of pure, self-present intuition, a *hic et nunc* of temporal experience which would then constitute an axial point for the interpretation of past and future modalities.

In fact Husserl's analysis is rather more complex, since he conceives of the conscious 'present' not so much as a punctual (indivisible) point in time but as a moment that partakes of both 'retentive' and 'protentive' vectors. The former have to do with those traces of just-past experience – those vivid after-images or items of immediate recollection – that impinge so directly on our present awareness as to constitute an integral part of that awareness from a phenomenological viewpoint. And the same applies symmetrically to that forward-reaching ('protentive') movement of thought which anticipates the course of experience from one moment to the next, and which thus makes it possible for consciousness to achieve a unifying grasp – or transcendental synthesis – of the otherwise inchoate flux of sensory data. Both dimensions are characterized, so Husserl maintains, by their contrast with those longer-term (past and future) perspectives which lack such an intimate involvement with the act of self-present reflective thought. That is, they may be summoned to memory on the one hand, or figure in the realm of as yet unrealized (i.e., imagined or anticipated) perceptions and experiences on the other. But in each case there is a crucial difference to be marked – in phenomenological terms – between the 'living present' which contains elements of retentive or protentive consciousness and those receding vistas, past and future, which extend beyond that *de jure* privileged sphere. What Derrida brings out in his reading of Husserl is the 'absolute and principled' *im*possibility of holding these distinctions in place. And he does so by showing that they are undermined by the logic of Husserl's own argument, for instance, by his finding it necessary to resort to the language of 'representation' – which should properly belong to items of long-term memory – in order to characterize the process or structure of immediate 'retentive' consciousness. Thus where Husserl sought to establish a qualitative difference between the temporal orders of 'perception and non-perception' his text bears witness rather to 'a difference between two modifications of non-perception'.

It is at this point – in a highly specific context of argument often ignored by his exegetes – that Derrida first deploys the deconstructive key-term *différance*, a term whose play on the nominalized forms of the two French verbs 'to differ' and 'to defer' captures precisely the aporias generated by Husserl's thinking about language and time-consciousness.[70] Such is the process of differing-deferral that affects Husserl's argument wherever he attempts to make good his claims for the priority of 'expressive' over 'indicative' signs, or for temporal experience as centred upon a moment of pure, self-present intuition. For *on Husserl's own account* – and against his express declarations elsewhere – the present can be defined only in contrastive or differential terms, that is to say, as an empty locus or a virtual point of intersection between two vectors (the retentive and protentive) which affect that moment at source. What is thereby revealed is a counter-logic that runs athwart Husserl's overt professions of intent and which manifests – in Derrida's words – the 'movement of différance' that always inhabits 'the pure actuality of the now'. By the same token, '[w]henever the immediate and full presence of the signified is concealed, the signifier will be of an indicative nature.'[71] And this despite Husserl's reiterated claim that the project of phenomenological enquiry should *in principle* concern itself with those modes of authentic (primordial) intuition that are given to consciousness only in the form of 'expressive' signs and their intentional correlates, and which may be quite absent from the realm of mere 'indicative' (conventional or arbitrary) signification. For that claim is in effect countermanded – and that stipulative order of priorities thrown into doubt – by Husserl's on occasion conceding what amounts to a structuralist view of the conditions of possibility for thought and language in general. Quite simply, there is no conceiving of a language that would meet this requirement of expressive (self-present) transparency and intuitive grasp, or that would somehow communicate meaning from mind to mind through the privileged access to a realm of thought – a domain of 'absolute ideal objectivities' – exempt from the detour through indicative signs, linguistic conventions, or 'writing' in Derrida's extended sense of that term. On the contrary: it is only in so far as language partakes of these latter conditions that it can possibly be thought of as communicating *any* kind of meaning, whether utterer's intentions (spoken or written) or intuitions, ideas and concepts that aspire to the status of absolute (*a priori* self-evident) truth. And this by an order of structural necessity which announces itself clearly in Husserl's texts despite and against his repeated statements to a contrary effect.

Derrida's readings of Husserl have received a good deal of informed critical commentary, so I shall say no more about them here.[72] My point is that they have little or nothing in common with the loose talk of *différance*, undecidability, textual 'free play' and so forth which often passes for 'deconstructive' thinking among disciples and opponents alike. What such readings (or non-readings) standardly ignore is the particular context of argument – in this case the issue between structuralism and Husserlian phenomenology – from which those terms take rise and which alone provides an adequate philosophical justification for their usage as Derrida deploys them. Nor are these issues in any way confined to the currency of present-day 'advanced' theoretical debate. Thus when Derrida writes that 'a certain structuralism has always been philosophy's most spontaneous gesture' he is clearly not referring to developments in the wake of Saussurean linguistics or to that (erstwhile) vaunted 'revolution' in the human sciences that started out from Saussure's inaugural programme. Rather, he has in mind the commitment to those values of conceptual rigour, clarity and truth – to the quest for 'absolute ideal objectivities' in thought and language – which has been the single most characteristic feature of Western philosophical tradition from Plato to the present.

This is also why Derrida remarks more than once that transcendental phenomenology, whatever its aporias or localized blind spots of argument, none the less constitutes a simply *indispensable* stage in the uncovering of just those problems that confront the structuralist enterprise.[73] In fact, he interprets these two projects – phenomenology and structuralism – as engaged in a process of mutual interrogative exchange and critique which resumes some of the most central issues in the history of Western thought. Thus Husserl's insistence on the 'expressive' character of signs – on their irreducibility to the order of given, preconstituted sense – allows us to perceive, in Derrida's words, how structuralism lives on 'the difference between its promise and its practice', since there is always a signifying (intentional) surplus in language that eludes the best efforts of structural analysis. And conversely, what is entailed by a 'structuralist' approach – on Derrida's broad (but legitimate) construal of that term – is the kind of transcendental argument from the conditions of possibility for thought and language that constitutes philosophy's 'most spontaneous gesture'. It is in the aporetic movement between these opposed yet strictly inseparable orders of reflection that thinking discovers 'the principled, essential, and structural impossibility of closing a structural phenomenology'.[74]

IV

This is also – as I have argued – the point at which deconstruction bears most directly on issues in contemporary science and philosophy of science. For if there is one topic that has preoccupied commentators on quantum mechanics and related phenomena it is the question as to whether we could *possibly* succeed in representing, understanding or explaining those phenomena given their remoteness from received ('classical') conceptions of scientific thought and method. In his debates with Niels Bohr, Einstein took the view that such perplexities were owing to the limits imposed by our present inadequate means of observation or conceptualization.[75] That is to say, they resulted from the lack of a unified theory that would finally subsume quantum mechanics in the way that General Relativity had subsumed Newtonian physics, i.e., as a special case whose laws could be taken to hold under certain (local) conditions and for certain well-defined (but limited) purposes. Plotnitsky – no friend to this 'classical' picture – describes the position as follows:

> Einstein recognized that quantum mechanics or complementarity was a rigorous and effective theory in so far as both its mathematical formalism and its ability to account for experimental data are concerned. He thought that it would continue to have its place as a partial account when the true *complete* theory, free of uncertainty, was discovered. Einstein resisted, and in fact rejected, quantum mechanics and complementarity primarily on the grounds of their incapacity to offer a complete conceptual synthesis, as he understood it.[76]

For Bohr, conversely, any 'true complete theory' such as Einstein envisaged would perforce be expressed in a language that failed to capture the strangeness of quantum phenomena in so far as it adhered to the 'classical' requirements of excluded middle, non-contradiction, observational warrant, explanatory rigour, theoretical consistency, and so forth.[77] Or again: the very notion of some ultimate 'conceptual synthesis' involves a whole range of such values – including (as Plotnitsky interprets it) a Hegelian notion of truth at the end of enquiry – which simply don't hold in the context of quantum mechanics and complementarity. It was a kindred conviction that led Bohr to issue his well-known series of statements regarding the unbridgeable gulf that existed between quantum phenomena and the categories of received (everyday or scientific) discourse.

The following is a representative sample of what Heisenberg noted as the 'darkly metaphysical' quality of Bohr's thinking:

> [H]owever far the phenomena transcend the scope of classical physical explanation, the account of all evidence must be expressed in classical terms. The argument is simply that by the word 'experiment' we refer to a situation where we can tell others what we have done and what we have learned and that, therefore, the account of the experimental arrangement and of the results of observations must be expressed in unambiguous language with suitable application of the terminology of classical physics.[78]

This passage is also characteristic for the problems it poses with regard to the justificatory grounds – the reasons, arguments or evidence of whatever kind – that Bohr might have had for advancing such strongly counter-intuitive claims. After all, it is hard to see how anyone could be in a position to *know* (or to offer any adequate reason for supposing) that there existed phenomena – 'noumena' might perhaps be a better term here – whose very nature it was to exceed or elude the furthest capacities of human conceptual grasp. All the more so since Bohr, in that series of debates with Einstein, must clearly have accepted at least one tenet of ontological realism, i.e., the belief that such experiments could be used to determine how the wave or the particle hypotheses would actually fare under given (albeit idealized or imaginary) experimental conditions. Nor is this argument in any way disconfirmed by the fact that the upshot of Bohr's (and of Heisenberg's) experiments was to demonstrate the conditions of *im*possibility – the obstacles to any possible 'true complete theory' – that characterized the scientist's dealing with phenomena in the quantum domain. Thus it might be the case, as Bohr maintained, that there would always necessarily come a point where these phenomena were subject to different (incommensurable) modes of description and conceptualization, for instance, with the wave/particle dualism or the impossiblity of simultaneously measuring a particle's position and velocity. This is what united Bohr and Heisenberg in the view that such elusive or paradoxical properties must be taken as pertaining to quantum 'reality' rather than as products or artefacts of our limited knowledge concerning them. It was also what set them at odds with Einstein's more sanguine belief that those anomalies would eventually be laid to rest with the advent of some alternative, more satisfactory theory. Still, their very differences bore witness to the shared conviction that results could be achieved – or theories tested – by devising a relevant thought-experiment, adjusting its parameters to the case in hand, and then seeking to determine what would follow from this or that conjectural starting-point.

In other words, both parties adhered to at least one item of the 'classical' scientific world-view. This was the belief – espoused by many thinkers from Galileo down – that science can be justified in proceeding on the basis of hypotheses which exceed the present limits of physical experiment (or phenomenal cognition) but are none the less capable of yielding a knowledge with its own criteria of truth and falsehood.[79] At an early stage, such conjectures may as yet be so far underdetermined by the current state of scientific knowledge – or the best experimental results – as to constitute purely metaphoric or speculative ventures in thought. But their status changes – undergoes the crucial transition, in Bachelard's terms, from 'metaphor' to 'concept' – with the advent of more precise observational data or a more adequate theoretical-explanatory framework whereby to make sense of those data.[80]

Salmon puts the case most succinctly as follows, with reference to the history of atomist conceptions in physics and philosophy of science. 'As everyone knows,' he writes,

> a primitive form of atomism was propounded in antiquity by Democritus, Lucretius, and others, but no very strong empirical evidence in support of such a theory was available before the beginning of the nineteenth century. However, from the time of Dalton's work (early in that century) until the end of the century, the atomic hypothesis was the subject of considerable discussion and debate. During this period, well-informed scientists could reasonably adopt divergent viewpoints on the question. . . . Within a dozen years after the turn of the century, however, the issue was scientifically settled in favour of the atomic/molecular hypothesis, and the scientific community – with the notable exception of Ernst Mach – appears to have been in agreement over it.[81]

There are many such cases in the history of science – from nuclear physics and molecular biology to astronomy and other disciplines – where conjecture has often run ahead of discovery, or where the ground has been laid by speculative thinkers for the subsequent work that will serve to substantiate their claims.[82] As we have seen, Salmon is a lot more cautious about extending this argument to issues in the realm of quantum mechanics. Nevertheless, there is a sense in which, here also, such thinking involves a degree of ontological commitment, that is, the presumption that hypothetical results – no matter how baffling or counter-intuitive – may reveal certain features of the quantum domain that *necessarily* follow from a thought-experiment conducted with due respect to known laws of physics and consequential reasoning. The point can be made more simply by asking what purpose those experiments could possibly have served were it not for this presumed correspondence (or convergence) between the order of thought and the order of physical or causal necessity. And the argument holds even in cases – like those adduced by Bohr – whose upshot is to demonstrate the *impossibility* of carrying out certain sorts of measurement or deciding between different (incommensurable) paradigms of method, observation and theory. For this conclusion is still arrived at through a thought-experiment whose validity – or (for that matter) whose openness to subsequent falsification – rests on the premise that thought can have access to objects, processes or events in the quantum-physical realm that would otherwise lie beyond the reach of human enquiry.

Take, for instance, the following description by Plotnitsky of the way that Heisenberg set about proving his thesis with regard to the uncertainty conditions that obtained for all possible measurements of quantum phenomena. 'According to Heisenberg's reasoning,' he writes,

> in order to measure the position of an electron with sufficient accuracy one would need to illuminate it with light of a very short wavelength and to observe the reflected light in a microscope. Since the resolving power of the lens of the microscope is limited, the wavelength of the light would have to be as short as that of gamma radiation. The interaction of an electron with gamma rays, however, would discontinuously change the momentum of the electron, which would make our knowledge of the momentum of the electron uncertain. The strict causality in determining the electron's motion – its 'history' – would thereby be prohibited.[83]

My point about this passage is that it manages to state – to explain with admirable clarity and precision – just how Heisenberg derived his result and just why that result necessarily applied for experiments of the kind here envisaged. That is to say, his reasoning is no less rigorous for the fact that it issues in a generalized statement of the uncertainty conditions that attach to all quantum-physical observations or measurements. One could make this point in various ways, beginning with the simple remark that Heisenberg posits the existence of putative realia – electrons, wavelengths, gamma rays, etc. – whose nature or whose characteristic modes of behaviour science has already done much to explain, but whose status would clearly be called into doubt by any wholesale application of the uncertainty principle. Thus the very logic and probative force of Heisenberg's argument depends on his adopting a degree of ontological realism with regard to precisely those entities which – on a rigorous application of the principle – would figure only as artefacts of this or that chosen system of measurement, with the result that no genuine issue could arise between rival (e.g., wave or particle) theories. In short, there is a problem about how we construe such claims, caught up as they are in a form of self-referential paradox or a failure to reckon with their own implicit criteria of truth, method, consequential reasoning, ontological commitment, and so forth.

This is perhaps why the 'natural' attitude – or the appeal to some one or more of just those 'classical' criteria – so often tends to reassert itself in writings, like the passage from Plotnitsky cited above, which expressly reject such governing values. Peter Gibbins has noted this tendency in various (otherwise divergent) accounts of quantum-physical 'phenomena' and the changes they require in our normal descriptive or explanatory habits of thought.

> If the wave ontology seems natural for light, at least to the classical mind, then it seems anything but natural for matter. The common resolution of wave-particle has been its collapse in favour of the particle. One finds a preference for the particle view in a whole variety of quantum-mechanical realisms, from the naive Popperian view that quantum systems are particles, to the sophisticated quantum-logical view that wave-particle duality and the indeterminateness of the quantum world can be smoothed away with quantum logic.[84]

To which he might have added that even those thinkers (like Bohr) most determined *not* to 'smooth away' the irreducible strangeness of quantum phenomena – that is, their recalcitrance to theorization in 'classical' terms – are none the less obliged *on their own submission* to adopt such a classical logic and language when attempting to describe those phenomena. In part this has to do with a weak version of the argument from conditions of possibility, i.e., that in order to communicate at all these thinkers have no alternative but to utilize currently available linguistic or logico-conceptual resources. But there is also a stronger line of argument which holds that quantum-theoretical theories and conjectures entail certain basic ontological commitments which *cannot* be abandoned – whatever their problematic outcome – since to do so would render the exercise void of demonstrative or probative force. At very least it must be assumed that, in the quantum realm as elsewhere, what is possible (or impossible) as a matter of informed and rigorous conjecture will apply also to those various explananda – events, processes, particles, wave-forms, state-vectors or whatever – which constitute the relevant object-domain.

In this sense the case with quantum mechanics is not, after all, so utterly remote from that which has often obtained during periods of Kuhnian 'revolutionary' science when there existed (as yet) no adequate means of observation or physical measurement for hypotheses that were none the less strongly borne out by the best current theory. This will typically involve on the one hand an appeal to well-attested items of scientific discourse – a reference (say) to electrons, wave-forms, molecular structures, gamma rays, etc. – and on the other a passage beyond those items to a deeper level of enquiry at which they might yet turn out to figure as limiting cases or as products of a restricted (regional) ontology. There are many text-book examples of this process at work. They would include (most famously) the transition from Newtonian to Einsteinian concepts of the space–time co-ordinate system, the development of various (increasingly complex and detailed) accounts of the atomic structure of matter, and the passage from Euclidean to non-Euclidean geometries. Of course, this claim is precisely what is denied on the Kuhnian (ontological-relativist) view. Thus, for instance, it is argued that 'mass', 'space' and 'time' have entirely different (incommensurable) meanings as they figure in the discourse of Newtonian or Einsteinian physics; that the notion of atomic 'structure' has likewise been subject to an ongoing process of radical revision which prevents any straightforward comparison between rival theories; and again, that the advent of non-Euclidean geometries has brought about a wholesale shift in the sense of elementary terms like 'point', 'line' and 'plane'.[85] However, such arguments typically ignore the fact that even the most radical shift of allegiance from one to another paradigm will conserve at least some crucial elements from the earlier theory. For we should otherwise have no means of understanding just why those changes came about, or in what sense they constituted an advance – a more adequate or deeper explanatory account – as compared with previous work in the field.[86] As Ian Hacking puts it: 'one can see by direct inspection that knowledge has grown. . . . The point is not that there is knowledge, but that there is growth; we know more about atomic weights than we once did, even if future times plunge us into quite new, expanded, reconceptualizations of those domains.'[87]

There would seem to be two minimal requirements in order for the history of science to make any kind of sense. One is the assumption – as in Hacking's appeal to our knowledge of the growth of knowledge – that criteria exist whereby certain theories can be judged to have displaced (or at any rate improved upon) others by virtue of their greater explanatory yield. The second has to do with those observed or postulated entities – whether planets, tidal motions, electrons, atomic weights, DNA molecules, genes, viruses or whatever – whose *precise specification* may always be subject to change but whose ontological status (or existence as objects of scientific enquiry) is in no way open to serious doubt. Indeed, as Nicholas Rescher suggests, such doubt is the present-day equivalent, in philosophy of science, of Descartes's hyperbolic thought-experiment with the demon of epistemological scepticism. What it amounts to is a failure to see that we can properly and justifiably claim to have certain kinds of knowledge of certain kinds of object without going on to erect such claims into a cast-iron orthodox doctrine. On the contrary: 'It is the very limitation of our knowledge of things – our recognition that reality extends beyond the horizons of what we can possibly know or even conjecture about it – that betokens the mind-independence of the real.'[88]

References

1 See especially Rom Harré, *The Principles of Scientific Thinking* (Chicago: University of Chicago Press, 1979); *The Philosophies of Science* (London: Oxford University Press, 1972); Rom Harré and E. H. Madden, *Causal Powers* (Oxford: Blackwell, 1975); J. Aronson, R. Harré and E. Way, *Realism Rescued: how scientific progress is possible* (London: Duckworth, 1994); also Roy Bhaskar (ed.), *Harré and his Critics* (Blackwell, 1990).

2 Roy Bhaskar, *Scientific Realism and Human Emancipation* (London: Verso, 1986); also Bhaskar, *A Realist Theory of Science* (Leeds: Leeds Books, 1975); *The Possibility of Naturalism*, 2nd edn (Hemel Hempstead: Harvester-Wheatsheaf, 1989); *Reclaiming Reality: a critical introduction to contemporary philosophy* (London: Verso, 1989); *Philosophy and the Idea of Freedom* (Oxford: Blackwell, 1991); and *Dialectic: the pulse of freedom* (Verso, 1993).

3 Wesley C. Salmon, *Scientific Explanation and the Causal Structure of the World* (Princeton, NJ: Princeton University Press, 1984), p. 274.

4 See also Wesley C. Salmon, *The Foundations of Scientific Inference* (Pittsburgh: University of Pittsburgh Press, 1967); *Four Decades of Scientific Explanation* (Minneapolis: University of Minnesota Press, 1989); Salmon (ed.), *Hans Reichenbach: logical empiricist* (Dordrecht: D. Reidel, 1979); Adolf Grunbaum and W.C. Salmon (eds), *The Limitations of Deductivism* (Berkeley and Los Angeles: University of California Press, 1988).

5 See Thomas S. Kuhn, *The Structure of Scientific Revolutions*, 2nd edn (Chicago: University of Chicago Press, 1970); also Gary Gutting (ed.), *Paradigms and Revolutions* (Notre Dame, IN: University of Notre Dame Press, 1980); Ian Hacking (ed.), *Scientific Revolutions* (London: Oxford University Press, 1981); John Krige, *Science, Revolution and Discontinuity* (Brighton: Harvester, 1980).

6 See, for instance, Michel Foucault, *The Order of Things: an archaeology of the human sciences* (London: Tavistock, 1970); John Dupré, *The Disorder of Things: metaphysical foundations of the disunity of science* (Cambridge, MA: Harvard University Press, 1993); Gary Gutting, *Michel Foucault's Archaeology of Scientific Knowledge* (Cambridge: Cambridge University Press, 1989); Jean-François Lyotard, *The Postmodern Condition: a report on knowledge*, trans. Geoff Bennington and Brian Massumi (Manchester: Manchester University Press, 1984); also Kuhn, *The Structure of Scientific Revolutions* and W. V. Quine, 'Two Dogmas of Empiricism', in *From a Logical Point of View*, 2nd edn (Cambridge, MA: Harvard University Press, 1961), pp. 20–46.

7 Ferdinand de Saussure, *Course in General Linguistics*, trans. Wade Baskin (London: Fontana, 1974). For criticism of this standard post-structuralist take on Saussure see, for instance, Perry Anderson, *In the Tracks of Historical Materialism* (London: Verso, 1983); Valentine Cunningham, *In the Reading Gaol: post-modernity, texts and history* (Oxford: Blackwell, 1994); J. G. Merquior, *From Prague to Paris: a critique of structuralist and post-structuralist thought* (London: Verso, 1986); Christopher Norris, *Truth and the Ethics of Criticism* (Manchester: Manchester University Press, 1984); also – for a more extended historical perspective – Hans Aarsleff, *From Locke to Saussure: essays in the study of language and intellectual history* (Minneapolis: University of Minnesota Press, 1982).

8 See especially Michel Foucault, *The Order of Things; The Archaeology of Knowledge*, trans. A. M. Sheridan Smith (London: Tavistock, 1972); *Language, Counter-Memory, Practice*, ed. D. F. Bouchard and S. Simon (Oxford: Blackwell, 1977).

9 See notes 5 and 6 above; also W. V. Quine, *Word and Object* (Cambridge, MA: MIT Press, 1960); *Ontological Relativity and Other Essays* (New York: Columbia University Press, 1969); *Theories and Things* (Cambridge, MA: Harvard University Press, 1981).

10 Lyotard, *The Postmodern Condition*; also *The Differend: phrases in dispute*, trans. Georges van den Abbeele (Manchester: Manchester University Press, 1988).

11 See Quine, 'Two Dogmas of Empiricism' and *Ontological Relativity*.
12 Lyotard, *The Postmodern Condition*, p. 112.
13 Immanuel Kant, 'Transcendental Aesthetic', in *Critique of Pure Reason*, trans. N. Kemp Smith (London: Macmillan, 1964), pp. 65–91; *Critique of Judgement*, trans. J. C. Meredith (Oxford: Clarendon Press, 1978).
14 Lyotard, *The Differend*; also *The Inhuman: reflections on time*, trans. Geoff Bennington and Rachel Bowlby (Cambridge: Polity Press, 1991).
15 Christopher Norris, 'Kant Disfigured: ethics, deconstruction, and the postmodern sublime', in *The Truth About Postmodernism* (Oxford: Blackwell, 1993), pp. 182–256; also Norris, *Truth and the Ethics of Criticism*.
16 Martin Heidegger, *Kant and the Problem of Metaphysics*, trans. James Churchill (Bloomington: Indiana University Press, 1962).
17 Michel Foucault, *The Order of Things*; also 'What Is Enlightenment?', in *The Foucault Reader*, ed. Paul Rabinow (Harmondsworth: Penguin, 1986), pp. 381–90.
18 On this relationship between politics and philosophy with regard specifically to Heidegger's reading of Kant, see Pierre Bourdieu, *The Political Ontology of Martin Heidegger*, trans. Peter Collier (Cambridge: Polity Press, 1991).
19 Immanuel Kant, *Critique of Pure Reason*.
20 See Norris, 'Kant Disfigured'.
21 See Richard Rorty, *Consequences of Pragmatism* (Minneapolis: University of Minnesota Press, 1982); *Essays on Heidegger and Others* (Cambridge: Cambridge University Press, 1991); *Objectivity, Relativism, and Truth* (Cambridge University Press, 1991).
22 Wesley C. Salmon *Scientific Explanation and the Causal Structure of the World*, p. 273.
23 Ibid., p. 274.
24 Ibid., p. 272.
25 Ibid., p. 272.
26 Ibid., p. 271.
27 See, for instance, Richard B. Braithwaite, *Scientific Explanation* (Cambridge: Cambridge University Press, 1953); Rudolf Carnap, *The Logical Structure of the World* (London: Routledge and Kegan Paul, 1937); Carl G. Hempel, *Aspects of Scientific Explanation* (New York: Macmillan, 1965) and *Fundamentals of Concept Formation in Empirical Science* (Chicago: University of Chicago Press, 1972); Hans Reichenbach, *Experience and Prediction* (University of Chicago Press, 1938).
28 See note 27 above; also R. Carnap, *An Introduction to the Philosophy of Science* (New York: Basic Books, 1974) and Wesley C. Salmon (ed.), *Hans Reichenbach: logical empiricist*.
29 See Kuhn, *The Structure of Scientific Revolutions* and Quine, 'Two Dogmas of Empiricism'.
30 Bertrand Russell, *Mysticism and Logic* (New York: W. W. Norton, 1929), p. 180.
31 Bertrand Russell, *Human Knowledge, its Scope and Limits* (New York: Simon and Schuster, 1948), p. 459.
32 See, for instance, J. S. Bell, *Speakable and Unspeakable in Quantum Mechanics: collected papers on quantum philosophy* (Cambridge: Cambridge University Press, 1987); Arthur Fine, *The Shaky Game: Einstein, realism, and quantum theory* (Chicago: University of Chicago Press, 1986); Peter Gibbins, *Particles and Paradoxes* (Cambridge University Press, 1987); Tim Maudlin, *Quantum Nonlocality and Relativity: metaphysical intima-*

tions of modern science (Oxford: Blackwell, 1993); Michael Redhead, *Incompleteness, Nonlocality and Realism: a prolegomenon to the philosophy of quantum mechanics* (Oxford: Clarendon Press, 1987).

33 See note 32 above; also A. Einstein, B. Podolsky and N. Rosen, 'Can quantum-mechanical description of reality be considered complete?', *Physical Review*, ser. 2, 47 (1935), pp. 777–808 and Niels Bohr, 'Discussion with Einstein on Epistemological Problems in Atomic Physics', in P. A. Schilpp (ed.), *Albert Einstein: philosopher-scientist* (La Salle: Open Court, 1969), pp. 199–241.

34 See entries under note 32 above; also Erwin Schrödinger, *Letters on Wave Mechanics* (New York: Philosophical Library, 1967) and – for a good introductory read – John Gribbin, *In Search of Schrödinger's Cat: quantum physics and reality* (New York: Bantam Books, 1984).

35 See, for instance, P. Davies, *The Mind of God: the scientific basis for a rational world* (New York: Simon and Schuster, 1992); J. Trefil, *Reading the Mind of God* (New York: Charles Scribner's Sons, 1989); W. Yourgrau and A. D. Breck (eds), *Cosmology, History, and Theology* (New York: Plenum, 1977).

36 Russell, *Human Knowledge, its Scope and Limits*, p. 459.

37 See entries under notes 32 and 33, above; also Michael Audi, *The Interpretation of Quantum Mechanics* (Chicago: University of Chicago Press, 1973); D. Bohm and B. J. Hiley, *The Undivided Universe: an ontological interpretation of quantum theory* (London: Routledge, 1993); Peter Forrest, *Quantum Metaphysics* (Oxford: Blackwell, 1988); John Honner, *The Description of Nature: Niels Bohr and the philosophy of quantum physics* (Oxford: Clarendon Press, 1987); Josef M. Jauch, *Are Quanta Real? a Galilean dialogue* (Bloomington: Indiana University Press, 1973); Karl Popper, *Quantum Theory and the Schism in Physics* (London: Hutchinson, 1982); A. I. M. Rae, *Quantum Physics: illusion or reality?* (Cambridge: Cambridge University Press, 1986); F. Rohrlich, *From Paradox to Reality: our basic concepts of the physical world* (Cambridge University Press, 1987).

38 See entries under note 33 above; also – on scientific thought-experiments more generally – James R. Brown, *The Laboratory of the Mind* (London: Routledge, 1991) and *Smoke and Mirrors: how science reflects reality* (Routledge, 1994); Paul Davies, 'The thought that counts: thought experiments in physics', *New Scientist*, 146 (6 May 1985), pp. 26–31; Thomas S. Kuhn, 'A Function for Thought Experiments', in *The Essential Tension: selected studies in scientific tradition and change* (Chicago: University of Chicago Press, 1977), pp. 240–65; Roy Sorensen, *Thought Experiments* (New York: Oxford University Press, 1992).

39 See Brown, *Smoke and Mirrors* for a more detailed treatment of these and other cases.

40 See Gibbins, *Particles and Paradoxes*; also M. Gardner, 'Is quantum logic really logic?', *Philosophy of Science*, 38 (1971), pp. 508–29; Susan Haack, *Deviant Logic: some philosophical issues* (Cambridge: Cambridge University Press, 1974); Peter Mittelstaedt, *Quantum Logic* (Princeton, NJ: Princeton University Press, 1994); Hilary Putnam, 'How to think quantum-logically', *Synthèse*, 29 (1974), pp. 55–61 and *Mathematics, Matter and Method*, 2nd edn (Cambridge: Cambridge University Press, 1979).

41 See Bohm and Hiley, *The Undivided Universe*; also Bohm, *Quantum Theory* (New York:

Prentice-Hall, 1951) and D. Z. Albert, 'Bohm's alternative to quantum Mechanics', *Scientific American*, 270 (May 1994), pp. 58–63.

42 See, for instance, Niels Bohr, *Atomic Theory and the Description of Nature* (Cambridge: Cambridge University Press, 1934) and *Atomic Physics and Human Knowledge* (New York: Wiley, 1958); also Henry J. Folse, *The Philosophy of Niels Bohr: the framework of complementarity* (Amsterdam: North-Holland, 1985); Honner, *The Description of Nature*; Dugald Murdoch, *Niels Bohr's Philosophy of Physics* (Cambridge: Cambridge University Press, 1987).

43 Peter Gibbins, *Particles and Paradoxes*, p. 9.

44 Ibid.

45 Salmon, *Four Decades of Scientific Explanation*, p. 186.

46 Ibid., p. 173.

47 Lyotard, *The Postmodern Condition*.

48 See Paul K. Feyerabend, *Against Method* (London: New Left Books, 1975) and *Science in a Free Society* (New Left Books, 1978).

49 Arkady Plotnitsky, *In the Shadow of Hegel: complementarity, history and the unconscious* (Gainesville: University of Florida Press, 1993) and *Complementarity: anti-epistemology after Bohr and Derrida* (Durham, NC: Duke University Press, 1994); also Plotnitsky, 'Complementarity, Idealization, and the Limits of Classical Conceptions of Reality', in *Mathematics, Science, and Postclassical Theory*, ed. Plotnitsky and Barbara H. Smith (*South Atlantic Quarterly*, 64/2, Spring 1995), pp. 527–70.

50 See especially Jacques Derrida, *'Speech and Phenomena' and Other Essays on Husserl's Theory of Signs*, trans. David B. Allison (Evanston: Northwestern University Press, 1973); *Of Grammatology*, trans. G. C. Spivak (Baltimore: Johns Hopkins University Press, 1974); *Writing and Difference*, trans. Alan Bass (London: Routledge and Kegan Paul, 1978); *Dissemination*, trans. Barbara Johnson (London: Atholone Press, 1981); *Margins of Philosophy*, trans. Alan Bass (Chicago: University of Chicago Press, 1982).

51 Derrida, *Edmund Husserl's 'Origin of Geometry': an introduction*, trans. John P. Leavey (Pittsburgh: Duquesne University Press, 1978).

52 Derrida, 'From Restricted to General Economy: a Hegelianism without reserve', in *Writing and Difference*, pp. 251–77.

53 See, for instance, Georges Bataille, *L'Expérience intérieure* (Paris: Gallimard, 1943).

54 Derrida, 'The Double Session', in *Dissemination*, pp. 173–286; p. 220.

55 See, for instance, Lyotard, *The Postmodern Condition*.

56 Derrida, 'The Double Session', p. 219.

57 See Christopher Norris, *Derrida* (London: Fontana, 1987); *Deconstruction and the Interests of Theory* (Leicester: Leicester University Press, 1992).

58 See entries under note 50 above.

59 See note 51 above.

60 See note 50 above; also Derrida, 'Signature Event Context', *Glyph*, vol. 1 (Baltimore Johns Hopkins University Press, 1977), pp. 172–97.

61 See Edmund Husserl, *Ideas: general introduction to pure phenomenology*, trans. W. R. Boyce Gibson (New York: Collier Books, 1972); *Logical Investigations*, trans. J. N. Findlay (New York: Humanities Press, 1970).

62 Derrida, ' "Genesis and Structure" and Phenomenology', in *Writing and Difference*, pp. 154–68; p. 162.

63 Ibid., p. 158.

64 Ibid., p. 156.

65 Husserl, *The Phenomenology of Internal Time Consciousness*, trans. J. S. Churchill (Bloomington: Indiana University Press, 1964); also Derrida, *Speech and Phenomena*.

66 Derrida, *Edmund Husserl's 'Origin of Geometry'*, p. 35.

67 Derrida, *Speech and Phenomena*, p. 40.

68 Ibid., p. 31.

69 See note 65 above,

70 Derrida, *Speech and Phenomena*.

71 Ibid., p. 40.

72 Among the more critical commentaries, see J. Claude Evans, *Strategies of Deconstruction: Derrida and the myth of the voice* (Minneapolis: University of Minnesota Press, 1991).

73 See especially Derrida, *Speech and Phenomena* and 'Genesis and Structure'.

74 Derrida, *Writing and Difference*, p. 162.

75 See notes 32, 33 and 37 above.

76 Plotnitsky, *In the Shadow of Hegel*, p. 40.

77 See notes 33, 37 and 42 above.

78 Cited by Gibbins, *Particles and Paradoxes*, p. 54.

79 See entries under note 38 above; also Alexandre Koyré, *Galilean Studies* (Brighton: Harvester, 1978).

80 Gaston Bachelard, *La Formation de l'esprit scientifique* (Paris: Corti, 1938); *The New Scientific Spirit* (Boston: Beacon Press, 1984); also Mary Tiles, *Bachelard: science and objectivity* (Cambridge: Cambridge University Press, 1984).

81 Salmon, *Scientific Realism and the Causal Structure of the World*, p. 214.

82 See, for instance, Michael Gardner, 'Realism and instrumentalism in nineteenth-century atomism', *Philosophy of Science*, 46/1 (1979), pp. 1–34; Jauch, *Are Quanta Real?*; J. Perrin, *Atoms*, trans. D. L. Hammick (New York: Van Nostrand, 1923).

83 Plotnitsky, *In the Shadow of Hegel*, p. 67.

84 Gibbins, *Particles and Paradoxes*, p. 21.

85 Kuhn, *The Structure of Scientific Revolutions*; also entries under note 5, above.

86 For some strong arguments against the idea of radical meaning change between paradigms, see Hartry Field, 'Theory change and the indeterminacy of reference', *Journal of Philosophy*, 70 (1973), pp. 462–81; 'Quine and the correspondence theory', *Philosophical Review*, 83 (1974), pp. 200–28; 'Conventionalism and instrumentalism in semantics', *Nous*, 9 (1975), pp. 375–405; 'Logic, meaning, and conceptual role', *Journal of Philosophy*, 74 (1977), pp. 379–409; also Michael E. Levin, 'On theory-change and meaning-change', *Philosophy of Science*, 46 (1979), pp. 407–24; David Papineau, *Theory and Meaning* (London: Oxford University Press, 1979).

87 See Ian Hacking, *Representing and Intervening: introductory topics in the philosophy of natural science* (Cambridge: Cambridge University Press, 1983).

88 Nicholas Rescher, *Scientific Realism: a critical reappraisal* (Dordrecht: D. Reidel, 1987), p. 125.

5

Hermeneutics, Anti-Realism and Philosophy of Science

I

Much recent work in philosophy of science has been concerned with the theme of ontological relativity or the extent to which – in Wittgenstein's oft-cited phrase – 'the limits of my language mean the limits of my world.'[1] One major problem with such thinking is its failure to explain how we could ever gain knowledge of world-views (or scientific systems of belief) that differed from our own in some significant respects. Donald Davidson makes the point in regard to those varieties of cultural-relativist doctrine that often go along with vaguely defined talk of 'language games', 'paradigms', 'ontological frameworks', 'conceptual schemes', etc.[2] Thus, for instance, it is claimed by ethnolinguistic relativists like Whorf that certain languages (e.g., Hopi Indian) deploy a range of lexical and logico-grammatical resources so utterly different from ours that there may be no possibility of translating between their 'mental universe' and our own.[3] It is the same line of argument that Quine pursues to a point where the process of 'radical translation' can only be envisaged on the basis of a bare stimulus-response psychology that denies all recourse to shared criteria of meaning, logic, validity conditions, evidential warrant, ontological commitment, or whatever.[4] And one could multiply examples – from late Wittgenstein to Kuhn, Feyerabend, Foucault and Lyotard – of this widespread drift towards forms of extreme epistemic and ontological relativism which take for granted the absence of adequate criteria for translating between different ('incommensurable') language games, discourses or paradigms.[5]

However, as Davidson remarks, these thinkers are all caught up in some version of the same performative contradiction. That is, they assert the non-availability of trans-paradigm or inter-linguistic criteria of meaning and truth while also purporting to locate the points at which such problems arise or to treat them in a manner that somehow allows for meaningful comparison between rival paradigms. But this could only be done, so Davidson maintains, if in fact it were possible – all sceptical doubts aside – to achieve a fair measure of communicative grasp across the otherwise unbridgeable gulf of wholesale ontological relativity. From which he concludes that we should give up thinking in terms of paradigms, discourses, theoretical frameworks, conceptual schemes, etc. What is required is a 'principle of charity' which tells us that *more likely than not* other people are right (rationally justified) in the majority of their beliefs, and therefore that we have no choice but to interpret their utterances – or construe their scientific theories – in accordance with the best going standards of truth in our (and presumably their) communities.[6] And for this purpose we can make do with a minimalist theory of truth (the Tarskian or so-called 'disquotational' theory) which takes each candidate sentence – canonically 'snow is white' – and treats it as true *iff*, i.e., if and only if, the quoted sentence corresponds to some accredited real-world state of affairs, in this case the fact of snow being white.[7] For then we have enough, as Davidson sees it, to get over the sorts of problem thrown up by Quine's, Whorf's, Kuhn's, Feyerabend's or suchlike forms of hyperinduced sceptical doubt. In other words we can carry on interpreting other people – or other scientific cultures – on the simple supposition that neither they nor we could be 'massively in error' about so many things as to render communication impossible.

This argument is convincing so far as it goes, that is, as a knock-down riposte to various forms of extravagant (and self-refuting) ontological relativism. But it is still a weak rejoinder in the obvious sense that it operates at a high level of abstract generality and fails to specify just what kinds of truth have a claim to cross-cultural or trans-paradigm validity. Thus philosophers of a relativist bent – Rorty among them – can cheerfully endorse the Davidson/Tarski line while seeing no role for 'truth' except as a semantic place-filler, a term that conveniently drops out for all practical (or pragmatist) purposes.[8] On this view the truth-predicate is wholly redundant, or at most (as F. P. Ramsey was the first to argue) just a means of adding rhetorical emphasis to statements whose sense, content or validity conditions are otherwise unaffected by their being thus recast in a formal mode.[9] And the way is then open to a holist interpretation which takes each candidate sentence as 'true' only in so far as it hangs together with the entire range of sentences that currently enjoy credence within this or that language, interpretive community, scientifc world-view, etc.

I have argued elsewhere that this reading of Davidson can be shown to get him wrong on numerous points of exegetical detail.[10] All the same, it possesses a certain prima facie plausibility, as might be gathered from Davidson's shifts of tack in response to the attempt, by Rorty and others, to enlist him as a kind of foot-dragging convert to the postmodern-pragmatist view.[11] Nor is it hard to see why he should find himself awkwardly placed to resist their arguments. For the Tarskian disquotational theory of truth is a redundancy theory both in the technical sense (i.e., that it cancels through to produce a plain statement of fact such as 'snow is white') and also in the sense that it can readily be construed as rendering such truth-talk otiose apart from its usefulness as a matter of pragmatic or linguistic convenience. Thus, as Rorty sees it, there is no real difference – no difference that makes any difference – between Davidson's approach and the position of those, like Quine, who endorse a thoroughgoing version of the argument for meaning holism and ontological relativity. One might as well accept that 'truth' can be defined (again for all practical purposes) as 'truth relative to the entire set of sentences, propositions, beliefs, attitudinal dispositions (etc.) that happen to obtain at this or that time in some given interpretive community'. That is to say, it is merely the compliment we standardly pay to those utterances that pass muster according to the current conversational rules of the game. And this despite Davidson's well-known objection to that whole line of thought which starts out from talk of paradigms, discourses, language games or other such variants on the scheme/content dualism, and which then proceeds to argue – in relativist fashion – for the impossibility of translating between or across those notional constructs. For this would count decisively against the Rortian view only if Davidson had offered a substantive (as opposed to a formal and technically redundant) theory of the truth conditions that apply in such contexts.

Indeed he has lately sought to do just that in his 1989 series of Dewey lectures on 'The Structure and Content of Truth'.[12] Here Davidson starts out by acknowledging that a Tarskian approach is not of much use – that it begs all the important questions – if one seeks to do more than establish a purely circular definition of truth for each and every sentence of any object-language taken as a whole. However, he still makes some large (and I think damaging) concessions to the line of argument that leads, so to speak, via Quine to Rorty and thence to all the problems of a wholesale relativist approach. This is mainly on account of his continuing to suppose – along with many thinkers in the 'post-analytic' camp – that philosophy goes wrong if it transgresses the boundary between offering *descriptively adequate accounts* of our first-order languages, communicative practices, modes of scientific enquiry, etc., and on the other hand attempting to *define and explain* what it is in the nature of some given explanandum that can justify the claim to provide a more adequate understanding. For the logical empiricists this approach took the form of a doctrine of 'semantic ascent', that is to say, the idea – as in Carnap and Tarski – that philosophy of science was best employed in constructing a formal (meta-linguistic) discourse which could link observation sentences to higher-level statements via the ground-rules of logical inference.[13] Such an approach would eschew talk of material (*de re*) necessity and confine itself rather to the order of *de dicto* (logico-linguistic) truth. It could thus hope to offer generalizations of a covering-law or deductive-nomological type which avoided any 'metaphysical' commitment to claims about the intrinsic (e.g., causal) properties of this or that real-world object of enquiry.

Thus in Hempel's classic formulation it was a matter of finding some valid deductive argument whose conclusion is a statement to the effect that the event for which an explanation is sought did in fact occur.[14] However, this programme turned out to be short-lived, as Quine and others set about deconstructing the analytic/synthetic distinction and, along with it, the idea that any firm line could be drawn between a first-order language of observation statements borne out by empirical data and a second-order language of logical entailment relations.[15] For there was then, so it seemed, no halting the drift towards a holistic theory of meaning and truth that treated all beliefs as ontologically on a par in respect of either their empirical status or their claim to find warrant in the ground rules of logical inference. And from here it was no great distance to the Rortian neo-pragmatist view – supposedly endorsed by Davidson – which holds truth to be simply a matter of preferential language games, descriptions or 'final vocabularies', habits of construing the (so-called) 'evidence' according to a (so-called) 'logic' of enquiry which in fact simply answers to our own ideas of what is 'good in the way of belief'.[16] For there could be no effective argument against these later, more wholesale versions of the 'linguistic turn' once philosophy of science had renounced all interest in causal-explanatory modes of understanding and adopted a logical-empiricist approach to questions of truth, meaning and method.

II

For some commentators – Rorty among them – these developments are all to the good. What they enable us to see is the open (endlessly revisable) character of scientific theories, the fact that they are grounded in nothing more durable than our present-day cultural practices and beliefs, and hence that science (together with philosophy of science) should address itself directly to our social needs and aspirations, rather than to false ideals of objectivity and truth. This is why Rorty has no objection to the Tarski formalism – or to Davidson's deployment of it – just so long as we recognize that its provisions are wholly redundant except as a useful deflationary technique for dispensing with other (less pragmatically amenable) kinds of truth-talk.

Joseph Rouse adopts a similar view of the advantages that accrue from this minimalist conception. Chief among them, he argues, are the negative virtue of deflating 'truth' to the point where it might just as well drop out of our predicate system altogether, and the positive virtue of opening up a whole new range of interpretive possibilities for hermeneutic or depth-ontological understanding. These Heideggerian echoes are evident enough in the following passage from Rouse's book *Knowledge and Power*.

> 'Truth' adds nothing to the assertions to which it is applied. In a similar way, 'existence' is not a real property; it adds nothing to the determinations of the things said to exist. The predicate 'true' can be applied only to sentences in a language. So

> to this extent what is true (i.e., what *can* be true or false) depends upon what can be said in that language. But this constraint does not mean that the language has some determining interactions with the states of affairs to be spoken of. A language opens a field of possibilities for speaking . . . without in any way determining which are true and which are false. It connects assertions with truth-conditions but does not determine whether those conditions obtain. Similarly, what exists depends upon the field of meaningful interaction and interpretation within which things can be encountered. This configuration of practices (including, of course, linguistic practice) allows things to show themselves in a variety of respects. These showings are not independent of this configuration any more than what is true-or-false can be independent of a language. There cannot be things that cannot interact with the things disclosed within a meaningful world (as there cannot be truths or falsehoods that cannot be expressed in a language). . . . Just as what is not a sentence-in-a-language is not true-or-false, there is no fact of the matter about whether things that cannot intelligibly be encountered within a meaningful world do or do not exist.[17]

I have quoted this passage at length because it brings out very clearly the elective affinity that Rouse and others have perceived between a minimalist (logico-semantic) conception of truth in philosophy of language and science and a broadly hermeneutic – Heideggerian – idea of truth as what emerges or stands revealed against the horizon of a meaningful form of life.[18]

On the face of it, nothing could be more unlikely than this coupling of Heidegger with a movement of thought whose origins lie in the programme of logical positivism. It is a strange conjunction on various counts, not least when one recalls the view of the positivists – Carnap and Ayer in particular – that Heidegger's depth-ontological talk was just a species of empty 'metaphysics', a strain of chronically inflated verbalism decked out in a variety of pseudo-arguments from (often fake) etymological pretexts.[19] And there is a further irony in the fact that Heidegger regarded the entire course of Western science and technological achievement as the upshot of a deep-laid (on his own terms 'metaphysical') will-to-power over nature and humanity alike.[20] Thus his thinking has been most influential among those – postmodernists, antifoundationalists, neo-pragmatists, 'strong' sociologists of knowledge, proponents of 'deep ecology' and so forth – who tend to equate science with a destructive and humanly degrading form of means-end rationality, an instrumental reason wholly devoid of critical or emancipatory values.[21] This understanding of modernity and its discontents has created some otherwise strange bedfellows. Among them was Adorno, himself quick to diagnose the dangers and confusions of Heidegger's obscurantist 'jargon of authenticity', but who none the less subscribed to a similar blanket diagnosis of science (or technology – the distinction is barely maintained) as among the chief symptoms of our well nigh terminal malaise.[22] Elsewhere one finds numerous echoes of Heidegger's thought in critics of the strong artificial-intelligence programme, in those who reject a physicalist (or 'central-state materialist') approach to the mind/body issue, and – as I have said – in deep ecologists who tend, following Heidegger, to view all science as complicit with the drive to subjugate nature to the purposes of brute unthinking technological mastery.[23]

What these commentators share is the perspective on science that Heidegger develops in his essays on 'The Question of Technology' and 'The Age of the World-Picture'.[24] This is the idea, simply put, that science 'does not think', that it inherits (and carries to a bad fruition) that purely instrumental relationship between mind and nature – or 'thought' and its 'objects' – which first entered Western philosophy with Plato and has since become entrenched in the entire tradition of post-Cartesian epistemological enquiry. Such enquiry is of its very nature incompatible with *thinking* – in Heidegger's depth-ontological sense of that word – to the extent that it supposedly works to exclude any consideration of the life-forms, the evaluative contexts or 'horizons of pre-understanding' within which all human projects – science included – discover their practical and existential bearings. So one can see why commentators like Rorty should locate a decisive turning-point for philosophy of science in the passage from logical empiricism and its various (equally problematic) offshoots to a pragmatist reading of Heidegger which stresses his talk of 'comportment', 'being-in-the-world', tools and equipment as 'ready-to-hand', etc., and which plays down his other (depth-ontological) talk of Being, *Dasein*, truth-as-unconcealment and suchlike – to their minds – otiose 'metaphysical' baggage. This reading falls in with the pragmatist idea that truth, whether in science or philosophy, is simply what works for practical purposes or what is currently and contingently 'good in the way of belief'.

However, it also has two rather large disadvantages. On the one hand, it comes up with a version of Heidegger which implausibly treats him as a pragmatist in the native American grain, and which ignores – or tactfully elides – all those other (*echt* Heideggerian) elements in his thinking which led him to denounce American pragmatism as the worst, most vulgar manifestation of philosophy in its latter-day technocratic guise. On the other, it cannot help but reproduce at least something of Heidegger's deep-laid antipathy to science, technology and all their works. Thus the same general picture tends to emerge of science as a project that first took rise from the epochal Western 'forgetfulness' of Being, and whose progress has since borne melancholy witness to the deepening rift between Being and beings, between the 'ontological' and the 'ontic', or again, between the 'historial' as a mode of thinking turned back towards the sources of authentic truth and the 'historical' (or merely 'factical') domain where genuine thinking can have no place.

These problems are everywhere apparent in Rorty's attempt to coax Heidegger down – or at any rate coax his readers around – to a sensible pragmatist interpretation of all that depth-ontological talk. And they are rarely far from the surface in other, more extended treatments of the theme, like Hubert Dreyfus's commentary on *Being and Time* and Mark Okrent's book *Heidegger's Pragmatism*.[25] For the trouble with all such half-way Heideggerian approaches is that they take just enough from Heidegger to obscure the most important issues about science and the history of science, while maintaining a residual (pragmatist) sense that science has after all come up with some rather impressive achievements, none of which can figure – to Heidegger's way of thinking – except as further stages in the epochal declension from Being to the modern (technologico-pragmatic) concealment or forgetfulness of Being. And this tension is no doubt sharpened by their awareness of the massive ethical and political confusions that went along with Heidegger's levelling of the difference between science and technology, as well as his crass failure to distinguish between the death-camp technology of mass extermination and those other developments (mechanized agriculture, factory production-lines, the hydroelectric dam on the Rhine, etc.) which he viewed as – 'essentially' – just so many symptoms of the same epochal malaise.

One can sympathize with Dreyfus, Okrent and Rorty in their wish to come up with a pragmatist reading of Heidegger that would attach the least possible weight to such pronouncements. But the question remains as to just how much is left of 'Heidegger' once this process has been carried through. And it is not, I think, an overly simplified or a crudely polemical view which would hold that this saving tactic leaves as much in doubt as the process of de-Nazification that Heidegger underwent in the immediate post-war period, or indeed his subsequent (less painful) rehabilitation at the hands of disciples or well-disposed exegetes. For these endeavours have at least one thing in common. That is, they underestimate the deep kinship that exists between Heidegger's thinking about language, Being and truth, his thinking about politics and the emergent high destiny of the German nation-state, and his attitude to science (or technology) as a phenomenon *indissociably* bound up with those other ubiquitous symptoms of decline – from 'Western metaphysics' to liberal democracy – whose overcoming he envisaged with the advent of National Socialism.[26] That Heidegger at one time fervently endorsed all this, and that he never went so far as to repudiate his belief in the 'inner truth and greatness' of the Nazi movement, is a fact that has been copiously documented in recent critical and historico-biographical studies.[27] However, these critics have had less to say about the role of science in Heidegger's philosophy and the way that the 'question of technology' so often stands in for those other values – of enlightened, liberal-democratic, or social-emancipatory thought – which were likewise (on his account) mere symptoms of our present destitute condition.

One notable exception is Jürgen Habermas, whose critique of Heidegger (and of this whole postmodern or counter-Enlightenment strain of thought) focuses precisely on his failure to distinguish between the various forms or modalities of reason that constitute the philosophic discourse of modernity.[28] What is thereby concealed (or summarily revoked) is that critical-emancipatory impulse which alone makes it possible for reason to reflect upon – and thus to resist – its own more reductive or narrowly instrumental uses. Hence the idea, orthodox among postmodernist thinkers, that the Enlightement was somehow directly to blame for those various pathological distortions of rationality whose effects range from environmental pollution to the kinds of death-dealing bureaucratic efficiency brought to their hideous perfection at Auschwitz.[29] To state the argument in this form is to reveal both its sheer moral crassness and its proximity to Heidegger's thinking in essays like 'The Question Concerning Technology'. It also recalls his contention that the greatness of National Socialism lay in its resisting the pressures upon Germany – upon the heartland of European culture – exerted by America and the Soviet Union, those twin avatars of the will-to-power in its modern techno-scientific guise. All of which suggests that Habermas is justified in linking Heidegger's massive failure of moral and political intelligence with his mystified conception of language and national destiny, his reductive (indeed caricatural) account of the legacy of Enlightenment thought, and – not least – his fixed habit of equating 'science' with the worst excesses of a purely instrumental reason. For if there is one thing that Heidegger's example should teach us, it is the need to keep faith with those values of reflective, self-critical thought which prevent the slide into a 'jargon of authenticity' premised on the notion of truth as vouchsafed to some few choice spirits by virtue of their national language or culture. And there is nothing more clearly indicative of Heidegger's contempt for such values than his demotion of science – at its best, a realm of shared (trans-cultural) thought and endeavour – to the level of brute 'unthinking' instrumentality.

Of course I am not suggesting – absurdly – that there are always such violent things at work behind the hermeneutic turn in contemporary philosophy of language and science. More often, as I have said, it enters at the point where 'truth' has been subjected to a deflationary (i.e., Tarskian) account which defines it in purely extensionalist and formal semantic terms as the set of equivalence relations holding between sentences in a given language and those (putative) real-world or factual states of affairs in virtue of which such sentences possess an assignable truth-value. Thus what counts as 'true' can only be determined by a process of recursive definition whereby sentences in the object language (like 'snow is white') are matched with some equivalent truth of observation (in this case, the fact that snow is indeed white) and then simply taken out of their quotation marks to stand as plain statements of fact.[30] In which case, according to Rouse, '[t]ruth plays no explanatory role, and the Tarski definitions do not need to be supplemented by a correspondence, coherence, or any other theory of truth.'[31] But this also means – as others like Rorty are quick to remark – that truth-talk is rendered pretty much redundant and might just as well drop out in favour of a straightforward pragmatist appeal to our normal (scientific or everyday) habits of belief. For the effect of adopting this Tarskian approach – that is to say, a disquotational theory of truth coupled with a holistic theory of meaning – is to empty 'truth' of all determinate content except in so far as it happens to fit with the entire set of beliefs held true at any given time. As Rouse puts it, '[t]he semantics of truth do not say anything about the material conditions under which the assertion of *p* is justified, or about the truth conditions of *p* (except in the limited sense that those conditions obtain whenever *p* holds). They only say that "*p* is true" and "*p*" are materially equivalent.'[32] Thus the whole business amounts to no more than an exercise in circular definition, an elaborately formalized way of making the point that 'truth' is in the end just a product of semantic convenience, a predicate that serves no purpose aside from its role in a meta-linguistic theory devoid of explanatory content.

So one can see why commentators on Davidson – Rouse among them – very often reach a stage in the argument where there seems no way to go except in the direction of a depth-hermeneutic approach that would somehow redeem this otherwise quite empty, trivial or tautological definition of 'truth'. The situation is summed up with pithy irony by Charles Taylor. 'Old-guard Diltheyans,' he writes, 'their shoulders hunched from years-long resistance against the encroaching pressure of positivist natural science, suddenly pitch forward on their faces as all opposition ceases to the reign of universal hermeneutics.'[33] J. E. Malpas has grasped the Heideggerian nettle most firmly in his book *Donald Davidson and the Mirror of Meaning*. After much patient and detailed explication in the broadly 'analytic' mode – aimed mostly to vindicate Davidson's position on these issues – he then has to acknowledge, with disarming candour, that almost nothing has been achieved by way of explaining the structure or content of 'truth' beyond its purely technical (Tarskian) sense. At which point Malpas, like Rouse, proposes that we cast academic prejudice aside and invoke Heideggerian depth-ontology as a means of providing what the theory self-evidently lacks. Thus, according to Malpas,

> Heidegger's rejection of representationalism and of the subject-object dichotomy mirrors Davidson's own rejection of those notions. In the case of both thinkers, it is a rejection accompanied by a similar rejection of standard theories of truth. Moreover, both Davidson and Heidegger emphasise the centrality of truth, and its presuppositional character. As presuppositional, truth is tied to the notion of horizonality. . . . A project is structured within a horizon, and with respect to some intention. It is, moreover, only within a horizon and with respect to an intention that things themselves can be encountered. A preliminary characterization of truth, in a horizonal sense, might be in terms of our having access to things as they really are – that is, our being able to encounter them. But our having access to things is only possible given the intentional-horizonal structure of projects. Only within that structure can things appear. . . . Such truth cannot, of course, have the structure of agreement in the sense of correspondence, since truth as appearing or

> 'Being-uncovering' must underlie the possibility of agreement. [For] the idea that a sentence might correspond to the world, or portions of the world, already presupposes that sentences can be distinguished, but this is only possible, on the Davidsonian account, if we already have mostly true beliefs – in other words, if the world is already apparent to us. It is not that such appearing is necessary for us to be able to verify correspondence, but that without it there is nothing to correspond with.[34]

This passage catches precisely the point of transition between a disquotational theory of truth and a depth-ontological account which seeks to make good the inadequacies of that theory by adopting a Heideggerian language of 'Being', 'uncovering', 'horizonality', primordial 'encounters' with things as they appear and so forth. For Malpas this is the only way forward once thought has run up against the inbuilt limits of an approach whose bankruptcy is evident – he thinks – in the history of failed analytical endeavours from logical positivism to its various present-day successor movements.

His position finds support in many quarters of present-day debate, for instance, in Hubert Dreyfus's Heideggerian reflections on the limits of scientific reason (narrowly conceived), and of course in Richard Rorty's more hybrid brand of 'post-analytic' cultural hermeneutics, where Heidegger rubs shoulders – incongruously enough – with Dewey, William James and the American pragmatists.[35] It is also, as we have seen, the position arrived at – and by much the same route of argument – in Rouse's book *Knowledge and Power*. Thus for Rouse the Tarski–Davidson approach has the negative virtue of demonstrating the inherent circularity of all such arguments, while on the other – more constructively – it points the way towards a depth hermeneutics of the natural and human sciences that would interpret their various methods and truth-claims always in the context (or against the 'horizon') of a meaningful form of life. As he puts it: 'One cannot understand any particular task or item of equipment (including a sentence) without some grasp of the whole configuration of things that makes up an intelligible world.'[36] And again: 'It is only through purposeful interaction with the world and the patterns of success and failure that emerge from it that our interpretations acquire meaning and the world becomes determinate.'[37] All of which follows, on this Heideggerian account, from the premise that nothing could emerge into the field of human purposive intent – neither 'things', 'tasks', 'items of equipment' or 'patterns of success and failure' – were it not for that pre-given world of situated practices, values and beliefs which constitutes their ultimate horizon of intelligibility.

This turn towards depth-hermeneutical models of scientific enquiry is also, as I have argued, a result of what is commonly perceived as the failure of those various alternative programmes – from logical empiricism to the covering-law or deductive-nomological approach – which have seemed to issue in a series of philosophic dead-ends. In its broad outline this story is familiar enough, as recounted by Rorty and other exponents of the jointly hermeneutic and post-analytical view. Where it begins is with the logical-empiricist doctrine that philosophy of science cannot do more than give a formalized (logically regimented) account of the methods, procedures, observation languages, theoretical constructs, modes of inferential reasoning, etc., that characterize scientific practice.[38] Where it ends – on this telling of the story – is with the advent of a thoroughly holistic conception of meaning and truth, coupled with a Quinean doctrine of ontological relativity, which between them cut away all the grounding assumptions of that original project. Or again, in so far as truth-values are preserved at all, they figure only as a formalized system of equivalence statements whereby to pair off sentences in the observation language with their counterparts in a second-order (meta-linguistic) schema quite devoid of explanatory content. From which point the way is open to a depth-ontological approach where the 'ontology' in question – and the operative notion of 'depth' – are construed, after Heidegger, in hermeneutic terms and hence as giving no hold for merely 'ontic', 'factical', or causal-explanatory modes of understanding.

III

It is worth remarking on the parallel movement that occurred within those branches of the human and social sciences influenced chiefly by developments in French theory. Here also there was a first stage of high structuralist abstraction, as witness the various disciplines – from anthropology to narrative poetics, historiography and political science – which invoked the pilot discipline of Saussurean linguistics as a means of achieving new standards of methodological rigour.[39] But this programme rapidly gave way to an extreme reactive movement, a repudiation of all such (quasi)-scientific aims and ideals in favour of a rhetoric that promoted the values of difference, discursive multiplicity, ontological relativism, the demise of meta-languages, the obsolescence of truth (or its textualist proxy the 'transcendental signified') and suchlike post-structuralist notions.[40] Thus Roland Barthes famously took to denouncing his own early quest for a universal grammar of narrative-poetic structures and devices that would point the way towards a genuine science of literature conceived after the Saussurean model. Henceforth, he proclaimed, we should abandon such delusory (authoritarian) procedures and should read each text – as Barthes sought to demonstrate in the case of Balzac's *Sarrasine* – with a view to releasing its maximum yield of plural significations, its irreducible *difference* from every other text and from every last precept of critical method.[41]

Foucault's work took a similar turn from the broadly structuralist 'archaeology' of knowledge to a Nietzsche-inspired genealogy of values that treated all truth-claims – all past and present disciplines of scientific, historical and philosophic enquiry – as so many products of the epistemic will-to-power vested in language or discourse.[42] Nowhere was this shift more dramatically apparent than in the waning fortunes of Althusserian Marxism as its critics – many of them erstwhile adherents – scrambled to vacate the exposed high ground of 'theoretical practice', conjunctural formations, economic determination 'in the last instance', the science/ideology distinction, and so forth.[43] What took their place, once again, was a rhetoric of multiple decentred 'subject positions', of reality as a wholly discursive – narrative or textual – construct, and of truth as a species of operative fiction sustained by the current (juridico-linguistic) status quo. In so far as 'the real' (or our knowledge thereof) retained any place in this theory, it was only when subject to redefinition in suitably Lacanian terms, i.e., as that which returns within discourse – or from the imperious order of the Symbolic – to disrupt the delusive imaginary realm of the subject-presumed-to-know.[44]

I should not wish to press too far with this comparison between developments in Anglo-American philosophy of science over the past four decades and changes in the currency of avant-garde thinking among cultural and literary critics. For these projects have little in common, it will surely be argued, aside from their each having started out with certain strong methodological commitments, and having then abandoned those commitments in the face of accumulating problems that eventually led to such doctrines as ontological relativity, semantic holism, the 'linguistic turn', the discursive construction of reality and so forth. Beyond that there is a great difference – I would readily concede – between the aims and objectives of philosophy of science in the broadly analytic mode and the kinds of thinking that claimed 'scientific' warrant during the period of high structuralist theoretical endeavour. Where they differ most sharply is on the issue of language, discourse or representation, that is to say, the extent to which language may be thought of as affording referential access to a domain of real-world (extra-discursive) objects, processses and events. For structuralists – and even more so for post-structuralists – the notion of our having such access can only be a product of those current (perhaps deeply naturalized) signifying codes and conventions that constitute 'reality' so far as we can possibly know it. Hence Foucault's structuralist premise, most explicit in his early book *The Order of Things*: that all kinds of knowledge in the natural and the human sciences alike, along with their various epistemological and ontological assumptions, can be shown to take rise from some particular (purely 'arbitrary') arrangement of signs or discursive representations.[45] Clearly, such an argument is worlds apart from anything envisaged by philosophers of langage and science in the Anglo-American analytic tradition. And this applies even to those – Quine among them – whose outlook of sturdy commonsense realism sits oddly with their otherwise wholesale talk of ontological relativity.[46]

Nevertheless, having entered these necessary caveats, I would still maintain that structuralism and analytic philosophy of science have both run up against similar problems and for much the same reason. That is to say, they each yield ground to the sceptic's arguments at precisely that point where (in Quine's terminology) a gulf opens up between 'word' and 'object', or again – following Saussure – where 'signifier' and 'signified' are conceived as standing in a purely arbitrary relationship, and the referent simply drops out as a notion devoid of theoretical or explanatory content. For it is then a very short distance to those other, more extreme variants of the linguistic, hermeneutic or textualist 'turn' which reject the appeal to any meta-language – any higher-level theory or mode of understanding – that would somehow place limits on the possible range of first-order signifying systems. This is what happens with the passage from structuralism to post-structuralism. It is also the starting-point of Foucauldian genealogy, of Lyotard's radically nominalist approach to issues of meaning and truth, and of Rorty's idea – drawing upon these and a promiscuous variety of other sources, from Wittgenstein to Heidegger, Quine, Kuhn and Feyerabend – that 'truth' *just is* whatever we take it to be according to our current interpretive lights.[47] Indeed, the very fact that Rorty can mix these sources with at least some show of justification is enough to suggest that they do, after all, have certain premises in common.

Those premises can be stated very briefly as follows. The legitimate business of theory – whether in philosophy of science or the various extensions of structuralist method – is *not* to give a causal-explanatory account of how knowledge comes about or how truth-claims are properly justified. Rather, it is to analyse the languages, discourses, narratives, modes of inferential reasoning, etc., which make up the field of accredited 'knowledge' at any given time. From which it follows, on the sceptical (post-structuralist or ontological-relativist) view, that in principle there can exist as many such schemes as there exist languages or ways of picking out objects in accordance with this or that Rortian 'final vocabulary'. For there is (so it is claimed) no fact of the matter that could decide which objects correspond to which statements in our observation language, nor again any core set of logical entailment relations or putative 'laws of thought' which could ultimately serve to adjudicate the issue between different (incommensurable) theories. After all, on this account, statements are not true in virtue of how things stand in reality, but only – as Quine and Foucault both argue – in virtue of their fit with the entire range of statements (discourses, language games, conceptual frameworks, etc.) that determine what shall count as a veridical utterance in relation to this or that ontological scheme. Hence the demise of those erstwhile 'strong' methodologies – like classic structuralism or the deductive-nomological programme in philosophy of science – which equated theoretical rigour with the move to a higher (formalized or meta-linguistic) level of enquiry where first-order statements could be subject to analysis without raising questions of their material truth or their explanatory power *vis-à-vis* objects in the physical domain. For at this stage there appeared to be nothing – no ground rules of method, no logical constraints, observational checks, or empirical validity conditions – that could save those programmes from the more extreme consequences of their own ontological-relativist or anti-realist position.

Hence also the currently widespread idea that the only alternative to a minimalist (or deflationary) account of truth is the turn towards hermeneutic models of depth-understanding that dispense altogether with such outworn conceptions of logic, method and truth. 'Another way to put this,' Rouse suggests, 'is that for there to be things of any particular kinds, there must be a world to which they belong. But the reality of that world is not a hypothesis to be demonstrated; it is the already given condition that makes possible any action at all, including posing and demonstrating hypotheses.'[48] This claim is borne out, so he believes, by the way that Tarski's theory manages to dispense with all those otiose ideas of truth to the facts, 'correspondence', or of science as a project supposedly aimed towards deeper (more adequate) explanatory hypotheses which can then be tested against the evidence. For such talk simply fails to make sense once we grasp the point that 'truth' is a redundant term – merely a formal convenience – which can always be made to cancel through by applying the standard disquotational procedure. This in turn suggests to Rouse – as likewise to others like Rorty and Malpas – that there must be something more to the issue of truth, and that the best place to look for that 'something more' is in the region of depth-ontological enquiry opened up by Heideggerian hermeneutics.

Thus (to repeat): 'There cannot be things that cannot interact with the things disclosed within a meaningful world (as there cannot be truths or falsehoods that cannot be expressed in a language).'[49] And again, in a passage that brings together echoes of Wittgenstein, Quine, Davidson and Heidegger:

> The conditions that must obtain for a sentence to be true constitute its meaning, and this depends upon the circumstances in which the sentence is appropriately used. What it is for an *x* to exist (as an *x*) is constituted in a similar way by the ways it can be encountered in the course of intelligibly dealing with the world. Both determinations are holistic. Understanding the meaning of a sentence requires understanding many other sentences along with it. Being able to recognize an *x* for what it is requires knowing how to recognize and deal with many other entities. An understanding of such appropriateness is sustained by the actual use of or other practical encounter with the things in question, in order to achieve the ends that call for such dealings.[50]

All the same, Rouse argues, we should not be misled into thinking this a charter for anti-realism in philosophy of science. Nor is it in any sense a pretext for those forms of wholesale ontological-relativist doctrine that treat the objects of scientific enquiry as so many optional contructs out of this or that discourse, language game, conceptual scheme, Rortian 'final vocabulary' or whatever. On the contrary: 'the language we speak does not determine which of its sentences are true.'

Moreover, '[t]he practices that constitute our "world" likewise do not determine which things exist, with what properties.'[51] The realist may suspect – when confronted with passages like the over above – that what is involved here is a thoroughgoing version of the linguistic or cultural-conventionalist 'turn', combined with a depth-ontological (hermeneutic) jargon that counts 'reality' a world well lost for the sake of such profound revelations. But he or she need have no such worries, according to Rouse. Quite simply, '[w]hich things there are, what properties they have, and what relations they enter into are determined by the things themselves and "how things stand" with them.'[52]

So one can carry on talking about the world and all its furniture – from cabbages and kings to planets, DNA molecules, electrons, wave-forms and particles – just so long as one takes the Heideggerian point, i.e., that those 'things' can only emerge against the horizon of intelligibility that constitutes the 'world' of meaningful practices and life-forms. However, this argument will appear less than convincing if one asks how the properties of x can be determined *both* by its intrinsic nature – as revealed through scientific investigation – *and* by its role within some given context of linguistic, cultural or 'world-disclosive' pre-understanding. For in the latter case nothing could count as an object – or as evidence for that object's nature and properties – except in so far as it showed up among the range of currently accepted practices and beliefs. That is to say, there could be no escaping the 'hermeneutic circle' which Heidegger – and Gadamer after him – have raised to a high point of interpretive principle.[53] Rouse is quite happy to embrace this idea, on the one hand because he thinks that we have no choice in the matter, and on the other because it involves, in his view, not a 'vicious' circularity, but a way of rendering science more accountable to the range of human needs, values, interests and social priorities. For if indeed it is the case, following Quine, that 'there is no nonlinguistic, pre-theoretical fact of the matter to which we could appeal to resolve disagreements about how the world is,' and if moreover – as he takes Tarski and Davidson to have shown – 'truth is a metalinguistic predicate' devoid of substantive content, then surely nothing is lost (and a great deal gained) by acknowledging the hermeneutic circle.[54] What we lose is just a set of mistaken, residually positivist ideas about the logic of scientific enquiry and its bearing on first-order observation statements. What we stand to gain, as Rouse sees it, is a much enhanced sense of science's social and ethical responsibiltiies, along with a suitably scaled-down conception of its claim to determine how things stand 'in reality'.

However, this Heideggerian line of argument sits awkwardly with Rouse's attempts elsewhere to placate his typecast 'realist' opponent by adopting a liberal attitude with regard to the various entities turned up in scientific research or in our everyday dealings with the world. More precisely: it brings out the strain of implicit anti-realism which attaches to such talk of 'things' (or entities) once subject to reinterpretation in the depth-ontological or hermeneutic mode. Thus when Heidegger asks 'What is a Thing' – in the essay of that title – his purpose is to question both our everyday ('ontic' or 'factical') notions of thinghood and that entire history of philosophico-scientific thought which has always been captive to 'metaphysical' ideas of knowledge, truth and representation, this latter conceived on the model of a perfect correspondence (*homoiosis*) between the knowing mind and the objective order of things.[55] This is why, in Heidegger's estimation, 'science does not think.' What it fails to think is the 'question of Being' (or the 'ontological difference' between Being and beings) which first found expression in those fragmentary texts that have come down to us from the pre-Socratics, and whose subsequent withdrawal – with the advent of 'philosophy' as we know it from Plato to Descartes, Kant, Husserl and beyond – is coterminous with the epoch of 'Western metaphysics'.

Hence Heideggger's treatment of science and technology as the *predestined* outcome of an age-old 'forgetfulness' of Being whose effects are everywhere apparent in the world-view of an epoch increasingly given over to the exploitation of natural resources, or the idea of nature as a 'standing reserve' for the purposes of human command and control. Such is his argument in the well-known passage – from 'The Question Concerning Technology' – about the hydroelectric power-plant built on the River Rhine. It is worth citing this passage at length since it helps to focus everything I have said concerning the problems with a depth-ontological or hermeneutic approach in philosophy of science. The power-plant, he writes,

> sets the Rhine to supplying its hydraulic pressure, which then sets the turbines turning. This turning sets those machines in motion whose thrust sets going the electric current for which the long-distance power station and its network of cables are set up to dispatch electricity. In the context of the interlocking processes pertaining to the orderly disposition of electrical energy, even the Rhine itself appears to be something at our command. The hydro-electric plant is not built into the Rhine River as was the old wooden bridge that joined bank with bank for hundreds of years. Rather, the river is dammed up into the power plant. What the river is now, namely, a water-power supplier, derives from the essence of the power station. In order that we may even remotely consider the monstrousness that reigns here, let us ponder for a moment the contrast that is spoken by the two titles: 'The Rhine', as dammed up into the *power* works, and 'The Rhine', as uttered by the *art*-work, in Hölderlin's hymn by that name. But, it will be replied, the Rhine is still a river in the landscape, is it not? Perhaps. But how? In no other way than as an object on call for inspection by a tour group ordered there by the vacation industry.[56]

We can see here all the major themes of Heidegger's brooding on the 'question of technology' and, beyond that, on the epoch of Western post-Hellenic 'metaphysical' thought. That epoch is defined – in brief – by its somehow 'predestined' forgetfulness of Being, by its exclusive concern with the beings (or objects of cognitive and technologico-scientific grasp) that henceforth lend themselves passively to treatment in this fashion, and by its fall into a mode of epistemological (or representationalist) thought which sets a gulf between subject and object that cannot be bridged but only 'dammed up' by the ever more exploitative uses of nature as a 'standing reserve' or an 'object on call'. Such is indeed the very 'essence' of technology, an essence whose manifestations include – as Heidegger notoriously listed them – mechanized agriculture, factory production-lines, the dam on the River Rhine, and the mass slaughter of human beings in the death-camps at Auschwitz and elsewhere.

I have already suggested that Heidegger's moral obtuseness in drawing these analogies – a stupidity noted, more in sorrow than anger, by Herbert Marcuse in their post-war exchange of letters – was deeply bound up with his attitude to science and his undifferentiating blanket diagnosis of the ills bought about by modern technology.[57] For if there is, as he claimed, some 'essence' of technology whose origins lie far back in the history of Western metaphysical thought, and which was thus predestined to manifest itself in this range of latter-day forms, then it becomes quite pointless to moralize on the subject or suggest that there is a difference – an all-important difference – between Auschwitz on the one hand and on the other such 'monstrosities' as mechanized agriculture or the dam on the Rhine. What these all have in common is their belonging to the 'age of the world-picture', an age of metaphysico-technological 'enframing' when everything is reduced to the dead level of instrumental or calculative reason. Mere 'correctness' (or accurate representation) henceforth becomes the sole aim of thought, which in turn gives rise to that epistemological – or technocratic – will-to-truth whose manifestations are everywhere around us. Thus: 'The correct always fixes upon something pertinent in whatever is under consideration.'[58] Whence the current reign of a purely instrumental (means-end) concept of rationality, joined to a causal-explanatory paradigm in the natural sciences which finds no room for that mode of revelatory thinking – that contemplative openness to Being – which Heidegger offers by way of contrast in his allusion to Hölderlin's poem 'The Rhine.' For '[w]herever ends are pursued and means are employed, wherever instrumentality reigns, there reigns causality.'

And yet, he continues, 'in order to be correct, this fixing by no means needs to uncover the thing in question in its essence.'[59] For to think the 'essence' of technology is also to open a space where technology appears in its true character as the simply inescapable destiny of thought for an epoch whose horizon has so long been configured by those same (metaphysical or representationalist) ideas of knowledge and truth. In which case the ontic/ontological difference has its equivalent in two ways of thinking about technology. On the one hand there is the kind of superficial denunciation which never gets beyond protesting its effects upon this or that aspect of our contemporary life-world. On the other there remains the possibility for an authentic (depth-ontological) thinking-back into the 'question of technology', a thinking that would reveal those various stages in the epochal declension – the 'concealment' of Being – which began with the theory of truth-as-correspondence (*adaequatio intellectus ad rem*), and which has issued in our present destitute condition, in the reign of technocratic reason and nature conceived instrumentally as a 'standing reserve'. Thus:

> When we consider the essence of technology we experience enframing as a destining of revealing. In this way we are alreading sojourning within the free space of destining, a destining that in no way confines us to a stultified compulsion to push on blindly with technology or, what comes to the same thing, to rebel helplessly against it and curse it as the work of the devil. Quite to the contrary, when we once open ourselves expressly to the *essence* of technology, we find ourselves unexpectedly taken into a freeing claim.[60]

This apparently more relaxed, open-minded and receptive approach to the 'question of technology' goes along with the famous 'turn' (*Kehre*) in Heidegger's later work, his belief that thinking can best comport itself through a kind of wise passivity (*Gelassenheit*), an attitude of unforced acceptance in response to whatever may yet be revealed of Being and truth. Still there is the danger, as he warns, that 'in the midst of all that is correct the true may withdraw.'[61] However, we shall not be any better equipped to avert that danger if we suppose that simply by denouncing its particular causes or effects – say, industrial pollution, soil erosion, the destruction of animal species, or the depletion of the ozone layer – we have thereby confronted the 'essence' of technology as a challenge to thought.

Not that Heidegger altogether ignores these specific matters for concern. 'Agriculture is now the mechanized food industry. Air is now set upon to yield nitrogen, the earth to yield ore, ore to yield uranium, for example; uranium is set upon to yield atomic energy, which can be unleashed either for destructive or for peaceful purposes.'[62] Yet such concerns are of secondary importance when set against the question of technology and its demand for an authentic (depth-ontological) thinking that would address that question in its very essence. 'What is dangerous is not technology,' Heidegger writes. 'Technology is not demonic; but its essence is mysterious.'[63] And again, in a passage that offers at least some degree of clarification:

> The threat to man does not come in the first instance from the potentially lethal machines and apparatus of technology. The actual threat has already afflicted man in his essence. The rule of enframing threatens man with the possibility that it could be denied to him to enter into a more original revealing and hence to experience the call of a more primal truth.[64]

What this passage clarifies, it seems to me, is not so much the logic or the justifying grounds of Heidegger's argument but the sheer amount of deep-laid verbal mystification – 'bewitchment by language', in Wittgenstein's phrase –

which befogs his thinking about the 'question' (or the 'essence') of technology. For technology has no such 'essence', only a diverse range of applications and purposes, some of them beneficial, others harmful, and others again – like genetic engineering or certain branches of 'peaceful' sub-atomic or particle research – as yet very largely untested as to their social and ethical consequences. To think of them all as possessing some 'mysterious' (though not necessarily 'demonic') essence is just another version of that moral obtuseness which led Heidegger to compare mechanized agriculture with the gas-chambers at Auschwitz, or – in his correspondence with Marcuse – to equate the treatment of the Jews in Nazi Germany to the post-war suffering of displaced populations in Soviet-occupied Eastern Europe.[65]

Nor is there much improvement to be noted in his habits of moral judgement when Heidegger undergoes his mid-life 'turn' to a non-assertive ethos of *Gelassenheit*, waiting upon truth, or letting be. What this change amounted to in ethico-political terms was a switch from the language of 'resoluteness', *Dasein*, 'being-unto-death', 'the self-assertion of the German university' (with Heidegger as its self-appointed philosopher-Führer), etc., to a language of wise acquiescence in the given (predestined) occurrence of events which effectively absolved him of all responsibility for his words and actions during that earlier period.[66] And there is a parallel shift in his thinking about the 'question of technology', one that emerges most clearly when Heidegger writes about technology as 'no mere human doing' but as a 'revealing that orders', that 'sets upon man to order the actual as standing-reserve in accordance with the way it shows itself'.[67] For in this case the question would be posed at a level far deeper – and historically much further back – than could be grasped by any merely 'ontic' or 'factical' dealing with particular problems in the nature of our current technologically oriented life-world. That is, it would require us to contemplate the 'essence' of technology as having somehow been set in place from the ancient Greek beginning, from that moment when philosophy and science took their destined turn towards an epistemological (or representationalist) mode of being-in-the-world. On this view it is the merest of anthropocentric illusions to think that we might come up with particular remedies for particular aspects of our current technological fix, or indeed to suppose that *we* – as individuals or coordinated action groups – could affect its outcome in any significant way. This presumption errs on the one hand by mistaking accident (or localized symptom) for essence, and on the other by attributing to human agents a power of judgement and choice over questions of ultimate (primordial) import.

IV

I shall cite one further passage from 'The Question of Technology' which brings out this close relation, in his later work, between Heidegger's thinking about techno-science and his attitude to issues of ethical, historical and political responsibility. 'The essence of modern technology,' he writes,

> starts man upon the way of that revealing through which the actual everywhere, more or less distinctly, becomes standing-reserve. 'To start upon a way' means 'to send' in our ordinary language. We shall call the sending that gathers (*versammelnde Schicken*), that first starts man upon a way of revealing, *destining* (*Geschick*). It is from this destining that the essence of all history (*Geschichte*) is determined. History is neither simply the object of written chronicle nor merely the process of human activity. That activity first becomes history as something destined. And it is only the destining into objectifying representation that makes the historical accessible as an object for historiography, i.e., for a science, and on this basis makes possible the current equating of the historical with that which is chronicled.[68]

One could find no better example of that *echt*-Heideggerian 'jargon of authenticity' whose effect is to obfuscate those very issues – of science, technology, historical understanding, the ethics (and the politics) of environmental concern – which it purports to address at a level of depth-ontological enquiry beyond the merely 'ontic' or 'factical'. Thus history in the authentic sense of that word should be thought of as proceeding from (or answering to) the summons of a destiny whose 'sending' may be heard – for those wise enough to hear – in the etymological link between *Geschick* and *Geschichte*. In so far as we persist in ignoring that summons (for instance, by seeking to ascertain the facts of history or consulting 'written chronicles') we thereby demonstrate our incapacity for 'thinking' at anything like the requisite depth. All the more so – presumably – if we carry this refusal to the point of contrasting Heidegger's ersatz mystique of origins, destiny, truth-as-unconcealment, etc., with the prime imperative to get things right as a matter of ethical and socio-historical justice. This merely shows that we are still in the grip of that 'objectifying' (representational) mode of thought that 'makes history accessible' (in Heidegger's words) 'as an object for historiography, i.e., for a science'. But if so, then this error is itself predestined and in no sense a matter of 'our' responsibility. For it belongs to the horizon (or the 'world-picture') of an epoch that has witnessed the withdrawal of Being and the giving-over of thought to the reign of metaphysics, of instrumental reason, and of modern science as the technocratic will-to-power in its latest – possibly terminal – phase.

All of which suggests that philosophers like Rouse should at least pause before enlisting Heideggerian depth-hermeneutics (along with Foucault's Nietzschean genealogies of power/knowledge) as an alternative resource in thinking about present-day science and its discontents. In both cases there is a failure to conceive how ethical values could possibly be reconciled with cognitive or knowledge-constitutive interests. That is to say, what Foucault takes over uncritically from Heidegger is his notion of the subject – especially the Kantian transcendental subject – as a mere figment of the humanist imaginary, a point of intersection between (on the one hand) the order of objectifying representations and (on the other) a sphere of 'suprasensible' precepts, dictates and values which lies altogether outside and beyond the realm of phenomenal cognition. Hence Foucault's well-known description, in *The Order of Things*, of the insoluble antinomies to which this illusory conception supposedly gave rise. Thus Kant's chief bequest to the nineteenth-century natural, human and social sciences was the idea of 'man' as a curious 'empirical-transcendental doublet', a hybrid creature whose existence was purely an artefact of this short-term (soon to be mended) rift in the fabric of discursive signs and representations.[69]

Foucault follows Heidegger very closely in his reading of Western metaphysics as the history of an error. It is an error that issues most visibly in Kant and the aporetic 'discourse' of a project which strives to reconcile the claims of understanding and reason, epistemology and ethics, determinism and free will, phenomenal cognition and reflective judgement, etc. For Heidegger it is manifest in that entire tradition of thought – from Plato to Husserl – whose history bears witness to the forgetfulness of Being and the retreat of thinking into a realm of merely 'ontic' (as opposed to depth-ontological) concern.[70] This tradition had substituted notions of truth-as-correspondence (*homoiosis*, *adaequatio* etc.) and of the knowing subject as its privileged locus for that originary wisdom whose truth is obscurely to be glimpsed in the few surviving fragments of the pre-Socratics. For Foucault it is less a matter of some long-lost primordial truth than of the sheer contingency – the range of always shifting discursive or epistemic formations – that come to light through an applied archaeology of 'knowledge' in its manifold guises to date. But in one crucial respect – as Foucault acknowledged – his thinking was deeply and lastingly indebted to Heidegger. It had to do with Heidegger's questing back into the history of that root 'metaphysical' illusion which opened up the cleft between subject and object, or which conceived of truth on the order of an adequate correspondence between ideas (or accurate representations) and the various real-world objects, entities, or events to which those ideas had reference.

In Heidegger's account this notion takes hold with the passage of a metaphor from the Greek *hypokeimenon* ('substance'; 'foundation'; 'that which supports from beneath') to the Latin *subiectum* and thence – via Descartes – to the range of modern post-Kantian variants on the mind as a more or less clear or distorting mirror held up to reality. At which point the way appeared open for Foucault to dissolve this entire problematics of language, truth and representation into a field of open-ended discursive possibility where subject and object are likewise conceived as products of a specular (imaginary) 'fold' in the fabric of knowledge. To thinkers of a more pragmatic mind – Rorty, Dreyfus and Okrent among them – Heidegger has likewise seemed to offer a welcome escape route from the travails of old-style epistemological (and new-style analytic) thought. Thus all one need do, according to Rorty, is play down Heidegger's portentous jargon of 'Being', 'ontological difference', 'Western metaphysics' and so forth, and interpret him as headed towards the sensible pragmatist conclusion that truth is pretty much what we make of it at this or that stage in the ongoing cultural conversation. From which it follows that the truth-claims of the natural as well as of the human and social sciences are best regarded as so many language games, discourses, 'final vocabularies' or elective metaphors, adopted for no better (and for no worse) reason then their happening to fit with the current self-images of the age.

So Rouse is not alone in turning to Heidegger (and also to Foucault) as allies in his project for a new hermeneutically informed philosophy of science. From Heidegger he seeks an approach to science that would treat it as an enterprise, a practice or a cultural life-form whose significance can be grasped only through questioning in the depth-ontological mode. From Foucault – along with other (e.g., feminist) critiques of scientific theory and method – he derives the idea that 'truth' is nothing more than a transient product of those power/knowledge differentials that determine what shall count as an authorized discourse at any given time. Moreover, he can claim to have arrived at this position not only by drawing on 'Continental' sources outside the dominant analytic (i.e., Anglo-American) mainstream, but also by reflecting on the problems and the limits encountered within that latter tradition of thought. Thus for Rouse there must always come a point where the analysis of truth conditions in scientific discourse reduces to a matter of purely circular definition, that is, some version of the Tarskian (formalized or 'disquotational') theory according to which truth is just an honorific predicate attaching to sentences held true in any given system of belief.

I have already discussed this topic at some length and will not repeat the details here. Sufficient to say that analytic philosophy has beaten a rapid and somewhat dishevelled retreat from its early (verificationist) idea of truth as consisting in a one-to-one match between truthful propositions and observationally warranted states of affairs. In so doing it has espoused a whole range of contextualist strategies, among them – most notably – the Wittgensteinian appeal to 'language games' or cultural 'forms of life'; Quine's idea of truth as a predicate capable of redistribution in as many ways as there exist alternative ontologies or conceptual schemes; and various forms of so-called 'internal realism' (like Hilary Putnam's latest candidate) which seek to head off the more disabling consequences of a full-fledged relativist approach but which still define truth as immanent – or relative – to this or that particular context of enquiry.[71] What these arguments all have in common is their assumption that philosophy of science has to do with the linguistic structures (whether sentences, statements, propositions or entire 'fabrics' of belief) which alone make it possible to analyse the various orders of scientific truth-claim. In other words they all adopt some version of Carnap's doctrine of 'semantic ascent', i.e., the idea that philosophy (as a formal, second-order or 'meta-linguistic' mode of analysis) should concern itself *not* with questions of *de re* truth or causal explanation but with questions in the realm of meaning, logic and strictly *de dicto* necessity. And this despite what has mostly been perceived – by Quine, Davidson, Putnam and others – as the collapse of Carnap's logical-empiricist programme and the hopelessness of any such attempt to devise a formal meta-language for the analysis of empirical (first-order) statements of scientific truth.[72] All that is left, it then appears, is a Tarskian semantic or formalized conception of truth whose function is exhausted – or which cancels right through – when applied to every sentence of the canonical form: "Snow is white" is true if and only if snow is white.'

So one can see why recent commentators (Rouse and Malpas among them) should argue that there must be *something more* to scientific truth – more substantive, interesting or profound – than is allowed for anywhere on the Tarskian account.[73] Davidson expresses a similar conviction in his 1990 series of lectures on 'The Structure and Content of Truth'. 'My own view,' he writes, 'is that Tarski has told us much of what we want to know about the concept of truth, and that there must be more. There must be more because there is no indication in Tarski's formal work of what it is that his various truth predicates have in common, and this must be part of the content of the concept.'[74] But if this sounds promising, then it has to be said that Davidson's lectures don't live up to their promise. In fact, like so much of his work, they simply veer across at a certain point from acknowledging the problem – the apparent vacuity of Tarski's semantic formula – to proposing a 'solution' that consists in little more than the unreconstructed empiricist idea that what makes our sentences true or false is the pattern of sensory stimuli that impinge upon our nerve-ends from time to time and thus (supposedly) decide the issue quite apart from all talk of paradigms, languages, theoretical frameworks, 'conceptual schemes' or whatever. It is the same sort of strategy – a fall back to 'commonsense' empiricism as a counter to their own more sceptical arguments – that one finds in certain passages of Quine's 'Two Dogmas' and in Kuhn's attempt to placate his critics on the issue of paradigm incommensurability.[75] Its *locus classicus* is the closing sentence of Davidson's essay 'On the Very Idea of a Conceptual Scheme'. 'In giving up the dualism of scheme and world,' he writes, 'we do not give up the world, but re-establish unmediated touch with the familiar objects whose antics make our sentences and opinions true or false.'[76] One need not be a card-carrying Hegelian to detect in this passage what Hegel diagnosed as the hallmarks of naïve sense certainty. All the more so, given its occurrence at the close of an essay which has gone some lengthy and elaborate ways around to address precisely the issues that are here brushed aside with such breezy assurance.

So there would seem good reason to suppose that the 'something more' is not to be had by recourse either to a formalized semantic conception of truth in the Tarskian mode nor again – at the opposite extreme – to a radical empiricism adopted in default of more adequate conceptual resources. At this point, as Rouse and Malpas would argue, the only way forward is to seek some alternative conception of truth that doesn't reduce to mere tautology on the one hand or, on the other, to Davidson's kind of bluff no-nonsense empiricism. What is required is a Heideggerian (or depth-ontological) approach that preserves the basic structure of Tarski's theory – that is to say, its purely formal system of notation for capturing the intuitive idea of truth across each and every sentence held true in a given language – while at the same time introducing a new dimension of truth as 'unconcealment', *aletheia*, 'opening', 'horizonal pre-understanding', etc. For Malpas, indeed, this is the realization towards which – did they but know it – Davidson, Quine, Putnam, Kuhn, Rorty and company have long been travelling. Just as Rorty sees Deweyan pragmatism at the end of every philosophical road (including Heidegger's, once shorn of its grandiose ontological pretensions) so Malpas sees Heidegger as the one philosopher to have worked his way through and beyond the vexing antinomies of present-day analytic thought.

Of course, Malpas has to recognize the deep resistance that such an argument is likely to encounter. After all, '[s]o radical is this shift that we may well wonder why we should use the term "truth" to refer to this fundamental opening.' And again: 'Why call this opening, this unconcealing/concealing, truth? Why should we even pay attention to such opening?'[77] Especially – one might add – since Heidegger has so long been held up by philosophers in the 'other' (analytic) tradition as a cautionary instance of what goes wrong when thinking is seduced by the fake profundities of a language that trades upon obscurantist rhetoric and an ersatz jargon of authenticity.[78] Malpas, however, is undeterred by the odds stacked against him in making this argument. Had the objectors come up with anything more useful or substantive in the way of 'truth' then their criticisms might have carried some weight. As it is, he finds nothing bar a fixed aversion to Heidegger's style and a range of place-filler substitutes – Tarski's included – which satisfy only the barest requirements for a formal theory.

Hence, he argues, the signal importance of Davidson's work as an enterprise that presses all the way with this currrent 'post-analytic' trend in philosophy of mind and language. What Davidson brings out is the sheer *impossibility* of giving any genuine (i.e., non-trivial or informative) content to the idea of truth so long as one remains within the terms laid down for such debate by the tradition that runs from logical empiricism to its various latter-day offshoots. For if these schools of thought have one thing in common, it is their failure to produce any viable alternative to the logical-empiricist dichotomy between first-order 'facts', observation statements or matters of empirical warrant and the second-order logic of scientific or philosophical enquiry. Where they differ is on the question as to whether such a programme can be carried through with success. Thus the holists and pragmatists concur in thinking that this project was foredoomed to failure since it rested on a number of untenable assumptions – like the two last 'dogmas' of Quine's essay or (in Davidson's variant) the 'very idea of a conceptual scheme'

– which collapsed under sceptical scrutiny. But the first and most crucial stage in this chapter of developments was that whereby philosophy took the turn towards a formalized language where issues of material (*de re*) causal or explanatory truth were supplanted by issues of metalinguistic (*de dicto*) veridical warrant. For the way was then open to other, more thoroughgoing versions of the 'linguistic turn' which relativized truth no longer to particular sentences, statements or propositions, but rather to the entire existing 'web' or 'fabric' of beliefs held true at some given time.[79] At which point the Heideggerians can claim that such conceptions of truth are manifestly either circular, redundant or trivial, and that only a depth-ontological approach can save thinking from this dead-end predicament.

V

It may be felt that I have skewed the issues here by taking just a few of Heidegger's more questionable statements and presenting them as typical or symptomatic instances of a much wider trend in present-day 'post-analytic' philosophy. After all, there would appear to be little enough in common between (on the one hand) Heidegger's notorious remarks about the death-camps, technology and 'Western metaphysics' and (on the other) the genial suggestion by various pragmatically minded liberal thinkers that philosophy should henceforth take its place as just one language game or cultural life-form among others, those others including science – or technology – once shorn of its domineering will-to-truth and restored to a decently scaled-down sense of its role in the ongoing cultural conversation. Nevertheless it seems to me that there remain some worrying implications about this current desire to demote science – along with philosophy of science and epistemology – to the point where it possesses no greater claim to truth than the language games of, say, religion or literary criticism. Where the confusion comes in, I have argued, is with the idea that *any* conception of science as aimed towards a better, more adequate understanding of real-world objects, processes or events must *always and inevitably* carry along with it a doctrine of absolute 'cognitive privilege' or a technocratic drive to exclude or devalue ethical, social and political concerns.

This idea takes various forms according to its various interdisciplinary contexts and sources. It finds expression in the 'strong' sociology of knowledge, an approach that avowedly collapses all distinctions between context of discovery and context of justification; in Lyotard's notion of a postmodern science given over to 'performative' rather than 'constative' criteria of suasive efficacy; in Feyerabend's kindred ('anarchist') idea that we should lift all constraints of method, consistency and truth, thus allowing a thousand speculative flowers to bloom and deciding between them on purely social and ethical grounds; in Rorty's neo-pragmatist view that science *just is* whatever counts as such by the lights of some given cultural community with its own preferred metaphors, language games, narratives of scientific 'progress', etc.; in Wittgensteinian resorts to language and social context as the furthest one can get by way of justifying scientific or other orders of truth-claim; and lastly, in the Heideggerian (depth-hermeneutical) appeal to a dimension of truth as concealment-revelation whose import precedes all mere determinations of 'ontic' or 'factical' concern.[80] To which might be added the more logically refined anti-realist contentions of philosophers such as Michael Dummett and the arguments of a 'constructive empiricist' like Bas van Fraassen who sees no reason to adduce 'laws of nature', intrinsic properties, causal dispositions or other such extravagant hypotheses when we can get by just as well on the modest assumption that scientific theories are justified solely by observational warrant and predictive yield.[81] Elsewhere, among postmodernists especially, there seems little sense that reality may impose limits – sometimes (if not always) non-negotiable limits – on the human freedom to redescribe nature in terms that answer to present conceptions of what is 'good in the way of belief'.

Roy Bhaskar has pursued some of the consequences of such thinking in his book *Philosophy and the Idea of Freedom*, most of which is taken up with a critique of Rorty's neo-pragmatist position.[82] His argument – in brief – is that Rortian talk of creative 'redescription' is misleading when applied to epistemology or philosophy of science in so far as those disciplines cannot be construed as creating or inventing the various objects that constitute their proper ontological domain. To adopt this extreme anti-realist line is to invite all manner of antinomies, paradoxes or aporias when it comes to explaining how science could possibly afford us knowledge of the world, or again, how we could ever have reason – cultural-linguistic preference aside – for counting certain theories more adequate (better borne out by argument or evidence) than others of notionally similar scope. Indeed, there is something decidedly myopic about Rorty's idea of freedom, construed as it is in such wholesale world-transformative terms as to leave no room for any irksome constraints upon the human will to redescribe 'reality' in response to an ever-shifting range of interests, values or desires. For there can then be no accounting – no making due allowance – for those various factors of a physical, causal, environmental or psychological nature that do place limits – whether we like it or not – on our freedom to refashion ourselves and the world in whatever way we choose.

Moreover, as Bhaskar remarks, this idea goes along with a way of drawing the line between 'private' and 'public' spheres such that strong-willed individuals are free to pursue their own projects of inventive self-fashioning just so long as they don't lay claim to authority in the wider (ethical or socio-political) domain. This may be wise counsel if applied to thinkers – such as Nietzsche or Foucault – whose primary values or ideas of autonomous selfhood are indeed far removed from any viable conception of the wider public good. However, it is a doctrine that leaves little room for the exercise of judgement and moral responsibility in matters of shared concern. Thus it speaks hardly at all to those issues of social, political and ethical conscience – environmental issues among them – that involve something more than the (notional) freedom to pursue one's idea of the life well lived in accordance with private-individual aims and inclinations. For this is to redefine the concept of 'autonomy' in much the same way – and with much the same drastic narrowing of scope – as affects the concept of 'liberalism' when used by proponents of free-market doctrine or by advocates of reduced welfare provision in the name of private enterprise. Or again – more to the point in this context – it goes along readily with versions of that argument which uphold the full liberty of persons to protest against what they regard as bad (antisocial or destructive) techno-scientific developments while treating such protest as strictly a matter of individual conscience and hence as irrelevant to policy decisions arrived at with a view to maximizing profit. In each case there is a severance between private and public spheres which leaves people free to follow their conscience – or engage in the process of creative self-fashioning – just provided they do it, in Rorty's phrase, 'on their own time' and without any claim to moralize or legislate in matters of collective concern.[83] For the rest, such issues are much better dealt with by those – the social, ethical and political theorists – whose discourse belongs to the public sphere (to the realm of *Wertfrei* adjudicative reason) and should therefore properly find no place for the voice of individual conscience or the expression of private values.

Of course, there is a large current literature on the antinomies of 'liberal' theory, thus construed, and its failure to acknowledge the complex relationship between those various value-spheres that Rorty so blithely puts asunder.[84] However, my main concern here is not so much with debates in present-day social and political theory as with their bearing on questions in philosophy of science, and more specifically the questions that science raises with regard to the scope and limits of human freedom. For on the Rortian (strong descriptivist) view, the best way to maximize that scope and to minimize those limits is to treat science as just another language game or cultural life-form, one that is always open to change or revision in keeping with the current self-images and values of the time. Such – he would persuade us – is the benefit to be had by removing science's false prestige as an 'objective', 'constructive', 'progressive' or truth-oriented discourse and henceforth treating its deliverances as strictly on a par with those of poets, novelists, philosophers and anyone else with a voice in the ongoing cultural conversation. Moreover, Rorty can call upon a wide range of sources – hermeneutics, narrative pragmatics, postmodernism, post-structuralism, post-Quinean talk of ontological relativity, post-Kuhnian philosophy of science, the 'strong' programme in sociology of knowledge – to support this view of scientific 'truth' as whatever best suits our current descriptive or socio-cultural purposes. For if the world and all its contents are pretty much what we make of them according to this or that preferential language game or elective 'final vocabulary' then of course there can be no restriction on our freedom to reinvent the world (and ourselves along with it) from one paradigm to the next. Certainly there is nothing 'in the nature of things' – no real-world factual, circumstantial or causal constraints – that could offer a check to these endless possibilities of creative redescription. To think that there might be is for Rorty just a sign that we have not yet broken with the delusory idea of science as a process of *discovering* those facts about the world that make our statements or theories true or false. Much better, he thinks, to view it as a process of *inventing* new languages – paradigms, images, 'metaphors we can live by' – whereby to provide as many novel perspectives on 'truth' as there may be social or cultural opportunities for changing the current topics of conversation.

Nothing could more clearly illustrate the point that I have been making with regard to the implications of anti-realism in epistemology and philosophy of science. The Heideggerian (depth-hermeneutic) variant is no doubt the most drastic in its claim upon our thinking about issues of truth, knowledge, responsibility and freedom. But there are also great problems with Rorty's more laid-back neo-pragmatist version if one asks what sense can possibly be made – in ethical and political as well as in epistemological terms – of a freedom whose limits in the matter of inventive redescription are set only by the limits of our current imagining or our choice among various (more or less novel) metaphors, language games, etc. Such 'freedom' is entirely nugatory – the merest of wishful fantasies – if it takes no account of the real-world conditions, both restrictive and potentially enabling, which bear upon the human quest for knowledge and truth. In Harry Frankfurt's words, 'There must be limits to our freedom if we are to have sufficient personal reality to exercise genuine autonomy at all. What has no boundaries has no shape.'[85] And again, in a passage from his aptly-titled essay 'On Bullshit' which will bear quoting at length:

> '[A]nti-realist' doctrines undermine confidence in the value of disinterested efforts to determine what is true and what is false, and even in the intelligibility of the notion of objective inquiry. One response to this loss of confidence has been a retreat from the discipline required by dedication to the ideal of *correctness* to a quite different sort of discipline, which is imposed by pursuit of an alternative ideal of *sincerity*. . . . Convinced that reality has no inherent nature, which he might hope to identify as the truth about things, he [the anti-realist] devotes himself to being true to his own nature.[86]

However, as Frankfurt remarks, there is something very odd – not to say 'preposterous' – about the notion that a self might be known for what it is (known in its authentic selfhood) more reliably than any knowledge to be had of real-world objects and events. Certainly Kant was under no such illusion, as witness those passages in the first *Critique* that tie the conditions for coherent first-person identity (the 'transcendental unity of apperception') to the conditions obtaining for our knowledge and experience of events in the spatio-temporal domain.[87]

Of course Kant was careful – and indeed went some long and tortuous ways around – to distinguish this order of subjectivity from any notion of the subject empirically or psychologically construed, that is to say, any attempt to derive substantive knowledge of the self from an argument in the purely transcendental (or 'conditions of possibility') mode. Only by maintaining this distinction, he thought, could philosophy be saved from the kinds of dilemma – the failure to connect sensuous or phenomenal cognitions with the concepts and categories of understanding – which had confronted empiricists like Hume on the one hand and rationalists like Descartes and Leibniz on the other. It is debatable whether Kant succeeded in his aim of reconciling transcendental idealism with empirical realism. However, his argument is surely valid to this extent at least: that we can make no sense of an idea of the subject (the knowing, willing or judging subject) whose 'world' would be entirely a construction out of its own sense data, language games, final vocabulary, conceptual scheme, or whatever. For such a sovereign disposer would encounter no resistance – no check upon its world-creating powers – from anything beyond or outside the domain of its own internal representations. It could therefore achieve no grasp of the distinction between waking and dreaming states, or again, between veridical (undistorted) perceptions and those brought about by various forms of perceptual illusion. Nor could the subject, thus conceived, be in any position to explain – in its own case or that of others – just how (through what kinds of causally explicable process) such illusions typically take rise. This is the main problem with strong anti-realist or descriptivist theories such as Rorty's: that in counting 'reality' a world well lost for the sake of inventive or imaginative self-creation they effectively dissolve any notion of the self to which that process could refer or apply. This is why, as Frankfurt says, there must be 'limits to our freedom if we are to have sufficient personal reality to exercise genuine autonomy at all'. And this 'personal reality' cannot be achieved without a sufficiently well-developed sense of that other (objective and mind-independent) reality that may always resist our best efforts of creative redescription. For there is otherwise nothing that can halt the drift towards, on the one hand, a wholesale anti-realism devoid of epistemic content, and on the other a kind of transcendental solipsism that views both the subject and its ambient world as just what we make of them according to this or that fictive or imaginary projection.

References

1 Ludwig Wittgenstein, *Tractatus Logico-Philosophicus*, trans. D. F. Pears and B. F. McGuiness (London: Routledge and Kegan Paul, 1961), Section 5.6.

2 Donald Davidson, 'On the Very Idea of a Conceptual Scheme', in *Inquiries into Truth and Interpretation* (Oxford: Clarendon Press, 1984), pp. 183–98.
3 B. L. Whorf, *Language, Thought and Reality: selected writings of Benjamin Lee Whorf*, ed. J. B. Carroll (Cambridge, MA: MIT Press, 1956).
4 See W. V. Quine, *From a Logical Point of View*, 2nd edn (Cambridge, MA: Harvard University Press, 1961); *Word and Object* (Cambridge, MA: MIT Press, 1960); *Theories and Things* (Harvard University Press, 1981).
5 See Ludwig Wittgenstein, *Philosophical Investigations*, trans. G. E. M. Anscombe (Oxford: Basil Blackwell, 1958); Thomas S. Kuhn, *The Structure of Scientific Revolutions*, 2nd edn (Chicago: University of Chicago Press, 1970); Paul K. Feyerabend, *Against Method* (London: New Left Books, 1975); Michel Foucault, *The Order of Things: an archaeology of the human sciences* (London: Tavistock, 1973); Jean-François Lyotard, *The Postmodern Condition: a report on knowledge*, trans. Geoff Bennington and Brian Massumi (Manchester: Manchester University Press, 1984).
6 See Davidson, 'On the Very Idea of a Conceptual Scheme'.
7 Alfred Tarski, 'The Concept of Truth in Formalized Languages', in *Logic, Semantics and Metamathematics*, trans. J. H. Woodger (London: Oxford University Press, 1956), pp. 152–278.
8 See especially Richard Rorty, 'Pragmatism, Davidson, and Truth', in Ernest LePore (ed.), *Truth and Interpretation: perspectives on the philosophy of Donald Davidson* (Oxford: Basil Blackwell, 1986), pp. 333–55; also Rorty, 'Is truth a goal of enquiry? Davidson *versus* Wright', *Philosophical Quarterly*, 45 (1995), pp. 281–300.
9 See F. P. Ramsey, *Philosophical Papers*, ed. D. H. Mellor (Cambridge: Cambridge University Press, 1990).
10 Christopher Norris, 'Reading Donald Davidson: truth, meaning and right interpretation', in *Deconstruction and the Interests of Theory* (Leicester: University of Leicester Press, 1992), pp. 59–83.
11 See, for instance, Davidson, 'A Coherence Theory of Truth and Knowledge', in LePore (ed.), *Truth and Interpretation*, pp. 307–19.
12 Donald Davidson, 'The structure and content of truth', *Journal of Philosophy*, 87 (1990), pp. 279–328.
13 See Tarski, 'The Concept of Truth in Formalized Languages'; also Rudolf Carnap, *The Logical Structure of the World* (London: Routledge and Kegan Paul, 1937) and Carl G. Hempel, *Fundamentals of Concept Formation in Empirical Science* (Chicago: University of Chicago Press, 1972).
14 See Hempel, *Fundamentals of Concept Formation*.
15 W. V. Quine, 'Two Dogmas of Empiricism', in *From a Logical Point of View*, pp. 20–46.
16 See, for instance, Richard Rorty, *Consequences of Pragmatism* (Brighton: Harvester, 1982) and *Objectivity, Relativism, and Truth* (Cambridge: Cambridge University Press, 1991).
17 Joseph Rouse, *Knowledge and Power: toward a political philosophy of science* (Ithaca, NY: Cornell University Press, 1987), p. 160.
18 See also Hubert L. Dreyfus, *Being-in-the-World: a commentary on Heidegger's Being and Time, Divison One* (Cambridge, MA: MIT Press, 1991); J. E. Malpas, *Donald Davidson*

and the Mirror of Meaning (Cambridge: Cambridge University Press, 1992); Stephen Mulhall, *On Being in the World: Wittgenstein and Heidegger on seeing aspects* (London: Routledge, 1990); Mark Okrent, *Heidegger's Pragmatism: understanding, being and the critique of metaphysics* (Ithaca, NY: Cornell University Press, 1988); Richard Rorty, *Essays on Heidegger and Others* (Cambridge University Press, 1991).

19 See especially Rudolf Carnap, 'The Elimination of Metaphysics through Logical Analysis of Language', in A. J. Ayer (ed.), *Logical Positivism* (New York: Free Press, 1959), pp. 60–81.

20 See, for instance, Martin Heidegger, *The Question Concerning Technology and Other Essays*, trans. William Lovitt (New York: Harper and Row, 1977).

21 See entries under notes 16–18 above; also Don Ihde, *Technology and the Lifeworld: from garden to earth* (Bloomington: Indiana University Press, 1990); John Loscerbo, *Being and Technology: a study in the philosophy of Martin Heidegger* (The Hague: Nijhoff, 1981); Michael E. Zimmerman, *Heidegger's Confrontation with Modernity: technology, politics, art* (Indiana University Press, 1991) and 'Re-thinking the Heidegger–deep ecology relationship', *Environmental Ethics*, 15/3 (1993), pp. 195–224.

22 See T. W. Adorno, *The Jargon of Authenticity*, trans. T. Tarnowski and F. Will (London: Routledge and Kegan Paul, 1973); also T. W. Adorno and Max Horkheimer, *Dialectic of Enlightenment*, trans. John Cumming (New York: Seabury Press, 1972).

23 See entries under note 21 above; also Hubert L. Dreyfus, *What Computers Can't Do: on the limits of artificial intelligence*, rev. edn (New York: Harper and Row, 1979).

24 See note 20 above.

25 See note 18 above.

26 See, for instance, Pierre Bourdieu, *The Political Ontology of Martin Heidegger*, trans. Peter Collier (Oxford: Polity Press, 1991); Hugo Ott, *Heidegger: a political life*, trans. Alan Blunden (London: Harper-Collins, 1993); Tom Rockmore, *On Heidegger's Nazism and Philosophy* (London: Harvester-Wheatsheaf, 1992); Hans Sluga, *Heidegger's Crisis: philosophy and politics in Nazi Germany* (Cambridge, MA: Harvard University Press, 1993); James F. Ward, *Heidegger's Political Thinking* (Cambridge, MA: University of Massachusetts Press, 1995); Richard Wolin, *The Politics of Being: the political thought of Martin Heidegger* (New York: Columbia University Press, 1990); Richard Wolin (ed.), *The Heidegger Controversy: a critical reader* (Cambridge, MA: MIT Press, 1993).

27 See especially the discussion in Rockmore, *On Heidegger's Nazism and Philosophy*, pp. 239ff; also other entries under note 26 above.

28 See especially Jürgen Habermas, *The Philosophical Discourse of Modernity: twelve lectures*, trans. Frederick Lawrence (Cambridge: Polity Press, 1987).

29 See, for instance, Zygmunt Bauman, *Postmodern Ethics* (Oxford: Blackwell, 1993).

30 See note 7 above.

31 Rouse, *Knowledge and Power*, p. 147.

32 Ibid., p. 150. On this deflationary theory see also Paul Horwich, *Truth* (Oxford: Blackwell, 1990).

33 Charles Taylor, 'Understanding in human science', *Review of Metaphysics*, 34 (1980), pp. 25–38; p. 26.

34 Malpas, *Donald Davidson and the Mirror of Meaning*, pp. 264–5.
35 Dreyfus, *Being-in-the-World*; Richard Rorty, *Essays on Heidegger and Others* (Cambridge: Cambridge University Press, 1991) and *Consequences of Pragmatism*.
36 Rouse, *Knowledge and Power*, p. 155.
37 Ibid., pp. 155–6.
38 See notes 13 and 19 above.
39 See Ferdinand de Saussure, *Course in General Linguistics*, trans. W. Baskin (London: Fontana, 1974); also Frederick Jameson, *The Prison-House of Language: a critical account of structuralism and Russian formalism* (Princeton, NJ: Princeton University Press, 1972); Michael Lane (ed.), *Structuralism: a reader* (London: Allen Lane, 1970); Richard Macksey and Eugenio Donato (eds), *The Structuralist Controversy* (Baltimore: Johns Hopkins University Press, 1972); Thomas G. Pavel, *The Feud of Language: a history of structuralist thought* (Oxford: Blackwell, 1990).
40 See, for instance, the essays collected in Roland Barthes, *Image, Music, Text*, ed. Stephen Heath (London: Fontana, 1977); also Derek Attridge, Geoff Bennington and Robert Young (eds), *Post-Structuralism and the Question of History* (Cambridge: Cambridge University Press, 1987); Catherine Belsey, *Critical Practice* (London: Methuen, 1980); Colin MacCabe, *James Joyce and the 'Revolution of the Word'* (London: Macmillan, 1978).
41 Roland Barthes, *Image, Music, Text*; also *S/Z*, trans. Richard Miller (London: Jonathan Cape, 1975).
42 See especially Michel Foucault, *Language, Counter-Memory, Practice*, ed. and trans. D. F. Bouchard and S. Weber (Oxford: Blackwell, 1977).
43 See, for instance, Ted Benton, *The Rise and Fall of Althusserian Marxism* (London: Macmillan, 1984) and Gregory Elliott, *Althusser: the detour of theory* (London: Verso, 1987).
44 See, for instance, Ernesto Laclau and Chantal Mouffe, *Hegemony and Socialist Strategy: towards a radical democratic politics* (London: Verso, 1985).
45 Michel Foucault, *The Order of Things*; also *The Archaeology of Knowledge*, trans. A. M. Sheridan Smith (London: Tavistock, 1972).
46 Quine, 'Two Dogmas of Empiricism'; also *Ontological Relativity and Other Essays* (New York: Columbia University Press, 1969).
47 See Rorty, *Objectivity, Relativism and Truth* and *Essays on Heidegger and Others* (notes 16 and 18 above).
48 Rouse, *Knowledge and Power*, p. 160.
49 Ibid., p. 160.
50 Ibid., p. 161.
51 Ibid., p. 161.
52 Ibid., p. 161.
53 See Hans-Georg Gadamer, *Truth and Method*, trans. W. Glen-Doepel, ed. John Cumming and Garrett Barden (London: Sheed and Ward, 1979) and David C. Hoy, *The Critical Circle: literature and history in contemporary hermeneutics* (Berkeley and Los Angeles: University of California Press, 1978).
54 Rouse, *Knowledge and Power*, p. 162.
55 Martin Heidegger, *What is a Thing?*, trans. W. B. Barton and V. Deutsch (South Bend: Gateway, 1967).

56 Martin Heidegger, 'The Question Concerning Technology', in *Martin Heidegger: basic writings*, ed. D. F. Krell, 2nd edn (London: Routledge, 1993), pp. 311–41; p. 321. See also Heidegger, *The Question Concerning Technology and Other Essays*, trans. William Lovitt (New York: Harper and Row, 1977); *What Is a Thing?*; John Caputo, 'Heidegger's Philosophy of Science', in J. Margolis (ed.), *Rationality, Relativism and the Human Sciences* (The Hague: Nijhoff, 1986); Philip R. Fandozzi, *Nihilism and Technology: a Heideggerian investigation* (Washington, DC: University Press of America, 1982); Don Ihde, *Technology and the Lifeworld: from garden to earth*; Joseph J. Kockelmans, *Heidegger and Science* (University Press of America, 1985); John Loscerbo, *Being and Technology: a study in the philosophy of Martin Heidegger.*

57 For further discussion see Rockmore, *On Heidegger's Nazism and Philosophy* and other entries under note 26 above; also Hans Sluga, *Heidegger's Crisis: philosophy and politics in Nazi Germany* (Cambridge, MA: Harvard University Press, 1993).

58 Heidegger, 'The Question Concerning Technology', p. 313.

59 Ibid., p. 313.

60 Ibid., pp. 303–31.

61 Ibid., p. 331.

62 Ibid., p. 320.

63 Ibid., p. 333.

64 Ibid., p. 333.

65 See Rockmore, *On Heidegger's Nazism and Philosophy*; also Günther Neske and Emil Kettering (eds), *Martin Heidegger and National Socialism: questions and answers*, trans. L. Harries and J. Neugroschel (New York: Paragon House, 1990); Alan Rosenberg and Alan Milchman, *Martin Heidegger and the Holocaust* (Atlantic Highlands, NJ: Humanities Press, 1994); Richard Wolin, *The Politics of Being: the political thought of Martin Heidegger* (New York: Columbia University Press, 1990); Richard Wolin (ed.), *The Heidegger Controversy.*

66 See note 26 above; also Heidegger, 'The Rectoral Address' and 'The Rectorate 1933/34: facts and thoughts', in Neske and Kettering (eds), *Martin Heidegger and National Socialism*, pp. 5–14 and 15–32.

67 Heidegger, 'The Question Concerning Technology', p. 324.

68 Ibid., p. 329.

69 Foucault, *The Order of Things.*

70 Martin Heidegger, *Being and Time*, trans. John Macquarrie and Edward Robinson (Oxford: Blackwell, 1962).

71 See notes 4, 5 and 18 above; also Hilary Putnam, *Representation and Reality* (Cambridge: Cambridge University Press, 1988) and *Realism With a Human Face* (Cambridge, MA: Harvard University Press, 1990).

72 See entries under note 13 above.

73 Rouse, *Knowledge and Power* and Malpas, *Donald Davidson and the Mirror of Meaning.*

74 Donald Davidson, 'The Structure and Content of Truth', p. 284.

75 Quine, 'Two Dogmas of Empiricism' and Kuhn, 'Postscript – 1969', in *The Structure of Scientific Revolutions*, 2nd edn.

76 Donald Davidson, 'On the Very Idea of a Conceptual Scheme', in *Inquiries into Truth and Interpretation* (Oxford: Clarendon Press, 1984), pp. 183–98, esp. p. 198.

77 Malpas, *Donald Davidson and the Mirror of Meaning*, pp. 269–70.

78 See, for instance, Rudolf Carnap, 'The Elimination of Metaphysics through Logical Analysis of Language'.

79 For a critical survey of the field, see Jerry Fodor and Ernest LePore, *Holism: a shopper's guide* (Oxford: Blackwell, 1991).

80 See notes 5, 18 and 53 above.

81 See Michael Dummett, *Truth and Other Enigmas* (London: Duckworth, 1978) and Bas van Fraassen, *The Scientific Image* (Oxford: Clarendon Press, 1980); also Michael Luntley, *Language, Logic and Experience: the case for anti-realism* (Duckworth, 1988) and Crispin Wright, *Realism, Meaning and Truth* (Oxford: Blackwell, 1987).

82 Roy Bhaskar, *Philosophy and the Idea of Freedom* (Oxford: Blackwell, 1991).

83 See Richard Rorty, *Contingency, Irony, and Solidarity* (Cambridge: Cambridge University Press, 1989) and *Objectivity, Relativism, and Truth*.

84 See, for instance, Michael J. Sandel, *Liberalism and its Critics* (Oxford: Blackwell, 1984); Michael Walzer, *Liberalism and the Limits of Justice* (Cambridge: Cambridge University Press, 1982) and *Spheres of Justice* (Oxford: Blackwell, 1983).

85 Harry G. Frankfurt, 'On Bullshit', in *The Importance of What We Care About: philosophical essays* (Cambridge: Cambridge University Press, 1988), pp. 117–33.

86 Ibid., p. 133.

87 Immanuel Kant, *Critique of Pure Reason*, trans. N. Kemp Smith (London: Macmillan, 1974).

6

Anti-Realism and Constructive Empiricism: Is There a (Real) Difference?

I

Richard Rorty's 1985 essay 'Texts and Lumps' is probably his best-known statement of the case for an anti-realist (or strong descriptivist) approach to questions of meaning, method and truth.[1] Here I shall discuss the implications of that approach for epistemology and philosophy of science, both of which disciplines Rorty would regard as prime candidates for debunking treatment. After that I shall turn to some arguments mounted by Bas van Frassen in defence of an outlook ('constructive empiricism') which attempts to steer a more cautious path between the realist and anti-realist positions.[2] However, that path turns out to veer sharply towards anti-realism at just those points where van Frassen is keenest to mark his distance from Rorty-style strong-descriptivist talk. I thus conclude that he fails in his attempt to draw this line and that constructive empiricism is heir to all the problems that anti-realism confronts when required to offer some workable account of what constitutes scientific knowledge.

Rorty's method, here as elsewhere, is to set up two supposedly distinct or autonomous domains of enquiry, one of which (e.g., physical science or the *Naturwissenschaften*) is standardly taken as a paradigm of truth, objectivity, progress, constructive or problem-solving endeavour, etc., while the other (e.g., literary criticism or the *Geisteswissenschaften*) is standardly viewed as lacking those criteria and hence as involving a fall-back appeal to 'softer' notions such as meaning, intention, hermeneutic insight, cultural significance and so forth.[3] He then whittles away at that distinction through a series of purportedly borderline or anomalous cases to the point where it either collapses altogether or – the preferred Rortian pay-off – turns out to work much better in reverse. Thus '[t]hink of a paradigmatic text,' Rorty invites us,

> as something puzzling which was said or written by a member of a primitive tribe, or by Aristotle, or by Blake. Nonlinguistic artifacts, such as pots, are borderline cases of texts. Think of a lump as something which you would bring for analysis to

> a natural scientist rather than to someone in the humanities or social sciences – something which might turn out to be, say, a piece of gold or the fossilized stomach of a stegosaurus. A wadded-up plastic bag is a borderline case of a lump. Most philosophical reflection about objectivity – most epistemology and philosophy of science – has concentrated on lumps. Most discussion of interpretation has concentrated on texts. (TL, pp. 84–5)

However, it is Rorty's main object in this essay to make us think of lumps (under certain descriptions) as having a text-like or interpretable aspect, and conversely of texts – again from a certain viewpoint – as amenable to treatment in the sort of vocabulary normally reserved for lumps.

Thus he sets out two columns of multiple criteria pertaining to the text/lump dichotomy and ranging all the way, on a five-point scale, from a physicalist/inscriptionalist to a strong-hermeneutic or *Geisteswissenchaftlich* mode of understanding. Under description I, for instance, a parallel is drawn between 'the sensory appearance and spatio-temporal location of a lump' and 'the phonetic or graphic features of an inscription' (TL, p. 85). Where the former may involve – among other things – certain methods for explaining and avoiding perceptual illusion, so the latter may involve philological techniques that enable the more accurate dating or deciphering of texts quite aside from their meaning, value or cultural significance. Description II matches up a pair of non-starters so far as Rorty is concerned: an idea of 'the real essence of the lump which lurks behind its appearances – how God or Nature would describe the lump' and 'what the author would, under ideal conditions, reply to questions about his inscription which are phrased in terms which he can understand right off the bat' (p. 85). The former he takes to express the view of neo-realist philosophers of science or language (Boyd, Kripke, early Putnam) whose arguments Rorty considers just a throwback to old, quasi-theological notions of substance, essence and the like. The latter he associates chiefly with the 'strong' intentionalist stance adopted by a literary critic such as E. D. Hirsch. This is the idea that an author's original (intended) meaning can somehow be recovered and separated out from the various accretions of cultural 'significance' that constitute the history of a work's reception or its afterlife of changing construals and interpretations.[4] However, Rorty argues, there is simply no prospect of upholding this distinction once critics come around to the sensible pragmatist view that the meaning of a text *just is* whatever it happens to signify for this or that member of some given interpretive community.[5] To suppose otherwise is to think – deludedly – that criticism should have some *method* or *theory* for establishing an author's intent, a method that would fix textual meaning in much the same way that the scientific realist thinks to fix the nature of some candidate lump by adverting to its essence, its underlying nature, micro-structural properties, or whatever.

At this level there is not much difference – as Rorty sees it – between 'God' or 'nature' conceived as the last metaphysical guarantees behind notions of meaning, truth or objectivity. Thus a realist about science or philosophical semantics is in much the same position as a critic like Hirsch who defends a realist stance with regard to the existence and recoverability of authorial intention. And the same applies to those literary theorists (e.g., critics of a formalist or structuralist persuasion) who think that criticism can become more 'rigorous' or philosophically respectable by imitating the methods of the natural sciences. For this merely shows that their model of 'science' is the old-fashioned (no longer credible) idea of scientific knowledge as 'cutting nature at the joints', or as actually having something to do with the way things stand 'in reality'.

In Rorty's view, the best cure for these delusive Level-II beliefs is to move straight on up to Levels III and IV. This is the stage at which pragmatism starts to look good and where realist epistemologies have at last given way to a conventionalist paradigm incorporating Kuhn's famous distinction between 'normal' and 'revolutionary' periods in science. Thus at Level III philosophers of science will stop talking about the lump's 'essence' or 'real nature' and will henceforth think of 'the lump as described by that sector of *our* normal science which specializes in lumps of that sort' (TL, p. 85). Such would be, for instance, 'a routine analysis performed by a chemist' or 'a routine identification performed by a biologist' (ibid.). The same stage is reached in the human sciences when philosophers, anthropologists, literary critics and others give up on the quest for such illusory items as truth, original meaning or author's intention and instead rely on their own best ideas as to what constitutes a relevant idea in this or that context of debate. One criterion might be: 'What the author would, under ideal conditions, reply to *our* questions about his inscription – questions he would have to be reeducated to understand . . . but which are readily intelligible to a present-day interpretive community' (p. 85). Think, for instance, Rorty suggests, of 'a Cambridge-educated primitive' or 'an Aristotle who has assimilated Marx and Freud'.

Level IV works out – in Kuhnian terms – as the transposition of this conventionalist view from a 'normal' to a 'revolutionary' phase of scientific or cultural activity. Thus in science it gives us '[t]he lump as described by . . . somebody who wants to redo chemistry, or entomology, or whatever, so that the currently "normal" chemical analyses or biological taxonomies are revealed as "mere appearances"' (pp. 85–6). That is to say, they may wish to proclaim their revolution as a matter of 'discovering' some hitherto unknown truth, or in keeping with some notion of scientific method based on the realist paradigm. But we should still do better – Rorty thinks – to take such claims with a Kuhnian pinch of salt and regard them as just a useful (strategic or rhetorical) means of winning support for the new belief among members of the relevant interest group or wider interpretive community. Level IV has something of the same ambivalence when translated into literary-hermeneutic terms. Thus a typical case might be: '[t]he role of the text in somebody's revolutionary view of the sequence of inscriptions to which the text belongs (including revolutionary suggestions about which sequence that is)' (pp. 85–6). As examples Rorty invites us to consider 'the role of an Aristotle text in Heidegger or a Blake text in [Harold] Bloom'. Once again there is a tendency – unfortunately so, Rorty thinks – for critics to make some stronger claim, for instance, by purporting to reveal what previous critics had missed on account of their doctrinal attachments, pet fixations, ideologically induced blindnesses, inferior techniques of close reading, and so forth. But we can always discount for those claims – here as with the scientists – by construing them as just an enabling fiction, or as a form of suasive rhetoric designed to overcome the resistance standardly encountered by strong-revisionist readings.

Such readings acquire their maximum scope with the advent of Rorty's Level V, at which point – predictably – his descriptions take on a greater inventiveness and brio. In the human sciences it is a matter of lifting the various disciplinary and generic conventions that standardly decide what shall count as a serious, competent or good-faith contribution to debate. For we will then be better placed to appreciate '[t]he role of the text in somebody's view of something other than the "kind" to which the text belongs – for example, its relation to the nature of man, the purpose of my life, the politics of our day, and so forth' (TL, p. 86). Nor are such freedoms by any means denied to the natural scientists or philosophers of science who might be thought (from a realist or 'essentialist' viewpoint) to be working under more stringent demands of method, objectivity and truth. On the contrary, he argues: at Level V it is just as possible for science – 'revolutionary' science – to exploit the full range of creative redescription enjoyed by people in the arts and humanities. Later on in the essay Rorty cites Feyerabend as one of those 'silly relativists' who have spoiled a good case by overstating its claims and thus yielded ground to the current (as he thinks it) reactive and retrograde neo-realist trend.[6] Still, it is hard to see what separates Rorty from a thinker like Feyerabend, given his eagerness to push right through with the Level-V idea that any wished-for extension of our hermeneutic freedom with regard to texts will also apply, give or take a few adjustments, to our dealing with physical lumps. These latter may concern '[t]he place of the lump, or of that *sort* of lump, in somebody's view of something other than the science to which the lump has been assigned' (TL, p. 86). Thus if *gold* is in question then the range of alternative descriptions might include 'the role of gold in the international economy, in sixteenth-century alchemy, in Alberich's fantasy life, my fantasy life, and so forth, as opposed to its role in chemistry' (ibid.).

In Rorty's summation, 'levels II–V under "texts" can be thought of as four possible meanings of "meaning",' while 'the same numbers under "lumps" can be thought of as so many meanings of "nature"' (p. 86). The Level II definition of 'lump' is the only one for which he finds no use whatsoever, committed as it is – or as Rorty views it – to a wholly outmoded realist-essentialist ontology. Level I descriptions (the kind that are of interest to graphologists, philologists, uninquisitive lump-fanciers or students of perceptual illusion) may properly be counted a case apart since they afford little scope for inventive licence. Otherwise the general purport of Rorty's argument is to shift the emphasis so far as possible from 'lumps' to 'texts' – or from the natural sciences to the arts and humanities – as our best source of new and productive ideas in the ongoing cultural conversation. At the same time we should think of Level V descriptions as pointing a way towards the sort of culture from which scientists – as well as philosophers, anthropologists or literary critics – would surely benefit. For on the pragmatist view, as Rorty presents it, there is nothing in the nature of those disciplines or their objects of study that would warrant any strong version of the text/lump dichotomy or any argument for upholding the cognate discipline boundaries. This is not to deny some use for the varieties of special expertise that chemists or biologists deploy in their analysis of lumps, which will probably incline us to consult their opinion – rather than consulting, say, a literary critic – for certain sorts of advice. But it *is* to deny that there exists some specific lump-related reason (some fact about its chemical properties, molecular constitution, DNA structure or whatever) that would single them out as uniquely well-qualified to pronounce on the topic. Or again, more to the point: at Level V there is no kind of knowledge regarding any lump – no privileged description or expert discourse – that would have a better claim than any other to specify what makes it *that* kind of lump.

II

I have summarized Rorty's essay at length because it offers an unusually forthright statement of the anti-realist case. Also it has the merit of pushing right through with that case and not falling back on some ill-defined midway stance or else – like so many present-day cultural relativists – simply taking anti-realism for granted as an aspect of our current 'postmodern' condition.[7] For Rorty, on the contrary, what most needs explaining is why anyone should wish to defend any version of the realist (or 'essentialist') argument, given the insuperable problems that arise with all efforts to secure it against the pragmatist's standard counter-arguments. Thus from his Level-V point of view 'the source of realist, antipragmatist philosophy of science is the attempt, characteristic of the Enlightenment, to make "Nature" do duty for God – the attempt to make natural science a way of conforming to the will of a power not ourselves, rather than simply facilitating commerce with the things around us' (TL, p. 87). And it is here – rather oddly – that he finds room for the strong line on authorial intention taken by a theorist such as Hirsch. For if indeed, as Rorty argues, 'the *only* interesting difference between texts and lumps is that we know how to form and defend hypotheses about the author's intention in the one case but not in the other' (p.

87), then Hirsch can himself be 'reinterpreted' – Rorty's word – so as to weaken his claim to the point where it becomes quite compatible with the pragmatist case. That is to say, one has only to press a little further with Hirsch's Vico-derived insight that 'the human realm is genuinely knowable while the realm of nature is not.'[8] For the kind of 'knowability' in question here is not a matter of possessing some hermeneutic method or divinatory skill that would allow the interpreter privileged access to an author's actual or original intent. Rather, it makes the more modest pragmatist point: that for some activities (e.g., interpreting texts or figuring out the motives of historical agents) to understand *just is* to attribute intentions – to ascribe humanly intelligible meaning or purpose – while for other activities (e.g., analysing lumps) such questions scarcely arise. Thus Rorty's way with Hirsch involves nothing more than a simple application of this same idea, i.e., that one can give up the strong claim to know or respect some *specific* authorial intention while still profiting from Vico's (and Hirsch's) crucial insight as regards the prior significance of cultural over natural-scientific modes of understanding.

It is not hard to see why 'Texts and Lumps' has achieved wide currency among literary critics, cultural theorists, 'strong' sociologists of knowledge and others who approve its anti-realist line on the history and philosophy of science. However, I would contend that the source of this essay's appeal lies more in its suasive rhetoric – along with the existence of a readily persuaded readership – than in its merits as a piece of philosophical argument. For what are we to make of Rorty's claim that realism amounts to just a kind of secularized ('Enlightenment') version of older scholastic doctrines such as those of the divine substance or of God as the *ens realissimum*? If this argument is to work then it must be the case that there is no difference – in pragmatist terms, no 'difference that makes any difference' – between the sort of thing that theologians predicate of God and the sort of thing that scientists discover about atomic nuclei, molecules, DNA proteins, viruses, electromagnetic fields, planetary orbits, quasars, or (in aerodynamics) the transition from laminar to turbulent flow for a given airfoil section under given ambient conditions.

Of course, it may be said that these items compose such a heteroclite collection as to place severe strain on any realist ontology designed to accommodate them all. Moreover, they invite the standard anti-realist response: that what counts as 'real' will always be relative to some ontological scheme, some particular way of picking out objects, events, causal processes or higher-order theoretical-explanatory 'laws' that happen to conform with current ideas about the nature and structure of reality. Hence Quine's famous argument in 'Two Dogmas of Empiricism' that we possess no means – aside from certain entrenched habits of belief – for distinguishing in point of ontological status between (say) centaurs, Homer's gods, numbers, set-theoretical clases and brick houses on Elm Street.[9] However, as various critics have noted, this doctrine gets into conflict with

Quine's more sturdily realist pronouncements as soon as he allows the priority of the natural sciences (physics especially) as our best source of guidance on these matters.[10] At any rate, there is something deeply obscurantist about Rorty's idea that realism *vis-à-vis* the objects or domains of scientific enquiry can only be upheld through a kind of illicit transfer from theological doctrines of substance or essence. This soon becomes apparent if one asks (for instance) what in that case could distinguish scholastic speculation on the substance of angels – corporeal, spiritual or various degrees in between – from scientific speculation on the nature of elementary particles and their various modes of interaction.

Of course it would take a very hard-headed realist (and one with little grasp of developments in particle physics over the past hundred years) to assert that any current such candidates for 'elementary' status – say quarks or leptons – were real in the sense of 'cutting nature at the joints' or representing the furthest that science can go along this path of enquiry. That is to say, one need only extrapolate from the history of modern (post-Daltonian) atomic and sub-atomic physics in order to refute any positive claim for the finality or truth of our current best ideas with regard to such putative realia. Still the very fact that we can speak of advances in this field – or (not to beg that particular question) that we can speak intelligibly of changing ideas *with regard to* atoms, nuclei, electrons, quarks, leptons and so forth – is sufficient grounds for rejecting the 'strong' anti-realist or Rortian neo-pragmatist case.[11] For we should otherwise be driven to the sort of conclusion that Kuhn first espoused in certain well-known passages of *The Structure of Scientific Revolutions* and then felt obliged to retract or qualify in response to his various critics.[12] That is to say, we should have no reason to suppose that scientists from Dalton to Eddington, Bohr, Dirac, Heisenberg, Schrödinger et al. were in any sense referring to the same sorts of object or addressing issues of shared theoretical concern.[13]

The same would apply to any branch of scientific enquiry where thinking had changed – and where certain issues remained open – as regards the precise ontological status or the degree of descriptive-explanatory warrant accorded to its objects of enquiry. Thus molecular biologists or aerodynamicists would not be talking about anything 'real' when they referred on the one hand to viruses and DNA proteins, or on the other to airfoil sections, stalling speeds, laminar and turbulent airflow patterns, etc. Still less could it be claimed, in the latter case, that theoretical 'objects' like Reynolds numbers (used to calculate the stalling speed or point of transition for a given airfoil) might possess any kind of objective warrant or real-world validity.[14] In each case the most one can say – on Rorty's account – is that these notions are pragmatically 'true' (= valid for us) in so far as they help to 'facilitate our commerce with the things around us'. Just *how or why* they manage to do that is of course no part of Rorty's pragmatist tale. For he rejects any version of the naïve 'realist' idea that their usefulness might itself be a function of their getting certain things right with respect to (say) the sub-atomic or molecular structure of matter, the replicative pattern of genes and chromosomes, or the best way to calculate airfoil performance under certain specified conditions. Rather, we should think of these and other such 'discoveries' as holding good only in so far as they happen to fit with our current descriptive conventions or the sorts of argument that go down well with members of our own interpretive community.

Thus a critic like Hirsch is on the right track – headed towards a pragmatist view of things – when he takes the point of Vico's strong hermeneutic or anti-realist theory, that is, the idea that 'the human realm is genuinely knowable while the realm of nature is not.'[15] On this account, as Rorty puts it, 'our ideal of perfect knowledge is the sympathetic knowledge we occasionally have of the state of mind of another person' (TL, p. 87). Such knowledge is the sort that is typically acquired through the reading of a literary text – where that reading has to do willy-nilly with ascriptions of authorial intent – or again, through the kind of motive-based historical enquiry that interprets the actions of historical agents in some given context of humanly intelligible choices, decisions and commitments.[16] That this approach can hardly work for lumps – unless with regard to the sociocultural or psychological 'context of discovery' – is one good reason (so Rorty believes) why the human rather than the natural sciences are our best source of insight into questions of interpretive method and truth. Since lumps are 'just whatever it is presently convenient to describe them as' – since 'they have no "inside" in the way that persons do' – then it follows that any realist ('essentialist') approach in philosophy of science must amount to no more than an illicit transfer of *geisteswissenschaftlich* to *naturwissenschaftlich* modes of knowledge and enquiry. Up to this point Rorty is wholly in agreement with Hirsch. However, Hirsch goes wrong when he thinks he has found methods (philosophically respectable methods) for fixing authors' intentions or for holding the methodological line between text-immanent 'meaning' and culture-relative 'significance'. Thus when Hirsch puts forward his Husserlian claim – that 'if we could not distinguish a content of consciousness from its contexts, we could not know any object at all in the world' – Rorty suggests that he would do much better to push right through with the Viconian insight and drop *both* his talk about intentional 'contents of consciousness' *and* his talk about 'knowledge of objects in the world'.[17] For by adopting this line he is again giving in to a science-based (epistemological) paradigm, one that has recourse to unfortunate ideas such as that of 'validity in interpretation', and which thus pulls back from the liberating prospects of a full-fledged pragmatist or textualist stance.

Rorty sees this as just a shuffling or compromise position forced upon Hirsch by his residual sense that there must be something special about the natural sciences – their capacity to get things right – which the human sciences had best not challenge (and might even do well to emulate up to a point) so as to retain their intellectual dignity. On the contrary, Rorty declares:

> Whereas Hirsch wants to make realistic philosophy of science look good in order to make 'meaning' at Level II look good for texts, I want to do the opposite. I want to admit everything Hirsch says about the objective validity of enquiry into meaning (in that sense) in order to make realistic philosophy of science look bad. I want to insist that we *can* have what Hirsch wants at Level II for texts, just in order to ram home the point that we *cannot* have anything of the sort for lumps. (TL, p. 87)

In other words, if you take Hirsch's talk about objective 'validity' with the appropriate (pragmatist) pinch of salt then you can hang onto his Viconian point – the interpretive priority of the human over the natural sciences – and turn it back against *both* varieties of realism, i.e., realist philosophy of science and realism with regard to authors' (or historical agents') intentions. For in Rorty's view this latter idea is just another example of the unfortunate tendency among literary critics to suppose that their discipline can gain some extra kudos – or at any rate avoid the stigma of 'subjectivism', 'irrationalism' and the like – by hitching its waggon to some method or theory with vaguely scientific credentials. Far better if Hirsch could be brought to acknowledge first that there is no such method or theory to be had; second, that meaning (or author's intention) is just what we make of it in changing cultural contexts and with various interpretive ends in view; and third, that scientists and philosophers of science are in just the same boat, whatever their deluded objectivist beliefs or their realist persuasions to the contrary. This is manifestly the way things are going, Rorty believes, and none too soon, given all the wasted time and ingenuity that have been spent in the effort to uphold one or another version of the realist paradigm.

Thus 'recently both "analytic" and "Continental" philosophers have been suggesting that, once the holism common to Quine and Gadamer is pushed along a few more steps, it cannot confine itself to one side of the text/lump distinction, but blurs that distinction' (TL, p. 91). Once this lesson is taken to heart then there will be no reason for philosophers of science – or for literary theorists like Hirsch – to keep on working with realist notions, whether of lumps, properties, essences, structures, texts, meanings, intentions or whatever else. At this stage their 'ontological' status will become pretty much a non-issue, or simply a matter of their cultural salience for us as participants in the ongoing 'conversation of mankind'. For that conversation will at last have swung around to the viewpoint that science – like everything else – is a cultural construct whose various objects, theories, truth-claims, explanatory 'successes' and so forth can have no reality outside or beyond some particular context of humanly significant meanings, purposes and values. In which case there won't be any room for Level-II realists about lumps, those who stubbornly continue to believe in 'the real essence of the lump which lurks behind its appearances – how God or Nature would describe the lump' (p. 85). As for Level-II realists about texts (Hirsch among them) they can perhaps best be viewed as half-way converts with a useful point to make regarding the hermeneutic matrix of all understanding, scientific knowledge included, but with a sad penchant for obscuring that point through their misplaced attachment to notions of truth and method.

I shall cite another passage from 'Texts and Lumps' which catches the entire drift of Rorty's argument in conveniently shorthand form. It is, he suggests, 'the *anti*pragmatist in philosophy of science and philosophy of language . . . who thinks that there are such things as "real natures" or "real essences"'. (His chief examples here are Kripke and Boyd.) For the pragmatist, conversely,

> we can only distinguish better and worse nominal essences – more and less useful descriptions of the lump. For him, there is no need for the notion of a convergence of scientific enquiry towards what the lump really, truly, is in itself. So the pragmatist philosopher of science may be tempted to interpret Hirsch's distinction between 'meaning' and 'significance' as a distinction between meaning *in se* and meaning *ad nos*, and to dismiss Hirsch as a belated Aristotelian who has not yet got the word that all essences are nominal. (TL, p. 86)

To which the scientific realist might be tempted to respond in kind and remark that Rorty's 'pragmatist philosopher' sounds for all the world like a belated Lockean who has not yet got the word that things have moved along since Locke arrived at his sceptical conclusions. That is to say, one source of Locke's nominalist doctrine that 'real essences' must remain forever unknowable was his persuasion that the physical sciences of his day, however impressive their achievements, had come nowhere near defining just what it was *in the nature* of various substances that might be thought to underlie and explain their observed properties, causal dispositions, chemical affinities, etc.[18] To maintain that doctrine – as Rorty does – without taking note of the progress that physics, chemistry and biology have made towards providing just such depth-explanatory knowledge suggests, to say the least, a very marked degree of fixed anti-realist bias.

Nor is his case borne out by the claim that talk of 'real essences' takes us back to Aristotle and an ancient – presumptively obsolete – idea of what is involved in the process of scientific explanation. For one need only look to the large body of work in recent causal-realist philosophy of science (Salmon, Boyd, Bhaskar) and to allied developments in philosophical semantics (Kripke, early Putnam, Donnellan) to see that such talk is by no means bereft of adequate arguments and evidence.[19] Indeed, the main interest of Rorty's essay is in remarking just how – by what sorts of rhetorical strategy – it contrives to make anti-realism appear the natural or commonsense (default) option and realism a kind of perverse throwback to quaint 'animistic' or 'theological' habits of thought. At any rate, there seems little virtue in a theory that can offer no viable account of scientific knowledge and its modes of acquisition; that erases the line between science and pseudo-science, or between (say) astronomy and astrology; that relativizes truth to the currency of belief within this or that (past or present) 'interpretive community'; that allows for any number of alternative theories, descriptions, language games, etc., just so long as they keep the conversation going; and that carries ontological relativity to a point where the distinction between 'texts' and 'lumps' pretty much drops out except as a matter of pragmatic convenience or a way of flagging disciplinary interests which themselves have a merely conventional (in no sense intrinsic) relation to their various particular objects of study. Or rather: those 'objects' have no existence – no referential status or claim to 'reality' – apart from their role in some discourse, ontological scheme or whatever which happens to quantify over objects of just that conventionally accepted sort. Thus 'Hirsch and his opponents are both too preoccupied with the distinctive textuality of texts, just as Kripke and his opponents are too preoccupied with the distinctive lumpishness of lumps' (TL, p. 89). From which it follows, according to Rorty, that '[r]ather than trying to locate sameness, we should dissolve both texts and lumps into nodes within transitory webs of relationships' (ibid.).

III

This does seem to me a quite extraordinary doctrine for anyone – even a convinced anti-realist such as Rorty – to take on board without in some way hedging his bets or backing off from its more awkward implications. However, by the same token, it is useful to have the doctrine set out in this extreme form since it also helps to show what is wrong with other, more moderate versions of the argument. Thus, for instance, a philosopher of science like Bas van Fraassen clearly wishes to avoid the excesses of anti-realism in its full-blown (sceptical-relativist) guise while none the less rejecting any appeal to 'laws of nature' – or causal-realist arguments – and espousing instead what he calls an outlook of 'constructive empiricism'.[20] On this view it is simply naïve to suppose that scientific theories can or should give us a truthful, valid or explanatorily adequate account of what the world is really like. Nor should we assume – again naïvely – that there exists some privileged (scientifically warranted) ontology that would allow us to quantify securely over the various objects, processes and events that play some role in our current scientific theorizing. On the realist view, according to van Fraassen, '[s]cience aims to give us, in its theories, a literally true story of what the world is like; and acceptance of a scientific theory involves the belief that it is true.' For the constructive empiricist, on the other hand, '[s]cience aims to give us theories which are empirically adequate: and acceptance of a theory involves a belief only that it is empirically adequate.'[21]

On this account there is simply no room – realist prejudice aside – for the idea that scientific theories refer to objects, entities or events in the physical domain that would then determine the truth-value of statements, propositions or hypotheses concerning them. Rather such items are selectively imported (to adopt the Quinean parlance) into this or that currently accepted frame of reference and it is then just a matter – for the constructive empiricist – of having good reason to believe that the theory fits the observational evidence. What counts as 'good reason' (or adequate 'fit') in any given case can be decided only through the kind of instrumentalist or pragmatist trade-off that balances considerations of scope, economy, elegance, falsifiability, predictive power, counterfactual–supporting warrant and so forth. What *cannot* thus count is any dubious recourse to truth, reality, 'laws of nature', causal necessities or other such remnants of an otiose metaphyics. But the question remains as to whether van Fraassen is entitled to invoke all or any among that first range of criteria without presupposing some or all of the second as ultimate justifying grounds. And this question is itself within reach of the issue as to whether anti-realism in philosophy of science can possibly be advanced – in however subtle or qualified a form – without some covert reliance on assumptions at odds with its own professed outlook.

Van Fraassen's arguments have a clarity and power – as well as a range and depth of scientific knowledge – that clearly place them in a class apart from Rorty's casual variations on the theme. However, his examples also have a curious way of resisting their proposed (anti-realist) construal and thus offering unintended support for the causal-realist argument. One such instance is van Fraassen's treatment of a case discussed previously by Reichenbach, Hempel and others. This has to do with the pair of propositions: (1) 'All solid spheres of enriched uranium (U235) have a diameter of less than one mile', and (2) 'All solid spheres of gold (Au) have a diameter of less than one mile' (*Laws and Symmetry*, p. 27). For the causal realist – or the believer in 'laws of nature' – it is not hard to specify the important difference between these superficially similar orders of truth-claim. Thus the truth of item (1) is a necessary truth in so far as it concerns the critical mass of enriched uranium, i.e., the causally explicable fact that a huge explosion would have occurred long before the lump achieved that scale. The truth of item (2), on the other hand, can only be a matter of the way things contingently are with respect to the amount of gold that exists in our planet (or maybe in any planet) and the impossibility – that being the case – of any such object turning up. Van Fraassen takes the point of this distinction and concedes that it captures something crucial to our best scientific (as well as our everyday commonsense) notions of necessary truth. Still he denies that we could ever be justified in treating (1) as a 'law of nature' that holds universally, and (2) in rigorous contradistinction as a matter of contingent fact. That is to say, he finds problems with any argument that would go beyond the stage of asserting it as a truth borne out by all our findings, experience or empirical observations to date that there cannot exist a sphere of enriched uranium more than one mile in diameter. It is no use invoking the criterion of universality since this would apply equally to a whole range of other cases (such as: 'no rivers past, present, or future are rivers of Coca-Cola or of milk') whose exceptionless truth we can confidently assert but which we would scarcely regard as *necessary* truths or as in any sense constituting 'laws of nature'. In which case, van Fraassen concludes, it cannot be the universality of (1) that sets it apart from contingent truths of the 'no *x* is *y*' (or the 'all *x* are *y*') variety. For we can produce any number of 'parallel examples' which 'employ exactly the same categories of terms, and share exactly the same logical form, yet evoke different responses when we think about what could be a law' (*LS*, p. 27).

However, van Fraassen can be seen to equivocate here between two quite distinct conceptions of 'universality', only one of which falls to his argument from parallel instances. What that argument shows, sure enough, is that we cannot establish a criterion for causal necessity (or for the existence of some putative 'law of nature') by examining the purely *formal* structure of any law-like proposition or hypothesis. Thus the moral to be drawn is one to the effect that 'laws cannot be simply the true statements in a certain class characterized in terms of syntax and semantics' (p. 28). And again: 'there is no general syntactic or semantic feature' in respect of which the two analogous examples (i.e., the rivers of Coca-Cola or milk or the sphere of gold more than one mile in diameter) would differ decisively in point of necessary truth from the case of the sphere of enriched uranium. But this argument begs the obvious question as to why the analysis should be thus restricted to the formal (*de dicto*) constituent features of law-like statements or truth-claims. For on this account *of course* there is nothing to distinguish contingent truths that just happen to possess 'universal' warrant to the best of our empirical knowledge from statements whose truth is a matter of substantive (*de re*) causal necessity. Indeed the point is made with considerable force – though ironically against the whole drift of his argument – by van Fraassen's choice of examples here. That is, we have strong scientific evidence (empirical, theoretical, depth-ontological and causal-explanatory grounds) for the claim that there *does not and could not* exist a sphere of enriched uranium more than one mile in diameter. As regards the non-existence of the equiproportional sphere of gold and the rivers flowing with milk or Coca-Cola, we can safely reckon it a universal truth in just that sense of the term 'universal' that is covered by van Fraassen's restrictive usage. In other words, it meets the requirements of constructive empiricism as a truth borne out by all the observational evidence to date and as belonging to that class of statements characterized by certain specifiable syntactic and semantic features. However, there would seem to be little merit in a theory that collapses the difference between these two sorts of case for no better reason than its being a priori committed to the tenets of constructive empiricism.

Van Fraassen is well aware of these difficulties and the extent to which his doctrine tends to conflict with our normal (commonsense-intuitive) judgement in these matters. Thus 'we have no inclination to call it a law' that there have been, are and will in future be no rivers of the kind hypothetically proposed. Of course, he concedes, 'we can cavil at the terms "river", "Coca-Cola", or "milk"' since 'perhaps they are of earthly particularity' and thus allow us to speculate on alternative possible worlds whose physical constitution might differ so markedly from our own that the generalization turned out in fact not to apply. Still there is a good (empirically adequate) sense in which van Fraassen can say: 'I think that this is true; and it is about the whole world and its history.' All the same, 'we have no inclination to call this general fact a law because we regard it as a merely incidental or accidental truth' (*LS*, p. 27). At which point one might expect him to draw the fairly obvious lesson (again in conformity with our normal intuitions) that there exist *other* sorts of truth – like that with respect to the critical mass of uranium – which possess a more than contingent status in virtue of their stemming from features or properties intrinsic to the object concerned. However, van Fraassen rejects this claim, despite its powerful intuitive appeal, since he sees no adequate or exceptionless way of distinguishing between the two (i.e., the empirical-contingent and necessary) orders of universality. Thus '[t]he mere linguistic form "All . . . are . . ." is not a good guide because it does not remain invariant under logical transformations' (p. 28). For in standard logic one can replace the statement 'Peter is honest' with the universal statement 'Everyone who is identical with Peter, is honest' (ibid.). That is to say, one can take any given statement that obtains for this or that particular instance and produce a universal equivalent whose truth conditions are identical but whose law-like appearance is merely a function of its formalized logical structure. And the same applies – so van Fraassen would argue – to the law-like generalizations (or statements in terms of universal and necessary truth) that are typically invoked by realist philosophers of science. For here also there is no guarantee that the 'laws' in question are anything more than a logical construct out of various items of empirical or observational evidence. In which case – habitual 'inclination' apart – we could never be truly, philosophically justified in talking about 'laws of nature' or supposing an order of causal necessity beyond or behind observational appearances.

As I have said, van Fraassen is not altogether happy about abandoning intuitive (commonsense-realist) ideas as regards the distinction between contingent and necessary truths. Thus we may well agree – on intuitive grounds – that there *must* be some important difference between the examples of the sphere of uranium and the sphere of gold, as likewise between the kind of argument that would establish non-existence in the former case and the kind of argument which leads us to believe that there exist no rivers of milk or Coca-Cola. Such agreement can safely be relied on, he suggests, 'before any detailed analysis of universality'. Nevertheless, there is reason to think 'that this analysis will not be easy', and moreover that 'it is extremely difficult to make the notion precise without trivializing it' (*LS*, p. 28). However, this argument can be turned back against van Fraassen by remarking that his analysis of universality is such as to exclude a priori any treatment of the issue that would *not* turn out either trivial or at odds with his own (narrowly defined) conception of empirical adequacy. That is to say, it finds no room for the idea – the commonsense-intuitive and scientifically proven as well as philosophically refined and elaborated idea – that issues such as truth, universality and explanatory warrant have to do with the way things are in reality and not (or not solely) with the issue of their justification in formal or logico-semantic terms. In this respect van Fraassen shows himself an heir to the dilemmas of logical empiricism, or the problem of bridging that troublesome gulf between first-order observation statements and higher-level statements (or covering-law generalizations) taken to characterize the logic of scientific enquiry.[22] 'To define generality of content,' he writes, 'turns out to be surprisingly difficult. . . . [and] in semantics, and philosophy of science, these difficulties have appeared quite poignantly' (*LS*, p. 28). What is perhaps more suprising – and poignant – is van Fraassen's at times quite heroic struggle to maintain a constructive-empiricist stance that avoids the dead-end (scientifically insupportable) problems of wholesale anti-realism while steadfastly refusing the best alternative arguments.[23]

Thus his book contains numerous examples of what look very much like 'laws of nature' but which van Fraassen declines to treat as such, in keeping with his major thesis. They invite that description in the sense that they hold good *necessarily* for this physical world – the one that we inhabit – and for all worlds relevantly similar to this in respect of their various (causal, structural, micro- or macro-physical) constitutive features. These examples are also very often counterfactual-supporting in so far as they explain why things just *had* to be this way in order for the world to exist at all as we know it or for us to exist within that world as gatherers of knowledge about it. Hence, for instance, the 'familiar idea', as van Fraassen describes it, that 'there are many different ways the world could have been, including differences in its laws governing nature'. After all, '[i]f gravity had obeyed an inverse cube law, we say, there would have been no stable solar system – and we don't think we are contemplating an absolute impossibility' (*LS*, p. 30). Now one might wish to go along with van Fraassen's immediate qualification – that 'we could be wrong in this' – since of course there is no argument to a necessary truth from within our Newtonian (inverse square-law-governed) world that could settle the issue of whether such a world was indeed *absolutely* impossible. On the other hand, his counterfactual example would lack any point or purpose if it failed to allow for this crucial distinction between the kinds of law-governed physical necessity that obtain in our own world and the kinds of hypothetically possible alternative that just might operate elsewhere.

Van Fraassen may very well reject such arguments in keeping with his general scepticism about laws of nature, causal necessity and kindred 'metaphysical' postulates. But it is hard to see how philosophy of science could manage without the basic idea that there exists a subset of *naturally or physically* possible worlds whose constitution is sufficiently like our own to mark them off from the wider range of merely speculative worlds in thought. For we should otherwise have no means of distinguishing the realm of science-fiction fantasy from that of scientific thought-experiments (as conducted by thinkers from Galileo to Einstein, Bohr and J. S Bell) or again, from the kinds of causal explanation that involve counter-factual-supporting statements or hypotheses.[24] What is required – at very least – is an adequate criterion (or range of criteria) for deciding that certain such worlds fall within the bounds of natural possibility while others do not, since they turn out to contravene some well-established physical constraint. This much van Fraassen implicitly concedes by producing instances – like the sphere of enriched uranium and the inverse cube law example – whose point would be altogether lost were it not for their appealing to distinctions of just that kind. In which case his scepticism about 'laws of nature' would appear more a terminological quibble than a matter of substantive philosophical commitment. This is why, as I have said, van Fraassen's 'constructive empiricism' stops well short of the Rorty-style notion that scientific 'truth' *just is* whatever we make of it according to our currently preferred range of language games, descriptive vocabularies, preferential metaphors, and so forth. That is, he makes a point of testing his argument against cases (such as those mentioned above) which must entail some order of physical necessity, and which therefore cannot be construed in line with a radical conventionalist, a purely instrumentalist or a blanket anti-realist approach. But again there is a puzzle as to why van Fraassen should apply his self-denying ordinance – his studious avoidance of talk about 'laws of nature' – when these provide by far the best (most adequate and economic) means of inference to the best explanation.

This issue is posed most sharply with regard to the role of 'unobservables' in scientific discourse, i.e., the kinds of theoretical posit (like 'atoms' and 'electrons' at one time and 'quarks' and 'mesons' more recently) whose ontological status – or the question of whose real-world (theory-independent) existence – has been or remains a matter of dispute.[25] Van Fraassen's belief is that the realist errs in atttributing reality – or assigning determinate truth conditions – to any talk concerning such notional entitities. Rather, we should take the constructive-empiricist line and avoid all forms of ontological commitment beyond the purely pragmatic appeal to what is borne out by current observation or the best evidence to hand. Thus here as elsewhere '[t]he aim of science is not truth as such but only *empirical adequacy*, that is, truth with respect to the observable phenomena' (*LS*, p. 192). Or again, 'acceptance . . . involves the opinion that the theory is successful – but the criterion of success is not truth in every respect, but only truth with respect to what is actual and observable' (p. 193). Yet van Fraassen proceeds on the very same page to remark that '[w]hen we come to a specific theory, there is an immediate philosophical question, which concerns the content alone: *how could the world possibly be the way this theory says it is?*' (ibid.; italics in original). And one then has to ask what could possibly count as an answer to van Fraassen's question if not the sort of answer that involves truth-claims – and ontological commitments – with respect to just those entities proposed as underlying (and explaining) the observable evidence.

Van Fraassen thinks that this is a non-issue if one just takes the point about constructive empiricism finding room for everything that can possibly (or properly) be needed in order to construct an observationally adequate theory. It is a question, he suggests, 'whose discussion presupposes no adherence to scientific realism, nor a choice between its alternatives'. Moreover, it is one area in philosophy of science 'where realists and anti-realists can meet and speak with perfect neutrality' (p. 193). However, this ecumenical attitude turns out to have sharp limits, or anyway not to accommodate the realist on terms that she or he would find remotely congenial. Thus, according to constructive empiricism, 'acceptance of a theory involves a certain amount of agnosticism, or suspension of belief' (ibid.). So much the realist might happily endorse as entailing nothing more than a due recognition of the non-finality of science as we have it and the likelihood (indeed near-certainty) that our current range of candidate theories – along with their various laws, entities, unobservables and so forth – will at length be superseded in the course of more advanced research. Nor could they find much to quarrel with in van Fraassen's remark that '[s]o far as science is concerned, of course, an individual scientist may additionally believe in the reality of entities behind the phenomena' (p. 193). However, the proposed truce begins to show signs of strain as van Fraassen makes it clear that any such beliefs had better be regarded – from the constructive-empiricist standpoint – as a kind of enabling psychological fiction devoid of assignable truth-content or ontological warrant. And one suspects that it must break down altogether when the passage in question goes on to say: 'similarly a chess player may wear flowers or hum a madrigal while playing' (ibid.).

There is – I submit – something quite absurd in the idea that inference to the best causal explanation, whether with regard to 'laws of nature' or the existence of unobservable entities, should be viewed as on a par with somebody's belief in the merit of wearing flowers or humming a tune while engaged in a game of chess. Of course, this may seem a rather humourless and literal-minded way of taking van Fraassen's joky analogy. But there is a similar problem, as I have argued, in knowing how best to construe his various specific examples, among them – again – the sphere of enriched uranium and the inverse cube law fictive hypothesis. That is to say, van Frassen's arguments have the signal virtue (unlike Rorty's) of actually providing the sorts of substantive test case that put his anti-realism to the proof and which open up a large credibility gap between the doctrine and the putative evidence for it. There is a similar ironic boomerang effect about his various citations from other philosophers of science whom van Fraassen regards as espousing an unworkable (ontologically over-committed) approach. Such is, presumably, his point in quoting a passage from Richard Boyd as the epigraph to his chapter 'Towards a New Epistemology'. Boyd's argument is a defence of causal realism and, moreover, of inference to the best explanation – 'abduction' in C. S. Peirce's terminology – as the only theory that can make sense of science in philosophically adequate terms.[26] Thus: 'It is by no means clear that students of the sciences . . . would have any methodology left if abduction is abandoned.' And again: 'If the fact that a theory provides the best possible explanation for some important phenomenon is not a justification for believing that the theory is at least approximately true, then it is hard to see how intellectual enquiry could proceed.'[27] There is just enough room for revisionism here – via Boyd's talk of 'approximate truth' and the leeway for an alternative (pragmatist) construal – to make it just conceivable that van Fraassen cited the passage in support of his own views. However, this reading is completely at odds with the logic of Boyd's argument from the fact that some theory – or candidate law – provides the best possible explanation of some given phenomenon to the presumptive *truth* of that theory and, moreover, the *existence* of whatever it entails in the way of (as yet) unobservable entities. What the alternative gains in anti-realist (or constructive empiricist) acceptability it loses twice over as a means of explaining how science could ever have achieved such a measure of descriptive, predictive and causal-explanatory warrant.

IV

If these are problems for van Fraassen – as I think they are – then no such problems can possibly arise for a thoroughgoing anti-realist such as Rorty, one who rejects the whole idea of science as in any way concerned with getting things right or coming up with truthful explanations. On this view, 'the notion of reality as having a "nature" to which it is our duty to correspond is simply one more variant of the notion that the gods can be placated by chanting the right words' (TL, p. 80). In short there is no difference – no difference that makes any difference – between 'essentialism' about gods, about nature totemically conceived, about magical effects or ritual incantations on the one hand, and on the other hand 'essentialism' about atomic configurations, molecular structures, causal dispositions, laws of nature or suchlike primitive animist residues in the discourse of philosophy of science. The pragmatist can afford to indulge such talk just so long as s/he remembers that it is just that – a certain line of talk – and open to whatever sorts of level-shifting may be advantageous in this or that cultural context. When trying to win the scientific realist around to her way of thinking the best tactic, so Rorty advises, is one which 'construes the reputed hardness of facts as an artefact produced by our choice of language game' (TL, p. 80). And then the conversation might go as follows:

> In some Mayan ball game, perhaps, the team associated with a lunar deity automatically loses, and is executed, if the moon is eclipsed during play. In poker, you

> know if you've won if you're dealt an ace-high straight flush. In the laboratory, a hypothesis may be discredited if the litmus paper turns blue, or the mercury fails to come up to a certain level. A hypothesis is agreed to have been 'verified by the real world' if a computer spits out a certain number. The hardness of fact in all these cases is simply the hardness of the previous agreements within a ommunity about the consequences of a certain event. The same hardness prevails in morality or literary criticism if, and only if, the relevant community is equally firm about who loses and who wins. (TL, p. 80)

Thus in literary criticism some such communities 'will throw you out if you interpret "Lycidas" as "really" about intertextuality', while 'others will take you in only if you do so' (ibid.). And with the natural sciences likewise – Rorty maintains – it is not so much a matter of getting things right or making progress towards some limit-point regulative notion of truth at the end of enquiry. For this would amount to just the old (realist and truth-fixated) habit of assuming that the world comes pre-packaged into objects, entities, 'natural kinds' and so forth whose properties, essences or real attributes we are somehow capable of knowing. Rather, we should see that there exist as many possible descriptions as there exist cultural interests – or 'interpretive communities' – engaged in producing (and endlessly revising) such descriptions.

From which it follows, on Rorty's account, that we shall have to give up any notion of distinguishing between the natural and the human sciences except in so far as they happen to adopt different language games for different suasive purposes. Moreover, the differences will be apt to fall out not so much *between* (say) physics and literary criticism, or astronomy and ethics, or biology and sociology but *within* those various topics of debate as they shift back and forth between periods of Kuhnian 'normal' and 'revolutionary' activity. For there is nothing *in the nature* of their objects of study – whether lumps or texts – that could serve to fix them once and for all as objects of just that kind or as properly requiring just that sort of investigative treatment. Of course, Rorty knows that his extreme anti-realist stance is likely to provoke all the standard counter-arguments familiar at least since Aristotle. Thus it may well be thought – as he pre-emptively concedes – 'to confuse the causal, physical force of the event with the merely social force of the consequences of that event' (TL, pp. 80–1). And again:

> When Galileo saw the moons of Jupiter through his telescope, it might be said, its impact on his retina was 'hard' in the relevant sense, even though its consequences were, to be sure, different for different communities. The astronomers of Padua took it as merely one more anomaly which had somehow to be worked into a more or less Aristotelian cosmology, whereas Galileo's admirers took it as shattering the crystalline spheres once and for all. But the datum *itself*, it might be argued, is utterly real quite apart from the interpretation it receives. (TL, p. 81)

Rorty can see no difficulty for the pragmatist here just so long as she keeps a level head and resists any attempt to sidetrack the discussion onto ground not of her choosing. Thus she can (and should) acknowledge that 'there is such a thing as brute physical resistance – the pressure of light waves on Galileo's eyeball, or of the stone on Dr. Johnson's boot.' But she should then go on to say – contra the realist about scientific 'facts' or 'laws of nature' – that quite simply 'there is no way of transferring this nonlinguistic brutality . . . to the truth of sentences' (p. 81). That is, the realist will always be stuck for a plausible account of just how – by what species of mysterious transubstantiation – the 'hardness' of this or that physical datum impresses itself on the meaning, content, logical form or truth-value of sentences concerning it.

So the pragmatist need have no qualms about acknowledging the stubborn facticity of *something* out there that roughly corresponds to the realist's stubborn belief in the existence of an objective, mind-independent, non-language-mediated world. However, she is still entitled to remark that we can have no access to that world – no knowledge of its various (supposed) constituent features – except by way of those concepts and theories that make up our current best notions of scientific truth. In which case the realist will have won a hollow victory since his argument finally amounts to no more than a minor variation on the Kantian idea of a noumenal 'reality' forever beyond reach of our human cognitive, perceptual or epistemic grasp. To this extent Rorty is wholly in agreement with those other, on the face of it more cautious anti-realists – Michael Dummett pre-eminent among them – who mount their case on what they take to be the manifest impossibility of verification-transcendent truths.[28] For how could we ever be in a position to assert or maintain such truths if we lacked any definite criteria for deciding what should count as evidence for or against? Thus if the meaning of a statement – or the content of a theory – is given by its truth conditions then those conditions must in turn be specifiable with reference to shared standards of demonstrative or evidential proof.

Hence Dummett's anti-realist appeal to 'warranted assertability' – rather than truth – as the concept that best captures what is involved in our various transactions with language and the world. For this concept has the virtue of allowing us to avoid the typical realist fallacy, i.e., the strictly meaningless (nonsensical) idea that we can know something without at the same time knowing what would count as an adequate criterion for our claim to possess just that item of knowledge. This is also to say – despite his differences with Rorty – that Dummett's anti-realism can very easily be pushed towards the kind of neo-pragmatist (or cultural-relativist) conclusion that he would find altogether unacceptable. Such is indeed Rorty's point when he expresses surprise at the unwillingness of some philosophers – those with residual 'analytic' leanings – to take the proferred route out of all their puzzles and perplexities. After all, as he remarks, '[w]hat we know of both texts and lumps is nothing more than the ways these are related to other texts and lumps mentioned in or presupposed by the propositions which we use to describe them' (TL, p. 88). On this much Dummett could surely agree, given his view of truth as a matter of warranted assertability and his rejection of realist (verification-transcendent) grounds of knowledge or belief. That he fails to follow this argument through to its natural pragmatist conclusion would for Rorty be merely a sign of his attachment to an idiom – that of old-style analytic philosophy – which is now long past its sell-by date.

Nor is van Fraassen much better placed to resist Rorty's blandishments. To be sure, he retains a sturdy sense of that something 'out there' – the physical world – which constitutes the object-domain of scientific knowledge and which in practice imposes certain working constraints upon the range of applicable language games, vocabularies, theories, paradigms, models, metaphors and so forth. Thus he never goes so far as to suggest – like Rorty – that the lump/text distinction might as well drop out since we can 'redescribe' the world in whatever way we choose just so long as our descriptions hang together and provide some novel or interesting slant on things. More than that: he takes the point that anti-realism can be made to look plausible – whether in philosophy of science or philosophical semantics – only on the basis of some highly counter-intuitive proposals with regard to our knowledge of the growth of knowledge. These include Nelson Goodman's famous 'new puzzle' of induction, arrived at by inventing factitious (i.e., non-natural) predicates and deploying them in order to cast doubt on the validity of *all* inductive procedures in the natural sciences or the various contexts of everyday practical experience.[29] Thus, for instance, if the colour-term 'grue' is defined as applying to any green object – say, an emerald – observed before AD 2000 and to any object examined thereafter and turning out to be blue, then clearly there is a problem for inductive arguments based on the presumed possibility of referring to the same sorts of object by the same predicate from one context to the next. Needless to say, this problem arises also for its symmetrical counterpart 'bleen', as likewise for any number of kindred examples – such as 'all raveswans are blight' – designed to block the passage by reliable induction from observed instances to more or less well supported generalizations. On one (perhaps the most charitable) reading we should construe Goodman's arguments as an ingenious *reductio ad absurdum* whose purpose it is to defend proper, legitimate forms of inductive reasoning against just those varieties of hyper-induced sceptical doubt that result from the application of such non-natural predicates. However, there is strong evidence elsewhere – notably in his book *Ways of Worldmaking* – that Goodman is ready to push his case to a point where it is taken to entail the most extreme anti-realist and ontological-relativist conclusions.[30] Indeed some philosophers, Quine among them, have found those conclusions frankly embarrassing despite their own adherence to the same doctrines in a somewhat less extreme (or more circumspect) form.[31]

Van Fraassen rehearses various arguments against the idea that induction has been somehow discredited – or shown to rest on untenable assumptions – by way of Goodman-type factitious predicates. They include the most obvious line of criticism, namely that there is a real (not merely a conventional, arbitrary or nominal) distinction to be drawn between naturally occurring predicates on the one hand and, on the other, those which can always be devised with the purpose of throwing a sceptical paradox into the process of inductive inference. Thus, quite simply, '[t]he predicate "green" stands for a real property and the predicate "grue" does not' (*LS*, p. 51). This connects in turn with the realist (anti-nominalist) conception of natural kinds, that is to say, the idea – as van Fraassen puts it – that 'mice do, and humans do, constitute kinds (mice-kind and human-kind) but their sum does not (there is no mouse-or-humankind)' (ibid.). Then again there is David Lewis's similar suggestion that the appeal to laws of nature can be strengthened against sceptical assault by construing those laws as 'theorems of all the best theories formulated in a *correct* language', this latter defined as 'one whose predicates all correspond to real distinctions'.[32]

Van Fraassen sets these arguments out at considerable length and – I would judge – to convincing effect against the sorts of sceptical-nominalist approach that find no room for such 'essentialist' talk. Along the way he provides various items of supporting evidence from the progress of science in offering ever more adequate (e.g., depth-ontological) grounds for distinguishing real from nominal kinds, and sometimes for revising membership conditions in the light of such continuing research. In other words, he provides a strong statement of the case that 'laws of nature', so far from giving way under pressure from various sceptical claims, must rather be viewed as inescapably built into our best (indeed our only adequate) modes of scientific knowledge acquisitition or theory construction. And this despite the fact – as he is quick to point out – that 'the distinctions which we use so easily – green vs. blue, hard vs. soft, mouse vs. cat – do not at all belong to the basic categories of physical science' (*LS*, p. 52). That is to say, there is no good reason to suppose (indeed compelling reason to deny) that the sorts of distinction which the sciences have and will come up with are the sorts of distinction which typically count as a matter of everyday, commonsense-intuitive judgement. This argument is familiar from Quine's writings and is often taken – though not (or not straightforwardly) by Quine himself – as lending further strong support to the case for ontological relativity.[33]

However, there is also the alternative construal according to which it is *precisely in virtue* of science's capacity to revise, refine or replace such commonsense categories that we can speak of progress with respect to our knowledge of just those specific individuating features that characterize this or that candidate kind.[34] Thus it might be maintained 'that humans have a special insight into the difference between natural and unnatural classes, and that this insight is one of the guiding factors in science' (*LS*, p. 52). On the other hand, this argument quickly runs into the above-mentioned difficulty, i.e., that science has increasingly tended to produce taxonomies or systems of classification at odds with those thrown up by intuitive 'insight'. Hence van Fraassen's alternative suggestion: 'that, without any such insight, scientists will tend to end up with natural predicates due to the ruthless weeding out of theories by empirical and/or theoretical success or failure' (ibid.). This might be thought to have various merits, among them its appeal to a naturalized (evolutionary) epistemology which explains both our knowledge of the growth of knowledge and the complex – sometimes conflictual – relation between intuitive and non-intuitive (scientific) modes of enquiry.[35] At any rate, there is no obvious reason to reject it as a viable account of what typically transpires in the process of scientific knowledge acquisition. It also has the virtue of not leading on to those various dead-end doctrines – of ontological relativity, radical meaning variance, paradigm incommensurability and the like – which have so bedevilled recent debate in philosophy of science.

As I say, van Fraassen gives a fair hearing to this line of argument and indeed puts up what could easily be taken (out of context) as a strong case for the defence. However, he then turns around very sharply and proceeds to attack it with just those sorts of objection that are standardly brought up by adherents to the opposed (anti-realist or ontological-relativist) view. Thus:

> if there really is an objective distinction between natural classes and others, and if laws in the sense of Lewis are what science hopes to formulate in the long run, then the only possible evidence for a predicate being natural is that it appears in a successful theory. If that is so, then science can never be guided even in part by a selection of natural over unnatural predicates. For the judgement of inferiority of any terminology on such a basis can be made only in retrospect, on the basis of some other lack of success. But in the absence of any selection for natural predicates, in independent fashion, at the time of theory choice or evaluation, we can have *no* reason to expect that science will tend to develop such a 'correct' language. (*LS*, pp. 53–4)

At which point van Fraassen reverts to anti-realist form and argues from the non-availability of natural predicates as a basis for informed or progressive theory choice to the non-existence of laws of nature except in the weak (non-explanatory) sense of their merely covering the evidence to date or describing observed regularities.

However, there are several odd features of the above passage which should strike any reader not strongly predisposed towards scepticism on both counts. One is the way that van Fraassen adopts a kind of Lamarckian position with regard to the issue of theory choice and the suggestion that candidate theories may be selected by a process broadly analogous to that which operates in the evolutionary domain. Thus his counter-argument relies on the idea that any relevant 'choice' must thought of as occurring both *consciously* on the part of the scientist(s) concerned and *at the very time* – or within the same punctual context of discovery – when that and other options *can now be seen* (in the wisdom of scientific hindsight) to have occupied the field. From which he derives the lesson that no such criterion can possibly apply since it is only 'in the long run' – on Lewis's submission – that we can sort out successful from failed theories, along with their respective (natural and non-natural) predicates, kinds, classes, etc. This objection seems curiously wide of the mark if construed with reference to the standard (broadly Darwinian) idea of theory selection as a phenomenon explainable in evolutionary terms. For it is precisely the point that such long-run explanations are just what is required in order to avoid the Lamarckian notion of heritable traits that develop within the life-span of a single organism or which result from 'choices' that the organism makes – and passes on to succeeding generations – through a process of ontogenetic change.

Of course, van Fraassen is well aware of all this. Indeed it is a central plank in his argument against the idea that theories are 'selected' according to criteria – of truth, fitness, descriptive adequacy, explanatory power, etc. – which must have played a decisive role in the original context of discovery. Still less can it be thought that the scientists concerned must themselves have possessed and applied those criteria with a full knowledge of how they would fare under subsequent investigation. To take such a view is clearly to confuse the short-run and long-run perspectives, that is to say, the issue of a theory's 'success' as judged by the state of informed knowledge at its time of introduction and the issue of its truth as thereafter borne out through its survival of further (often more rigorous) modes of scientific enquiry. Again, van Fraassen makes just this point when he rejects the idea that there could exist any criterion for sorting out natural from non-natural predicates 'at the time of theory choice or evaluation'. But there is a large (and in my view unwarranted) leap of argument when he then goes on to say – as if following logically from this – that 'we can have *no* reason to expect that science will tend to develop such a "correct" language' (*LS*, p. 54). For this is to impose a standard of 'correctness' (of theoretical consistency, observational warrant, natural predicate-hood and the like) which van Fraassen has shown to be wholly unworkable but which he none the less demands – strangely enough – from any argument in favour of theory selection on just such naturalized epistemological grounds. For there is no good reason why the defender of inductive procedures (or, for that matter, the believer in 'laws of nature') should accept van Fraassen's exorbitant demand. This requires, to repeat, that a theory be accounted successful – along with its cognate object language, ontological commitments, range of associated predicates, etc. – only on condition that the long-run criteria for theory choice can plausibly be thought to have governed its acceptance in the first place or played a decisive role in the thinking of those who endorsed it. But the problem disappears if one takes his point that inductive procedures are precisely justified by their capacity *over time* to withstand criticism, to cope with apparent counter-instances or anomalies, and to prove their worth by enabling a process of inference to the best explanation. In which case van Fraassen's entire argument against 'laws of nature' can be seen to rest on a false premise, namely the idea that such laws (and the kinds of evidence typically adduced in their support) should or must conform to an unreal standard of punctual accountability.

It is worth pressing this point a bit harder since van Fraassen is so insistent on pushing his case in the opposite direction despite – as I have said – his explicitly endorsing the long-run criterion for matters of inductive warrant. In his view, quite simply, 'there is no plausible way to improve on this dismal picture' (*LS*, p. 54). That is, we can have no adequate grounds for counting a theory successful (or its predicates natural) since the only relevant criteria of 'success' and 'naturalness' are those provided by the theory itself and hence, *ex hypothese*, sure to confirm that the theory will turn out justified. It is at this stage that van Fraassen's sceptical argument may be seen to extend from the initial moment of 'theory choice' to the subsequent contexts of 'theory evaluation'. Moreover, it is here that he comes closest to the kinds of wholesale anti-realism – or full-fledged ontological-relativist doctrine – that have gained wide currency in recent discussion. For if indeed it is the case that theories *entirely dictate* what falls within the range of scientifically accredited objects, predicates, evidential grounds, reasons for accepting (or rejecting) some explanatory hypothesis and so forth, then the way is wide open – via Kuhn, Rorty et al. – to just those forms of extravagant sceptical doubt.

Van Fraassen sharpens the focus of argument by directing it specifically to the issue of natural versus non-natural predicates and posing that issue as follows:

> [I]f the *only* link we have is that a predicate is more likely to be natural if it occurs in a successful theory, then we shall never have warrant to think that any predicate is natural. This sounds paradoxical, but consider the following example. Suppose:
>
> (a) 1 per cent of all available predicates have feature *F*
> (b) 2 per cent of all predicates which appear in successful theories have feature *F*
> (c) feature *F* is not correlated with any independently checkable characteristic.
>
> Then it is clearly true that a predicate is more likely to be natural if it occurs in a successful theory – indeed, *twice* as likely. Yet we shall never have reason to have any but an extremely low opinion of any predicate's claim to naturalness. (*LS*, p. 554)

Items (a) and (b) are clearly plausible enough on any theory – whatever its philosophical commitments – which takes account of the relation (or probability weighting) between evidence and justificatory warrant. With item (c), however, van Fraassen introduces a clause that effectively prejudges the issue in favour of wholesale theory dependence and the idea that we could never have adequate (non-circular) grounds for distinguishing natural from non-natural predicates. For *of course* if feature *F* is 'not correlated with any independently checkable characteristic' then its 'naturalness' can only be a product of theoretical definition, or a status accorded solely by virtue of its happening to figure as a needed predicate in just that theory, hypothesis, paradigm, ontological scheme or whatever. In which case – *concesso non dato* – we could never have valid grounds for arguing *either* to the naturalness of a given predicate from the mere fact of its turning up often in successful theories, *or* to the inductive warrant for just such theories from the fact that they contain a higher than average proportion of natural (feature-*F*-related) predicates.

Thus van Fraassen again:

> To think that our opinion of such a claim could cumulatively improve would require something like this: every time a predicate survives theory change, we must raise our opinion of its claim to naturalness. But that is exactly what would be plausible if independent selection in favour of naturalness were going on in theory change – the opposite of our present hypothesis. (*LS*, p. 54)

However, this begs the whole question as to *why* we should suppose – philosophical predilections aside – that 'independent selection in favour of naturalness' should *not* be the main factor in explaining theory change and indeed the only adequate means of accounting for our knowledge of the growth of scientific knowledge. Which is also to say (as Hilary Putnam used to argue before his conversion to a fig-leaf variety of 'internal realism') that a realist and causal-explanatory account along these lines is the only account which does not render the success of the physical sciences a downright mystery or miracle.[36] For it is otherwise hard to see where exactly the line can be drawn between van Fraassen's brand of constructive empiricism and those various anti-realist doctrines – Rorty's among them – that count the world well lost for the sake of our endless freedom to reinvent or redescribe it.

References

1 Richard Rorty, 'Texts and Lumps', in *Objectivity, Relativism, and Truth* (Cambridge: Cambridge University Press, 1991), pp. 78–92.

2 See Bas C. van Fraassen, *The Scientific Image* (Oxford: Clarendon Press, 1980) and *Laws and Symmetry* (Clarendon Press, 1989).

3 See also Rorty, *Consequences of Pragmatism* (Brighton: Harvester, 1982); *Contingency, Irony, and Solidarity* (Cambridge: Cambridge University Press, 1989); and *Essays on Heidegger and Others* (Cambridge University Press, 1991).

4 E. D. Hirsch, *Validity in Interpretation* (New Haven: Yale University Press, 1967).

5 See also Stanley Fish, *Is There a Text in This Class? The authority of interpretive communities* (Cambridge, MA: Harvard University Press, 1980).

6 See, for instance, Paul K. Feyerabend, *Against Method* (London: New Left Books, 1975).

7 For some fairly representative pronouncements see Jean-François Lyotard, *The Postmodern Condition: a report on knowledge*, trans. Geoff Bennington and Brian Massumi (Manchester: Manchester University Press, 1984).
8 Hirsch, *Validity in Interpretation*, p. 273.
9 W. V. Quine, 'Two Dogmas of Empiricism', in *From a Logical Point of View*, 2nd edn (Cambridge, MA: Harvard University Press, 1961), pp. 20–46.
10 See Quine's later thoughts on this and related matters in *Pursuit of Truth*, rev. edn (Cambridge, MA: Harvard University Press, 1990); also Robert Barrett and Roger Gibson (eds), *Perspectives on Quine* (Oxford: Blackwell, 1989).
11 See, for instance, Michael Gardner, 'Realism and instrumentalism in nineteenth-century physics', *Philosophy of Science*, 46/1 (1979), pp. 1–34; Josef M. Jauch, *Are Quanta Real? a Galilean dialogue* (Bloomington: Indiana University Press, 1973); Mary Jo Nye, *Molecular Reality* (London: MacDonald, 1972); J. Perrin, *Atoms* (New York: Van Nostrand, 1923); Wesley C. Salmon, *Scientific Explanation and the Causal Structure of the World* (Princeton, NJ: Princeton University Press, 1984); A. Sudbury, *Quantum Mechanics and the Particles of Nature* (Cambridge: Cambridge University Press, 1986).
12 Thomas S. Kuhn, *The Structure of Scientific Revolutions*, 2nd edn (Chicago: University of Chicago Press, 1970); also Kuhn, *The Essential Tension: selected studies in scientific tradition and change* (University of Chicago Press, 1977).
13 See Hartry Field, 'Theory change and the indeterminacy of reference', *Journal of Philosophy*, 70 (1973), pp. 462–81; 'Quine and the correspondence theory', *Philosophical Review*, 83 (1974), pp. 200–28; 'Conventionalism and instrumentalism in semantics', *Nous*, 9 (1975), pp. 375–405; also Arthur Fine, 'How to compare theories: reference and change', *Nous*, 9 (1975), pp. 17–32; T. S. Kuhn, 'Rationality and theory-choice', *Journal of Philosophy*, 80 (1983), pp. 563–70; Michael E. Levin, 'On theory-change and meaning-change', *Philosophy of Science*, 46 (1979), pp. 407–24.
14 For a philosopher's approach to these matters, see Norwood Russell Hanson, 'A Picture Theory of Theory-Meaning' and 'The Theory of Flight', in *What I Do Not Believe, and Other Essays*, ed. Stephen Toulmin and Harry Woolf (Dordrecht: D. Reidel, 1971), pp. 4–49 and 333–90.
15 Hirsch, *Validity in Interpretation*, p. 273.
16 See, for instance, Quentin Skinner, 'Meaning and understanding in the History of ideas', *History and Theory*, 8 (1969), pp. 3–53.
17 E. D. Hirsch, *The Aims of Interpretation* (Chicago: University of Chicago Press, 1976), p. 3.
18 See Hilary Kornblith, *Inductive Inference and its Natural Ground: an essay in naturalistic epistemology* (Cambridge, MA:MIT Press, 1993).
19 Salmon, *Scientific Explanation and the Causal Structure of the World*; also D. M. Armstrong, *Universals and Scientific Realism*, 2 vols (Cambridge: Cambridge University Press, 1978); J. Aronson, R. Harré and E. Way, *Realism Rescued: how scientific progress is possible* (London: Duckworth, 1994); Roy Bhaskar, *A Realist Theory of Science* (Leeds: Leeds Books, 1975) and *Scientific Realism and Human Emancipation* (London: Verso, 1986); Michael Devitt, *Realism and Truth* (Oxford: Blackwell, 1984); Clifford Hooker, *A Realistic Theory of Science* (Albany, NY: State University of New York

Press, 1987); Saul Kripke, *Naming and Necessity* (Oxford: Blackwell, 1980); Nicholas Rescher, *Scientific Progress* (Blackwell, 1979) and *Scientific Realism: a critical reappraisal* (Dordrecht: D. Reidel, 1987); Stephen Schwartz (ed.), *Naming, Necessity, and Natural Kinds* (Ithaca, NY: Cornell University Press, 1977); J. J. C. Smart, *Philosophy and Scientific Realism* (London: Routledge, 1963); Peter J. Smith, *Realism and the Progress of Science* (Cambridge: Cambridge University Press, 1981); M. Tooley, *Causation: a realist approach* (Oxford: Blackwell, 1988); Roger Trigg, *Reality at Risk: a defence of realism in philosophy and the sciences* (London: Harvester-Wheatsheaf, 1989).

20 Bas C. van Fraassen, *The Scientific Image* (Oxford: Clarendon Press, 1980) and *Laws and Symmetry* (Clarendon Press, 1989).

21 Van Fraassen, *Laws and Symmetry*, pp. 192–3. All further references indicated by *LS* and page number in the text.

22 See, for instance, R. Carnap, *The Logical Structure of the World* (London: Routledge and Kegan Paul, 1937); C. G. Hempel, *Fundamentals of Concept Formation in Empirical Science* (Chicago: University of Chicago Press, 1972); Hans Reichenbach, *Experience and Prediction* (University of Chicago Press, 1938); also – from a critical standpoint – Wesley C. Salmon, *Four Decades of Scientific Explanation* (Minneapolis: University of Minnesota Press, 1989) and Adolf Grunbaum and W. C. Salmon (eds), *The Limitations of Deductivism* (Berkeley and Los Angeles: University of California Press, 1988).

23 See PSILLOS, 'On Van Frasssen's critique of abductive reasoning: some pitfalls of selective scepticism', *Philosophical Quarterly*, 46 (1996), pp. 31–47.

24 See, for instance, J. S. Bell, *Speakable and Unspeakable in Quantum Mechanics: collected papers on quantum philosophy* (Cambridge: Cambridge University Press, 1987); James R. Brown, *The Laboratory of the Mind: thought experiments in the natural sciences* (London: Routledge, 1991) and *Smoke and Mirrors: how science reflects reality* (Routledge, 1994); Paul Davies, 'The Thought that Counts: thought experiments in physics', *New Scientist*, 146 (6 May 1995), pp. 26–31; David Lewis, *Counterfactuals* (Oxford: Blackwell, 1973); Roy Sorensen, *Thought Experiments* (New York: Oxford University Press, 1992).

25 For a different (qualified realist) view of these matters, see Ian Hacking, *Representing and Intervening: introductory topics in the philosophy of science* (Cambridge: Cambridge University Press, 1983).

26 See, for instance, Charles Sanders Peirce, *Essays in Philosophy of Science* (New York: Bobbs-Merrill, 1957); *Reasoning and the Logic of Things*, ed. K. L. Ketner (Cambridge, MA: Harvard University Press, 1992).

27 Cited by van Frassen, *Laws and Symmetry*, p. 151. See also R. Boyd, 'The current status of scientific realism', *Erkenntnis*, 19 (1983), pp. 45–90; 'What realism implies and what it does not', *Dialectica*, 43 (1989), pp. 5–29; 'Realism, Conventionality and "Realism About"', in G. Boolos (ed.), *Meaning and Method: essays in honour of Hilary Putnam* (Cambridge: Cambridge University Press, 1990), pp. 171–95.

28 See Michael Dummett, *Truth and Other Enigmas* (London: Duckworth, 1978); also Michael Luntley, *Language, Logic and Experience: the case for anti-realism* (Duckworth, 1988); N. Tennant, *Anti-Realism and Logic* (Oxford: Clarendon Press, 1987); Crispin Wright, *Realism, Meaning and Truth* (Oxford: Blackwell, 1987); Gerald Vision, *Modern Anti-Realism and Manufactured Truth* (London: Routledge, 1988); Kenneth P. Winkler, 'Scepticism and anti-realism', *Mind*, 94 (1985), pp. 36–52.

29 Nelson Goodman, *Fact, Fiction and Forecast* (Cambridge, MA: Harvard University Press, 1955).

30 Goodman, *Ways of Worldmaking* (Indianapolis: Bobbs-Merrill, 1978).

31 W. V. Quine, 'Goodman's *Ways of Worldmaking*', in *Theories and Things* (Cambridge, MA: Harvard University Press, 1981); see also Joseph Margolis, *Texts Without Referents: reconciling science and narrative* (Oxford: Blackwell, 1987).

32 Cited by van Fraassen, *Laws and Symmetry*, p. 51.

33 This tension is evident if one compares Quine's 'Two Dogmas of Empiricism' with his essay 'Natural Kinds' in Schwartz (ed.), *Naming, Necessity, and Natural Kinds*, pp. 155–75. See also Philip Hugley and Charles Sayward, 'Quine's relativism', *Ratio*, 3 (1990), pp. 142–9 and Bas C. van Fraassen, 'Relative Adjustments' (review of Quine's *Pursuit of Truth*), *Times Literary Supplement*, 10 Aug. 1990, p. 853.

34 See especially Kornblith, *Inductive Inference and its Natural Ground*; also Schwartz (ed.), *Naming, Necessity and Natural Kinds* and other entries under note 19 above.

35 This evolutionist approach is one (but not the only) source of arguments for our knowledge of the growth of scientific knowledge. See, for instance, Peter J. Smith, *Realism and the Progress of Science*; also D. W. Hamlyn, *Experience and the Growth of Understanding* (London: Routledge and Kegan Paul, 1978); Imre Lakatos and Alan Musgrave (eds), *Criticism and the Growth of Knowledge* (Cambridge: Cambridge University Press, 1970); Peter Lipton (ed.), *Theories, Evidence and Explanation* (Aldershot: Dartmouth, 1994); E. McMullin, 'Explanatory Success and the Truth of Theory', in N. Rescher (ed.), *Scientific Enquiry in Philosophical Perspective* (New York: University Press of America, 1987), pp. 51–73; K. Moser, *Knowledge and Evidence* (Cambridge: Cambridge University Press, 1989); Alan Musgrave, *Common Sense, Science and Scepticism: a historical introduction to the theory of knowledge* (Cambridge: Cambridge University Press, 1993); W. H. Newton-Smith, 'Realism and inference to the best explanation', *Fundamenta Scientiae*, 7 (1987), pp. 305–16; John M. Ziman, *Reliable Knowledge: an explanation of the grounds for belief in science* (Cambridge: Cambridge University Press, 1978).

36 Compare, for instance, Putnam, *Philosophical Papers*, vols 1 and 2 (Cambridge: Cambridge University Press, 1975) with his various writings over the past decade, among them *The Many Faces of Realism* (La Salle: Open Court, 1987), *Realism With a Human Face* (Cambridge, MA: Harvard University Press, 1990), and *Renewing Philosophy* (Harvard University Press, 1992).

7

Ontology According to Van Fraassen: Some Problems with Constructive Empiricism

I

In this chapter I shall resume my critical account of Bas van Fraassen's 'constructive-empiricist' approach to issues in philosophy of science.[1] I shall chiefly be discussing his professed anti-realist stance with regard to 'unobservables', that is, his unwillingness to grant the reality of those various hypothetical items – from electrons to quarks, gluons and muons – that have often played a crucial role in the development of scientific theories. More precisely, it is a matter of van Fraassen's refusing to quantify over object domains – e.g., the more advanced or speculative reaches of micro-physics – where the 'objects' in question cannot (as yet) be observed but are none the less treated as necessary postulates if the theory is to explain what would otherwise lack any adequate descriptive or causal-explanatory account. Of course, this follows consistently enough from van Fraassen's scepticism about 'laws of nature' and his Humean reluctance to allow that such hypotheses may be justified on the basis of an inference to the best explanation from currently available evidence plus theoretically informed conjecture.[2] Thus he takes it as a ground rule of good scientific method – and good philosophy of science – that we should so far as possible avoid the appeal to laws, unobservables, hypothetical entities and so forth which have no place in an object language framed in proper (constructive-empiricist) terms.

However, this self-denying ordinance cannot be applied without creating large problems for philosophers and historians of science alike. That is to say, it becomes hard (if not impossible) to explain how knowledge has so often advanced from the stage of theoretical conjecture concerning the existence of various particles to the stage of demonstrative proof, for instance, by observing their tracks in a Wilson cloud-chamber or through the advent of electron microscopes with higher powers of resolution.[3] For otherwise we should have no warrant for the claim that our present-day knowledge of such entities has progressed far beyond that available to thinkers – like the ancient atomists – who proceeded on the basis of inspired guesswork from purely metaphysical premises. Moreover, this scepticism would have to extend to the modern (post-Daltonian) history of atomic and sub-atomic physics along with the various, increasingly refined or depth-ontological theories that have also, most often, started out as conjecture and then been subject to more rigorous forms of elaboration and critique.[4]

Of course van Fraassen is not denying the self-evident truth that such conjectural 'objects' have often played a role in scientific theories and that they have often turned out – in the case of 'successful' theories – to warrant eventual admission to the range of putative physical realia. All the same he sees no reason to accept this as an argument for (long-run) realism with regard to such entities or as an argument against the constructive-empiricist doctrine that would exclude them, on principle, from the object domain of an empirically adequate theory. Thus: 'Scientific models may, without detriment to their function, contain much structure which corresponds to no elements of reality at all' (*LS*, p. 213). And again: 'The part of the model which represents reality includes the representation of actually observable phenomena, and *perhaps* something more' (ibid.). This is why van Fraassen makes allowance for the fact that some 'individual scientist[s]' may 'believe in the reality of entities behind the phenomena', whether laws of nature or objects corresponding to their various preferred theories, ontologies, hypothetical constructs, etc. All the same, such beliefs are best regarded as a kind of enabling psychological fiction, on a par – van Fraassen suggests – with the chess-player's belief that his game is improved if he wears flowers or hums madrigals while playing (p. 213). But this is to reduce the whole argument to manifest absurdity. What counts as valid (long-run) scientific warrant for the existence of certain elementary particles or of certain laws of nature concerning them has nothing to do with the psychology of belief or indeed with any factor – any motivating impulse or circumstantial detail – that may have influenced their first being espoused by this or that 'individual scientist'. To suppose that it does – even (as here) by way of a joky analogy – is to court just the kind of vulgar reductionist error that typifies much recent work in the so-called 'strong' sociology of knowledge or in the wilder reaches of scientific psychobiography.[5] Not that van Fraasen's constructive empiricism really bears comparison with these or other such manifestations of the current anti-science vogue. However, it does leave a door open to just those forms of sceptical-relativist argument that begin from the (supposed) triumph of anti-realism in philosophy of science, and which then go on to draw the most extreme – scientifically and philosophically insupportable – conclusions.

I shall not attempt to survey the large and varied literature that exists on the topic of 'unobservables' (or theoretical entities) and their role in the discourse of the natural sciences.[6] Suffice it to say that anti-realism is far from having conquered the field and that there are, in fact, some strong counter-arguments which receive less than adequate attention in van Fraassen's various writings on this topic. Among the earliest was F. P. Ramsey's well-known account of the way that a theory of reference might be extended from the macroscopic domain to the range of those (then) unobservable entities – such as electrons – whose role in existentially quantified statements was borne out by the best (observational and theoretical) evidence to hand.[7] This connects in turn with Ramsey's well-known formulation of scientific 'laws' as 'consequences of those propositions which we should take as axioms if we knew everything and organized it as simply as possible in a deductive system'.[8] Van Fraassen cites this passage, sure enough, but in the context of a sceptical argument from the existence of many such possible axiomatic systems, all of them compatible with our best current state of knowledge, to the failure of any such theory when it comes to establishing the case for laws of nature.

Nor can it be saved, he argues, by a modal-realist approach – such as that of David Lewis – which seeks to narrow the range of relevant (this-world-applicable) predicates, theories, laws, etc., to those which most simply and informatively explain what is known to be the case respecting that world which we actually inhabit.[9] Van Fraassen's objection has to do partly with his scepticism about modal ('possible worlds') talk and partly with his argument against the idea that one could ever arrive at a satisfactory trade-off between 'simplicity' and 'informativeness' – along with 'balance' and explanatory 'strength' – as criteria of truth in scientific theory construction. Thus:

> Strength must have something to do with information; perhaps they are the same. Simplicity must be a quite different notion. Note also that we have here *three* standards of comparison: simplicity, strength, and balance. The third is needed because there is some tension between the first and second, which cannot be jointly maximized. Sometimes a simple theory is more informative. But if we have a simple theory, and just add more information to it, so as to make it stronger, we will almost always reduce its simplicity. These are intuitive considerations that surely everyone shares. As soon as we reflect on balance, however, the certainty of our intuitions dwindles fast. A person who weighs 170 pounds is overweight if he is five foot and underweight if he is six foot three – there we have a notion of balance. But how shall we gauge when a gain in simplicity is well bought for a loss in information? When does gain equal loss for two such disparate virtues? (*LS*, pp. 41–2)

I have cited this passage at length since it shows just why van Fraassen is so sceptical about Ramsey's account of scientific laws, as well as his proposal for treating unobservables – theoretical entities of various sorts – as proper candidates for inclusion in existentially quantified statements. For clearly there are various ways of interpreting what might be involved in Ramsey's claim. Thus his statement can be read as making no more than the trivial (self-evident) case that it is always possible to devise a purely logico-deductive 'system' that derives its criteria for the framing of valid scientific laws from whatever 'propositions' we currently accept as a matter of observational warrant plus whatever 'axiomatic' status they acquire within that particular system. This is how van Fraassen understands it, given his constructive-empiricist approach and his belief that any putative laws thus arrived at can only take the form of generalized (law-like) statements subsuming various observed regularities.

In short, van Fraassen here espouses something very like the logical-empiricist disjunction between first-order items of empirical fact belonging to this or that observation language and second-order (meta-linguistic) statements whose validity or truth is purely a matter of their providing a framework – a logico-deductive system – which covers or incorporates those first-level items.[10] This seems to be the main reason why he can find no room for genuine (causal-explanatory) 'laws of nature', as distinct from the kinds of instrumentalist fiction that commonly claim that title. It is also why van Fraassen's constructive empiricism very often comes close to a Quinean position as regards ontological relativity, or the lack of any adequate (non-framework-relative) criteria for distinguishing real-world entities or objects from the range of other items – mathematical sets and classes, centaurs, Homer's gods, etc. – thrown up by different ontological schemes.[11] This may appear oddly out of keeping with van Fraasen's anti-realist refusal to admit theoretical constructs (or unobservable posits) to the object language of the physical sciences. However it follows logically enough from his belief that such a language can only be based on criteria of good empirical warrant along with the sorts of covering-law statement that merely subsume – rather than explain – various occurrent regularities. For on this account there is simply no bridging the gulf between observed phenomena and whatever it is *in the nature* of those same phenomena by which one might hope to explain or specify their causally generative powers, dispositions, properties, etc.

That Ramsey's formulation can be read in this way – pushed towards a logical-empiricist construal – is chiefly the result of his appeal to what would be the case about our best theories and well-supported statements of scientific law if we knew everything and organized it as simply as possible in a deductive system.[12] For this leaves open the obvious question as to how such knowledge might be attained and whether 'knowing everything' would involve knowing more than could properly figure as legitimate 'knowledge' on van Fraasen's constructive-empiricist account. Of course Ramsey's is an ideal-case hypothesis, a regulative idea (in the Kantian–Peircean sense) of how science might achieve such an order of perfected harmonious adjustment between truths arrived at by investigative means and the order of logico-deductive thought wherein those truths would assume their necessary place. What sets it apart from any pure-bred rationalist metaphysics in the grand style is clearly the conditional clause ('*if* we knew everything') and the implicit proviso that no such system either is or could ever be attainable so long as there remain – as there always will – truths unknown to science. However, this need not be seen as lending support to the anti-realist case or as providing further justification for van Fraassen's sceptical-agnostic stance with regard to laws of nature. That is to say, there is still room for the long-run argument which holds those laws (and the forms of inductive inference on which they are typically based) to be the best – indeed the only – means of explaining how science has made such remarkable progress in its quest for causal-explanatory grounds beyond the mere enumeration of observed regularities in nature.[13]

At this point the sceptic will of course fall back on the standard Humean line of counter-argument. Thus any attempt to justify inductive procedures will always (so it is thought) involve some appeal to precisely those same premises – i.e., the constancy of natural processes and law-governed causal relations – which the inductivist takes for granted and which therefore reduce the whole case to vicious circularity. For Hume the only choice was that which fell out between 'truths of ideas' and 'truths of fact', the former purely analytic (or self-evident to reason) and hence quite devoid of substantive content, while the latter could provide no argument for supposing induction to rest on any grounds more secure than fixed prejudice or indurate habits of belief.[14] Logical empiricism may not have adopted so sceptical a view of the scope and limits of scientific enquiry. However, it did follow Hume in erecting its own version of that same dichotomy and in failing to acknowledge that causal explanations might have to do with the *intrinsic nature* of the things to be explained – their propensities, attributes, causal dispositions, micro-structural features, etc. – rather than their role as observational posits to be brought under some higher-level (e.g., deductive-nomological) scheme.[15]

To be sure, van Fraassen has his own differences with the programme of logical empiricism, as likewise with any sceptic doctrine that insists – after Hume – on the lack of justificatory grounds for our practices of inductive inference. All the same, his 'constructive' variant on that doctrine runs into comparable problems if it is asked just *why* we should regard some such inferences as carrying strong scientific or evidential weight while others are taken to possess no more than a provisional, heuristic or instrumental value in some given context of enquiry. What is required for this purpose is precisely what van Fraasen rejects: that is to say, a causal-explanatory approach which locates the truth of certain statements in their picking out just those features of the object domain that provide most support for the rational process of inference to the best explanation. And should the objection once again be raised, after Hume, that this claim is viciously circular – i.e., that it begs the question by seeking to justify inductive procedures on inductive grounds – then again there is a strong counter-argument from the impossibility of otherwise explaining how science could so far have managed to produce such a range of well-proven (explanatorily adequate and predictively reliable) theories.

Moreover, this argument gives reason to doubt that van Fraassen can be justified in his scepticism *vis-à-vis* 'laws of nature' and their role in scientific explanation. For it is just in virtue of those laws – of their holding good across various particular instances and observational contexts – that inductive procedures can be taken as warranted by something more than the mere existence of observed regularities, Humean 'constant conjunctions', or whatever. That is to say: it is the existence of deep further facts about (e.g.) the biochemical, molecular or sub-atomic structure of physical realia that enables scientists to proceed reliably on this basis of inference to the best explanation.[16] Thus anyone who, like van Fraasen, finds reason to doubt that such laws exist will by the same token be obliged – or strongly predisposed – to extend that doubt to the very existence of those underlying properties, structures, dispositions, etc., which provide the only possible (non-circular) warrant for inductive inference. Perhaps it is the case, as Peter Lipton suggests, that 'circularity is relative to audience.' One would then have to say that the issue allowed of no ultimate resolution either way since 'the inductive justification of induction is circular for an audience of sceptics, yet not among those who already accept that induction is better than guessing.'[17] However, there are still good reasons – many of them provided by Lipton himself – for regarding this as an overly agnostic conclusion, given the extent to which such forms of inductive inference play a crucial (indeed indispensable) role in the physical sciences and elsewhere. Thus the charge of circularity may be turned back against the sceptic by asking quite simply what better justification any theory or procedure could possess than that of explaining *on its own terms* – that is, on inductive grounds – why induction has succeeded in advancing so far beyond the level of mere observed regularities to the construction of scientific theories with ever-increasing scope, depth and causal-explanatory power.

Such is what Lipton calls the 'truth argument' in its simplest form: namely that 'we ought to infer first that successful theories are true or approximately true, since this is the best explanation of their success, and then that Inference to the Best Explanation is truth-tropic, since this is the method of inference that guided us to these theories.'[18] The sceptic – and indeed the constructive empiricist – may still want to say that these arguments are both circular in sense of presupposing what they claim to justify, i.e., the equation of 'truth' with 'success' and of 'success' in turn with just those kinds of inductive inference that are 'truth-tropic' (or knowledge-conducive) to the extent that induction holds good. However, he or she will then be obliged to deny that these issues are in any way affected by the kinds of evidence that might be supplied by adducing the progress of scientific knowledge in specific areas of enquiry. For it is evidence of this sort – ontologically grounded and justified in terms of its explanatory yield – that the inductivist can best make use of in order to give substance and precision to her otherwise question-begging talk of 'truth' and 'success'. That is to say, it can then be construed as involving something more than a merely pragmatic or instrumentalist conception of truth, one whose criteria for 'successful' application are dependent upon – or relative to – the interests, values or preferred self-image of some given interpretive community.

II

To this latter way of thinking, as urged by Richard Rorty in his essay 'Texts and Lumps', there is simply no reason (certainly nothing 'in the nature of things') why we should accept any limit on the range of possible interests and therefore the range of possible descriptions under which objects of whatever kind may plausibly or usefully be brought.[19] From which it follows that we might as well abandon the very notion of 'objects' as falling under 'kinds'. For they could do so only in virtue of some salient property – some structural, constitutive or depth-ontological feature – whose presence (and whose causal-explanatory power) would supposedly mark them out as meeting all the relevant criteria or membership conditions. On this view, as Rorty mischievously phrases it, 'the notion of reality as having a 'nature' to which it is our duty to correspond is simply one more variant of the notion that the gods can be placated by chanting the right words' (TL, p. 80). Or again: 'Kant's distinction between constituting gold by rule-governed synthesis and constituting texts by free and playful synthesis must be discarded' (p. 83). For the Rorty-style pragmatist these are just minor variations on an old ontologico-epistemological theme whose problems are pretty much the same whether the emphasis falls (as in Kant) on the mind's supposed power to 'synthesize' intuitions under inter-subjectively valid concepts and categories, or (as in Kripke) on natural-kind terms as a paradigm case of how language refers and how advances in scientific knowledge come about.[20] Give up those beliefs, so his argument runs, and the problems will simply disappear along with all the cramping disciplinary constraints – of 'truth', 'objectivity', 'method', 'rigour', 'correspondence to reality' and so forth – that have so far acted as a needless brake on our powers of creative redescription.

Rorty suggests the metaphor of the die and the blank (that is to say, of knowledge as somehow impressed upon the passively recipient mind) as a useful device for pointing us beyond these unfortunate habits of thought. For there is, he thinks, something intrinsically absurd about the notion that we must have 'respect for facts', as if 'facts' had the kind of objectivity – the indifference to us and our various descriptive purposes – that could make them somehow language-transcendent or capable of sorting out the true from the false in whatever range of descriptions we currently happen to favour. Thus the die-and-blank metaphor comes in as a handy reminder of Rorty's central point, namely (as Donald Davidson puts it) that 'causation is not under a description, but explanation is.'[21] In other words, one can be as realist as one likes about the ontical existence of a world 'out there' – along with all its objects, properties, causal processes, etc. –

while none the less failing to see any sense in which our descriptions could be verified (or our explanations borne out) by the way things stand 'in reality'. 'To say that we must have respect for facts is just to say that we must, if we are to play a certain language-game, play by the rules' (TL, p. 81). And again, linking up this Wittgensteinian message with the same metaphorical idea: 'To say that we must have respect for unmediated causal forces is pointless. It is like saying that the blank must have respect for the impressed die. The blank has no choice, neither do we' (ibid.).

However, one could just as well argue that this is a wholly inappropriate metaphor for describing how it is that scientisists achieve a more adequate, truthful or depth-explanatory knowledge of objects and events in the physical world. In short, it harks back to just the kind of 'blankly' empiricist doctrine – the idea of knowledge as somehow 'impressed' upon the mind through a passively receptive process – that has created all manner of intractable problems from Locke and Hume to the present. That is to say, it rejects the very idea that truth might have to do with with causal explanations whose validity involves both the nature of real-world (mind-independent) objects, processes and events and, on the other hand, whatever best explains the advancement of knowledge through a deeper understanding of those same objects, processes and events. For Rorty this whole line of argument is hopelessly beside the point. Thus '[t]he only way to get a noninstitutional fact would be to find a language for describing an object which was as little ours, and as much the object's own, as the object's causal powers' (TL, p. 84). Since this is quite simply not to be had – since indeed there exist as many descriptions or explanations as there exist alternative language games – then we might as well henceforth 'give up that fantasy' and accept that 'no object will appear softer than any other.' In short: 'What we know of both texts and lumps is nothing more than the ways in which these are related to the other texts and lumps mentioned in or presupposed by the propositions which we use to describe them' (p. 88). From which it follows, on this Rortian account, that any talk of descriptive 'truth' or explanatory 'depth' will always come down to just another language game adopted with a view to conjuring assent among this or that like-minded target community. All that is needed to make communication and persuasion, and thus knowledge, possible is the linguistic know-how necessary to move from level to level' (TL, p. 88). To suppose otherwise is merely to reveal one's persistent belief in some deep further fact – some truth, essence, structural feature, causal-explanatory nature or whatever – that would serve to distinguish lumps from texts, or different sorts of lump in respect of their real-world, physically constitutive (not merely descriptive or language-relative) attributes.

As I have said, this argument can be made to look plausible only through a narrowly empiricist construal of what counts as scientific knowledge. Thus Rorty's metaphoric talk of 'dies', 'blanks', 'impressions', etc., is just another version – albeit blunter than most – of the empiricist move whereby the *explanandum* is stripped of everything bar its surface, observable or passively registrant features and the *explanans* in turn prescriptively confined to a statement of observed regularities devoid of genuine explanatory depth. Such has been the story, with sundry variations, from Hume to the logical empiricists, that is to say, from a generalized scepticism *vis-à-vis* the status of causal laws to a widening divorce between the object language of scientific observation and the higher-level (meta-linguistic) discourse of logico-scientific enquiry.[22] Hence the emergence of those various 'post-analytic' alternatives – Wittgensteianian, hermeneutic, neo-pragmatist, 'strong' sociological and so forth – whose common feature is the notion of 'reality' as relative to (or constructed by) our current range of language games, cultural life-forms, ontological schemes, etc.[23] What they share is a failure to look beyond that sceptical-empiricist impasse to forms of realist and depth-explanatory argument in philosophy of science which don't thus succumb to the standard line of updated Humean objection. Furthermore, the kind of 'depth' in question here is not to be confused with those jointly Heideggerian and Wittgenstein-influenced doctrines that have lately exerted a growing appeal among post-analytic philosophers.[24] Such is the idea of those – like Rorty – who recommend the so-called 'linguistic turn' (or a suitably scaled-down, pragmatized version of the depth-hermeneutical approach) as our best hope for breaking the hold of old-style logical empiricism.

Hence Rorty's insouciant claim – purportedly with warrant from Nietzsche, Dilthey, Heidegger, Gadamer, Hirsch et al. – that '[r]ealistic interpretations of natural science' should be viewed as just 'hopeless attempts to make physical science imitate the *Geisteswissenschaften*' (TL, p. 87). For on this account there can be nothing more to our causal-explanatory theories than just the kind of 'depth' (hermeneutically construed) that the above thinkers have variously sought to provide by making language in some sense the ultimate horizon of all human enquiry. No matter that Dilthey and Hirsch are at pains to distinguish between the kinds of understanding appropriate to the natural and the human sciences, that is to say, between those languages specialized for the purpose of physical-causal explanation and those others kinds of language – e.g., literary criticism – where different criteria obtain.[25] For it is just this delusive idea, in Rorty's view, that leads a critic like Hirsch to ignore his own best pragmatist insight and thus to fall back on arguments – like that of 'validity' in interpretation – which betray his lingering attachment to science (or to science-based notions of method and truth) as a paradigm of human enquiry. With Dilthey likewise we can best get over the lump/text, explanation/understanding, or *natur-* versus *geisteswissenschaftlich* dichotomy by seeing that these are just handy – pragmatic or currently acceptable – ways of trying out different descriptive idioms in this or that context of enquiry.

Of course, as Rorty commonsensically concedes, '[t]here *seems* to be a difference between the hard objects with which chemists deal and the soft ones with which literary critics deal' (TL, p. 83). Thus even the pragmatist 'has to admit that there is a *prima facie* difference to be accounted for', despite his or her objection in principle to any such carving-out of disciplinary domains. After all, 'when chemists say that gold is insoluble in nitric acid, there's an end on it,' whereas 'when critics say that the problem of *The Turn of the Screw*, or *Hamlet*, or whatever, is insoluble with the apparatus of New, or psychoanalytic, or semiotic criticism, this is just an invitation to the respective critical schools to distil even more powerful brews' (p. 83). All the same, the pragmatist should not be too impressed by this argument from the seeming self-evidence of certain 'hard' scientific truths as compared with the 'soft' currency of literary-critical debate. For again, according to Rorty, there is simply no way that our descriptions can be thought of as somehow verified by whatever it is that constitutes that realm of 'hard', observer-independent or non-description-relative truth. Such descriptions are much better viewed as just stories, more or less persuasive or interesting accounts of how we and others have arrived at different ways of interpreting the 'scientific' evidence. Thus

> [t]he question of whether any of these stories *really* is appropriate is like the question of whether Aristotelian hylomorphism or Galilean mathematization is *really* appropriate for describing planetary motion. From the point of view of a pragmatist philosophy of science, there is no point to such a question. The only issue is whether describing the planets in one language or the other lets us tell stories about them which will fit together with all the other stories we want to tell. (TL, p. 82)

In which case the text/lump distinction can finally fall away, along with those various kindred dualisms (like that between the natural and the human sciences or the 'hard' and 'soft' disciplines) which may once have possessed a certain story-telling interest but are now – so the pragmatist hopes to persuade us – running out of narrative steam.

I have suggested that this swing towards language-based, narrative or depth-hermeneutic philosophies of science is largely a result of the failure of logical empiricism and its various successor doctrines, along with the absence – at least within mainstream analytic philosophy – of any adequate alternative approach. One such alternative is that of causal realism as developed with great sophistication and detail by proponents such as Wesley Salmon and Roy Bhaskar.[26] That it has not found acceptance among more philosophers of a (broadly) 'post-analytic' persuasion is perhaps best explained by the kinds of residual or qualified empiricist outlook which continue to exert a strong counter-influence among those trained up on the original programme. In Rorty's case, as we have seen, this takes the form of a breezily commonsense assurance that we can carry on believing in all those items which make up our everyday or received scientific ontology –

along with all their properties, attributes, causal dispositions, etc. – just so long as we refrain from counting them 'real' in any but a pragmatist ('real-for-us') sense. Beyond that, the only test we can possibly apply to some candidate item, theory, ontological scheme or whatever is 'how well it coheres with the best work currently being done in, for example, both biochemistry and literary criticism' (TL, p. 90). That is to say, there is nothing – no deep further fact – that would constitute the proper sort of object for this or that discipline, or the proper sort of discipline whereby to achieve a better (more detailed or accurate) knowledge of some given object domain. For, as Rorty sees it, this whole idea of matching objects to disciplines is one that involves a purely circular process of argument, a failure to take his simple point that neither 'objects' nor 'disciplines' have any existence outside the various language games, descriptions, interpretive communities, etc., which currently decide what shall fall within the range of accredited statements or beliefs.

Thus natural science is itself no more a 'natural kind' than those various, increasingly specialized sub-branches – from astrophysics to molecular biology or quantum mechanics – whose emergence, whose history, and (perhaps) whose eventual supersession is entirely bound up with their postulated objects of enquiry.[27] In each case, so Rorty would argue, we are talking about *one and the same thing* when we talk about the objects – construed under this or that description – and when we seek to construct a disciplinary profile of the research community or the set of presupposed (knowledge-constitutive) interests that takes those objects for its own. For then indeed there could be nothing – and most definitely nothing 'in the nature of things' – to rule out the prospect of some startling change in our whole conception of molecular biology brought about by the switch (say) from physics to literary criticism as the paradigm source of new models, metaphors or analogies. Hence Rorty's recommendation of 'text/lump parallelism' as a good way of thinking about the future of the disciplines once happily relieved of all the surplus ontological baggage that they have borne along with them for the past two millennia and more. 'My holistic strategy, characteristic of pragmatists (and in particular of Dewey), is to reinterpret every such dualism as a momentarily convenient blocking-out of regions along a spectrum, rather than as a recognition of an ontological, or methodological, or epistemological divide' (TL, p. 84). This goes equally for the text/lump and the humanities/natural sciences dualism, both of them products (so Rorty would maintain) of a short-term, contingent, natural for us but otherwise quite arbitrary divison of intellectual labour. So one might still decide that the chemist or physicist is the best person to visit if one wants an expert opinion on a lump of gold, or the biologist/chemist/archaeologist if the lump in question just happens to be the 'fossilized stomach of a stegosaurus' (p. 85). And again, if it is textual expertise that is required, then most likely a philologist or literary critic (or in borderline cases a palaeontologist) will tend to spring to mind as the best sort of person to consult for further guidance. But should one end up in the wrong faculty building – perhaps by inspired misdirection – there is at least a fair chance that one will learn something just as interesting and even (with luck) something that transforms a whole field of study and its objects. For 'when he is asked to interpret the felt difference between hard objects and soft objects, the pragmatist says that the difference is between the rules of one institution (chemistry) and those of another (literary criticism)' (p. 84). All that stands in the way of this wished-for *dénouement* is our stubborn attachment to certain ideas – reality, truth, natural kinds, 'laws of nature' and the like – which Rorty considers a world well lost in comparison with the new-found descriptive freedoms that are thus opened up.

III

It does seem to me quite extraordinary (and very much a sign of the times) that this argument, and others in a similar vein, have managed to exert such widespread appeal across a range of academic fields from cultural and literary criticism to 'science studies' and the 'strong' programme in sociology of knowledge.[28] On the other hand maybe it is not so surprising, given the perceived institutional stakes and the benefits to be had – supposedly – from adopting this rhetoric-of-science line and thus, at a stroke, both cutting the natural sciences down to size and accruing a measure of their old prestige for practitioners of the new disciplines. What is less understandable, I have argued, is the retreat by some philosophers (van Fraasen among them) to positions which claim to hold the line against any such wholesale anti-realist creed, but which still yield crucial argumentative ground at just that point where thoroughgoing sceptics such as Rorty bring maximum pressure to bear. Thus it is chiefly on the issue of inference to the best explanation – of whether such reasoning can be held to support a realist ontology and a causal-explanatory epistemology – that van Fraassen and Rorty both, for all their differences, turn out to inhabit the same (anti-realist) camp. Hence van Fraasen's steadfast refusal to countenance any form of inference to the best explanation as regards the existence – the ontological good standing – of theoretical entities or unobservable posits. For the effect of this self-denying ordinance is to open the way for more convinced anti-realists, such as Rorty, to extend that argument to the entire range of what we take as adequate (observational or inferential) grounds for maintaining the existence of objective realia that are not just a construct of some currently favoured discourse, paradigm, or language game.

This point is brought out in a passage from J. S. Mill where the argument is clearly intended to support a strong inductivist case but where the sceptic could easily turn it around to cast doubt on Mill's own premises. Thus: 'in almost every act of our perceiving faculties, observation and inference are intimately blended. What we are said to observe is usually a compound result, of which one-tenth may be observation, and the remaining nine-tenths inference.'[29] For Mill the reliability of induction was a lesson everywhere to be read in the everyday run of human experience and, most impressively, in the achievements of the natural sciences. But, of course, if one rejects that premise then the very fact to which this passage draws notice – the inference-laden character of all our perceptions, observation statements, ontological commitments, etc. – must count as a decisive reason for rejecting the entire inductivist case. That is to say, Mill's argument will fall to the charge of taking for granted what it sets out to prove, namely the existence of strong regularities (or laws of nature) that hold good from one observational context to another, and which thus provide a causal-explanatory link beyond any mere accumulation of discrete data. It is then a short step to those various sceptical doctrines – ontological relativity, meaning variance, the underdetermination of theory by evidence, paradigm incommensurability and so forth – which often take the form of a meta-induction from the problems with inductive inference.[30] Among these latter, so it is claimed, are (1) the theory-laden character of observation statements, (2) the non-existence (for all that we can know) of underlying causal laws, and (3) the inherent circularity of arguments that seek to justify induction on purely inductive grounds. For such arguments must always rest on some version of the continuity principle, that is, an assurance that the laws of nature will continue to operate as they have up to now and will thus give warrant for the same kinds of (hitherto successful) inference to the best explanation. But of course this principle goes by the board – along with Mill's case for the massive self-evidence of scientific progress to date – if one takes it that nothing could possibly count as sufficient guarantee for our presuming to extrapolate from past to present and future instances.

On the sceptic's construal nothing much has changed (give or take a few refinements of detail) since Hume first advanced his ingenious case against the validity of inductive reasoning. For there is clearly a sense in which the sceptic will always have the last word just so long as he or she is prepared to hold out and profess disbelief in whatever sorts of argument the realist cares to come up with. Nor can one get very far by adopting the so-called 'sceptical solution' to sceptical worries, that is, the Wittgensteinian line of response (more or less repeated by Kripke) which holds it nonsensical to doubt beliefs that are everywhere taken for granted in our everyday 'language games' or cultural 'form of life'.[31] For the hardline sceptic will surely reply, first, that this 'solution' is no such thing since it merely restates Hume's problem of induction along with his appeal to custom and commonsense as the sole recourse in such matters, and second, that Hume was perfectly able to communicate his sceptical doubts in a language intelligible to most readers (and indeed compelling to some of them). So there is little to be gained from the kind of Wittgensteinian, avowedly therapeutic approach which insists that these are unreal problems – mere artefacts of our 'bewitchment by language' – while simply evading the sceptic's question as regards our knowledge of an external world outside or beyond the various 'language games' commonly used to describe it. Hence the frequent appeal to Wittgenstein by anti-realists and cultural relativists of diverse persuasions – among them adherents of the 'strong' programme in sociology of knowledge – who are pleased to consider that world just a construct out of our various discourses, paradigms, conceptual schemes or whatever.

In short, any adequate answer to the sceptic will need to start much farther back, at the point where Hume's problem of induction first got a hold. That is to say, it will take the basic principle of inference to the best explanation and derive from it the massive – overwhelming – probability that Hume's scruples were indeed misplaced and that the strongest case for the validity of induction is precisely its having held good so far across numerous contexts of everyday and scientific reasoning. Quite simply, we could not have made a start in the process of knowledge acquisition were it not for the existence of physical realia, of law-governed regularities in nature, and also – crucialIly – of just those cognitive and inferential powers that allow human knowers to achieve progress in their dealing with various aspects of the world.[32] Thus there is no reason – sceptical prejudice aside – to agree with Hume as against Mill on the issue of inductive warrant. Least of all should one be over-impressed by the standard circularity argument, i.e., that induction is always involved in a *petitio principii* since any grounds for believing in the reliability of inference from past to present or future events must itself be reliant on inductive procedures and thus beg the whole question. For the question then arises what better evidence we (or the sceptic) could possibly have than the fact that such procedures have played an indispensable role in the entire process of cognitive advance that has led from the most primitive forms of sense certainty to the most elaborate – yet empirically testable – scientific theories. At very least the burden of proof should fall on the sceptic rather than the realist opponent who can point to a vast range of confirmatory findings from just about every field of human experience and knowledge to date.

This is, I submit, sufficient to refute the kinds of extreme anti-realist (or cultural-relativist) argument which find their *reductio ad absurdum* in Rorty's 'Texts and Lumps'. What remains to be shown is that a similar case can be made against van Fraasen's 'constructive empiricist' position, that is to say, his agnostic stance with regard to the existence of unobservable entities and his consequent refusal to admit such entities to any (scientifically valid) realist ontology. The best, most convincing source of counter-arguments here is the large body of detailed scholarship devoted to particular episodes in the history of microphysics and related fields.[33] They range from Roentgen's chance discovery of X-rays to the various sub-atomic particles – electrons, ions, neutrinos and so forth – whose existence was at first a matter of conjecture but was then borne out through a process of jointly theoretical and empirical research. Wesley Salmon provides one of the best brief rehearsals of the realist case. Thus 'we can, after all, make statements about unobservable entities without invoking a special theoretical vocabulary.' And, by way of illustration:

> In his work on Brownian movement, Jean Perrin created large numbers of tiny spheres of gamboge (a yellow resinous substance). He suspended them in water, observed their motions with a microscope, and inferred that many smaller particles were colliding with them. Without using any non-observational terms, I have just described the essentials of an epoch-making experiment on the reality of molecules, namely the ascertainment of Avogadro's number (the number of molecules in a mole of any substance). . . . The crux of the argument is this. From a series of physical experiments that are superficially extremely diverse, it is possible to infer the value of Avogadro's number, and the values obtained in all of these types of experiment agree with one another remarkably well. If matter were not actually composed of such micro-entities as molecules, atoms, ions, electrons, etc., this agreement would be an unbelievably improbable coincidence.[34]

On the strength of this reasoning Salmon concludes that 'the problem of the existence of entities not even indirectly observable was essentially settled for natural science as a result of the work of Perrin and others in roughly the first decade of the twentieth century.'[35] And there is plentiful evidence from other sources – case-studies in the history and philosophy of science – that would add weight to Salmon's conclusion.

Of course van Fraassen might still maintain, on basically Humean grounds, that any such inference (even 'to the best explanation') must always admit the possibility of error, or of future developments – maybe in the state of technological advance – which render that inference no longer tenable. Such, after all, has been the story of particle physics over the past hundred years as successive candidates (from atoms to gluons, muons, quarks and so forth) have presented themselves for the title 'ultimate constituent of matter'. However, there is no good reason to suppose that anti-realism is the inevitable upshot of this 'argument from error', i.e., from the fact that past scientific theories have so often proved inadequate, partial or subject to correction in the light of subsequent research. For it is just this process of increasing refinement and elaboration – of theories continually tested against the best evidence to hand – which supports the case for paradigm-transcendent truths, that is to say, for the existence of real-world entities that are *not* just constructs of our present range of language games, conceptual schemes, ontological frameworks or whatever. In Nicholas Rescher's well-chosen words, it is not so much an argument against scientific realism as an argument 'against the ontological finality of science as we have it'.[36] Thus the very possibility of our turning out wrong (or under-informed) with regard to those ultimate constituents is itself the best evidence for there being something in the nature of the physical world which can or could act as a future corrective to our present limited knowledge.

Rescher describes instrumentalism – and the same would apply to van Fraasen's 'constructive empiricism' – as 'a sort of deconstructionism of natural science'. That is, '[i]t tries to do for *theoretical* entities what Bertrand Russell tried to do for *fictional* entities – to reinterpret talk that is *ostensibly* about entities in terms that bear no ontological weight.'[37] (I have argued in previous chapters that it is wrong to equate 'deconstruction' with anti-realism, but let us leave that disagreement aside for now.)[38] His point is that the instrumentalist approach is one that quite simply fails to explain how scientific advances come about through a gradual, uneven, by no means preordained, but none the less real and intelligible process of disciplined knowledge acquisition. For if that process is to make any kind of rational sense – and not appear wholly miraculous – then there is simply no choice but to posit an objective (i.e., mind- and language-independent) domain towards which our theories must be thought to converge and in reference to which they progressively acquire whatever truth-value is conferred upon them by the way things stand in reality. Such was C. S. Peirce's well-known idea of 'truth at the end of enquiry', a belief that placed his 'pragmaticist' philosophy markedly at odds with Rorty's strain of full-fledged neo-pragmatist thinking.[39] It is also – I would argue – the basic supposition of scientific realism and, indeed, of science itself as a distinctive, self-critical, highly disciplined form of cooperative human endeavour.

Michael Devitt makes this point in a simple but telling passage from his book *Realism and Truth*. Thus: 'If scientific realism, and the theories it draws on, were not correct, there would be no explanation of why the observed world is as if they were correct; that fact would be brute, if not miraculous.'[40] It is worth remarking here that when the atheist Hume set out his case against the possibility of miracles – or against the idea that it could ever be rational to credit their occurrence – he did so on the ground that they required us to suspend all those other tried and tested beliefs (in causality, laws of nature, physical necessity and the like) which his own scepticism had elsewhere called into doubt.[41] Hume's basic error – like that of the current anti-realists – was to demand epistemic warrant (justification in terms of presently available knowledge) for ontological claims regarding the existence of a world and its constituent features, some of which are now unknown and others of which might forever elude the best efforts of human understanding. One powerful counter-argument, due to Michael Tooley, puts the case that there might have been laws of nature – operant in some other physically possible world – which ruled out the evolution of consciousness and hence of there existing any theories in respect of which truth-claims could be verified or falsified.[42] Besides, it must be self-evident to any but the hardline anti-realist (or transcendental solipsist) that the universe and much of its furniture existed for aeons before there were sentient creatures around to formulate theories concerning its nature and origins. All of which suffices to refute anti-realism as a doctrine that denies – what we surely know – that the measure of truth is how far our ideas correspond to reality, rather than (as Rorty would have it) how far 'reality' corresponds to our ideas.

It is perhaps not so obvious that this applies also to van Fraassen's case with regard to unobservables, given the greater margin of doubt concerning their precise ontological status. Nevertheless a similar argument can be made for the belief that science achieves genuine progress in these areas of fundamental physics and, moreover, that our knowledge of this growth of knowledge cannot be explained except on condition of its having to do with progressively more adequate (ontologically deeper) descriptions. Thus, to take an example from D. M. Armstrong, '[i]t is at least plausible that, as contemporary physics claims, the charge *e* on every electron is strictly identical.'[43] What marks such predicates out as strong candidates for truth – along with the existence of those entities (in this case the electron) which they are taken to define or to qualify – is their status as universals, that is to say, as properties which hold good of particulars in each and every context of description. Thus '[w]e should not assume,' Armstrong cautions, 'that any old predicate, which in all probability has come into existence for purely practical reasons, applies in virtue of such identities.' After all, '[i]t is very unlikely, as Wittgenstein perceived although he was no realist about universals, that all games have something identical which makes them all games.'[44] However, it is just as unlikely – indeed very much more so – that scientific truth-claims like that concerning the identical (negative) charge on all electrons are rather, as some Wittgensteinians would have it, meaningful only in the context of some given language game or cultural life-form, and hence on a par – ontologically speaking – with the postulated objects of religious, mystical or pseudo-scientific belief.[45] On the contrary, Armstrong maintains: 'just what universals there are in the world, that is, what (repeatable) properties particulars have and what (repeatable) relations hold between particulars, is to be decided *a posteriori*, on the basis of total science.'[46] This is not – I should add – the idea of 'total science' that leads Quine to adopt his radically holistic approach to issues of meaning, truth and interpretation.[47] Rather it is the argument that truth-claims are warranted – and laws of nature manifested – in just those cases (like the negative charge on electrons) where theory combines with accurate and repeated measurement to provide a strong basis for inference to the best causal explanation.

In its most important features this argument goes back to Aristotle's *Posterior Analytics* and the earliest stages of philosophical reflection on science as a disciplined and truth-seeking enterprise. When Wesley Salmon declares that 'the time has come to put the "cause" back into "because"' he is essentially restating what appeared self-evident to Aristotle, namely that genuine scientific explanations involved something more than the witness of repeated co-occurrent happenings or events.[48] Thus in Aristotle's words: 'We suppose ourselves to possess unqualified scientific knowledge of a thing, as opposed to knowing it in the accidental way in which the sophist knows it, when we think that we know the cause on which the fact depends, as the cause of that fact and of no other, and further, that the fact could not be other than it is.'[49] It is the sophist's 'accidental' way of knowing – a knowledge derived from enumerative induction or observed regularities in nature – which has struck many sceptics, from Hume on down, as the only way possible in matters of causal explanation. Nor have things fared much better with deductive-nomological approaches, that is to say, theories that avoid this (supposed) inductivist impasse only through the recourse to a covering-law account which applies at so abstract and generalized a level as to admit all manner of irrelevant premises and equally irrelevant conclusions.[50] What Aristotle clearly perceived – and what has dropped from sight in much recent debate – is the requirement that genuine explanations have a grounding in the nature of the given *explanandum*, or in properties (e.g., physical attributes, causal dispositions, micro-structural features, etc.) which offer a means of distinguishing valid from invalid modes of inference. For, lacking such means, we are back with all the problems that have plagued philosophy of science in the wake of logical empiricism.

This point can be made most simply with reference to the issue of priority in causal explanations of the deductive-nomological (or covering-law) type. That is, such theories will always run up against the difficulty that they provide no adequate decision procedure for determining the *order* of causal relations, or for establishing which is the cause and which the effect in any given pair of linked observations or events. This problem arises on account of their purely formal structure and the consequent reversability of terms between *explanans* (major and minor premises) and *explanandum* (some event falling under the relevant scheme). Thus, in Hempel's classic formulation, what is required in order for such reasoning to apply is 'a valid deductive argument whose conclusion states that the event to be explained did occur'.[51] But this leaves open the standing (logical) possibility that causal processes may be construed as working in reverse, i.e., on a covering-law model that allows for the ascription of past effects to present causes or which treats effects as causes of causes and causes as effects of effects. In this respect, be it noted, there is little to choose between deductive-nomological approaches and the mode of hypothetico-deductive reasoning espoused by Popper and his followers. For the latter is strictly isomorphic with the former and offers no more in the way of guidance on issues of causal priority.[52]

In short, philosophy of science goes wrong – fails to capture what is distinctive about genuine cases of scientific explanation – when it takes this turn from substantive (*de re*) to purely formal (*de dicto*) ideas of what counts as a valid causal-explanatory mode of reasoning on the evidence. David-Hillel Ruben puts the case most succinctly – again following Aristotle – when he remarks that these problems can best be avoided by requiring that the premises in any given case must always include something about the cause of the event to be explained. Thus: 'given the angle of the sun's elevation, it is the height of the flagpole that causes the length of the shadow, and not vice versa; the change in the atmospheric pressure that causes the rise or fall of the barometer, and not vice versa; the receding of the galaxies that causes the red shift, and not vice versa.'[53] Of course, in each case there might be hardline sceptics – disciples of Nietzsche maybe – who denied that we could have any grounds for such assurance, or who chose to maintain that our whole idea of causality is merely a species of consoling fiction, a mirage induced by our enslavement to forms of illusory (linguistically induced) 'commonsense' belief. Such arguments are not, after all, so remote from that strain of Humean scepticism – carried on by the logical empiricists – which effectively drives a wedge between first-order observation statements and higher-level covering-law principles. What they ignore, quite simply, is our frequent possession of knowledge (as in the above examples from Ruben) which permits us to assert *without serious or genuine doubt* that we are justified – scientifically warranted – in accepting certain orders of explanatory argument and rejecting others as ill-founded, uninformed or downright perverse. Of course, the sceptic may persist in his or her argument that this is just another piece of commonsense realist bluff which merely begs the question with regard to the truth-claims of science and our knowledge of the growth of knowledge. But in that case it seems only fair to conclude – with so much evidence to hand – that the burden of proof rests squarely with the sceptic rather than the realist.

References

1 Bas C. van Fraassen, *The Scientific Image* (Oxford: Clarendon Press, 1980).
2 Van Fraassen, *Laws and Symmetry* (Oxford: Clarendon Press, 1989). All further references indicated by *LS* and page number in the text.
3 See Ian Hacking, *Representing and Intervening: introductory topics in the philosophy of natural science* (Cambridge: Cambridge University Press, 1983).
4 For further discussion see Robert Ackerman, *Data, Instruments, and Theory: a dialectical approach to the philosophy of science* (Princeton, NJ: Princeton University Press, 1985); Michael Gardner, 'Realism and instrumentalism in nineteenth-century atomism', *Philosophy of Science*, 46/1 (1979), pp. 1–34; Josef M. Jauch, *Are Quanta Real? a Galilean dialogue* (Bloomington: Indiana University Press, 1973); Mary Jo Nye, *Molecular Reality* (London: MacDonald, 1972); J. Perrin, *Atoms*, trans. D. L. Hammick (New York: Van Nostrand, 1923).
5 For examples of the strong sociological approach see, for instance, Barry Barnes, *About Science* (Oxford: Blackwell, 1985); Harry Collins and Trevor Pinch, *The Golem: what everyone should know about science* (Cambridge: Cambridge University Press, 1993); Ludwik Flieck, *Genesis and Development of a Scientific Fact* (Chicago: University

of Chicago Press, 1979); Steve Fuller, *Philosophy of Science and its Discontents* (Boulder, CO: Westview Press, 1989); K. Knorr-Cetina and M. Mulkay (eds), *Science Observed* (London: Sage, 1983); Steve Woolgar, *Science: the very idea* (London: Tavistock, 1988).

6 See entries under note 4 above; also Niels Bohr, *Atomic Physics and Human Knowledge* (New York: Wiley, 1958); Michael Gardner, 'Realism and instrumentalism in nineteenth-century atomism'; Stephan Korner (ed.), *Observation and Interpretation* (New York: Academic Press, 1957); Richard E. Grandy (ed.), *Theories and Observation in Science* (Englewood Cliffs, NJ: Prentice-Hall, 1973).

7 Frank P. Ramsey, *Philosophical Papers*, ed. D. H. Mellor (Cambridge: Cambridge University Press, 1990).

8 Cited by van Fraasen, *Laws and Symmetry*, p. 40.

9 See David Lewis, *Counterfactuals* (Oxford: Blackwell, 1973) and *On the Plurality of Worlds* (Blackwell, 1986).

10 See A. J. Ayer, *The Foundations of Empirical Knowledge* (London: Macmillan, 1955); R. Carnap, *The Logical Structure of the World* (London: Routledge and Kegan Paul, 1937); C. G. Hempel, *Fundamentals of Concept Formation in Empirical Science* (Chicago: University of Chicago Press, 1972); Hans Reichenbach, *Experience and Prediction* (University of Chicago Press, 1938).

11 See W. V. Quine, 'Two Dogmas of Empiricism', in *From a Logical Point of View*, 2nd edn (Cambridge, MA: Harvard University Press, 1961), pp. 20–46; also Quine, *Ontological Relativity and Other Essays* (New York: Columbia University Press, 1969).

12 See note 7 above; also Ramsey, *The Foundations of Mathematics* (London: Routledge and Kegan Paul, 1931).

13 See, for instance, D. M. Armstrong, *What Is a Law of Nature?* (Cambridge: Cambridge University Press, 1983); John W. Carroll, *Laws of Nature* (Cambridge University Press, 1994); L. Jonathan Cohen, *The Implications of Induction* (London: Methuen, 1970); Rom Harré and E. H. Madden, *Causal Powers* (Oxford: Blackwell, 1975); Nicholas Rescher, *Induction* (Oxford: Blackwell, 1980); Wesley C. Salmon, *Scientific Explanation and the Causal Structure of the World* (Princeton, NJ: Princeton University Press, 1984); Brian Skyrms, *Causal Necessity* (New Haven: Yale University Press, 1980); Richard Swinburne (ed.), *The Justification of Induction* (London: Oxford University Press, 1974); M. Tooley, *Causation: a realist approach* (Oxford: Blackwell, 1988).

14 David Hume, *A Treatise of Human Nature*, ed. L. A. Selby-Bigge (Oxford: Clarendon Press, 1967).

15 See entries under note 10 above.

16 For further arguments to this effect, see Hilary Kornblith, *Inductive Inference and its Natural Ground: an essay in naturalistic epistemology* (Cambridge, MA: MIT Press, 1993); also Kornblith (ed.), *Naturalizing Epistemology* (MIT Press, 1985) and Salmon, *Scientific Explanation and the Causal Structure of the World.*

17 Peter Lipton, *Inference to the Best Explanation* (London: Routledge, 1993), p. 167.

18 Ibid., p. 167.

19 Richard Rorty, 'Texts and Lumps', in *Objectivity, Relativism, and Truth* (Cambridge: Cambridge University Press, 1991), pp. 78–92. All further references indicated by 'TL' and page-number in the text.

20 Saul Kripke, *Naming and Necessity* (Oxford: Blackwell, 1970); also Stephen Schwartz (ed.), *Naming, Necessity, and Natural Kinds* (Ithaca, NY: Cornell University Press, 1977).
21 Cited by Rorty, 'Texts and Lumps', p. 81.
22 See entries under note 10 above; also Alfred Tarski, *Logic, Semantics and Metamathematics*, trans. J. H. Woodger (London: Oxford University Press, 1956).
23 See note 5 above; also David Bloor, *Wittgenstein: a social theory of knowledge* (New York: Columbia University Press, 1983); Harry Collins, *Changing Order: replication and induction in scientific practice* (Chicago: University of Chicago Press, 1985); Andrew Pickering (ed.), *Science as Practice and Culture* (University of Chicago Press, 1992); Rorty, *Objectivity, Relativism, and Truth*; Joseph Rouse, *Knowledge and Power: towards a political philosophy of science* (Ithaca, NY: Cornell University Press, 1987); Steve Woolgar (ed.), *Knowledge and Reflexivity: new frontiers in the sociology of knowledge* (London: Sage, 1988).
24 See entries under note 23 above; also Martin Heidegger, *The Question Concerning Technology and Other Essays*, trans. William Lovitt (New York: Harper and Row, 1977); Stephen Mulhall, *On Being in the World: Wittgenstein and Heidegger on seeing aspects* (London: Routledge, 1990); Mark Okrent, *Heidegger's Pragmatism: understanding, being, and the critique of metaphysics* (Ithaca, NY: Cornell University Press, 1988).
25 E. D. Hirsch, *Validity in Interpretation* (New Haven: Yale University Press, 1967).
26 See Salmon, *Scientific Explanation and the Causal Structure of the World* and other entries under note 13 above; also Roy Bhaskar, *Scientific Realism and Human Emancipation* (London: Verso, 1986) and *Reclaiming Reality: a critical introduction to contemporary philosophy* (Verso, 1989).
27 See, for instance, Rorty, 'Science as Solidarity' and 'Is Natural Science a Natural Kind?', in *Objectivity, Relativism, and Truth*, pp. 35–45 and 46–62.
28 See notes 5 and 23 above.
29 Cited in Lipton, *Inference to the Best Explanation*, p. 181.
30 See entries under note 11 above; also Thomas S. Kuhn, *The Structure of Scientific Revolutions*, 2nd edn (Chicago: University of Chicago Press, 1970).
31 Saul Kripke, *Wittgenstein on Rules and Private Language* (Oxford: Blackwell, 1982).
32 See entries under notes 13 and 16 above; also J. Aronson, R. Harré and E. Way, *Realism Rescued: how scientific progress is possible* (London: Duckworth, 1994); Michael Devitt, *Realism and Truth* (Oxford: Blackwell, 1984); Jarett Leplin (ed.), *Scientific Realism* (Berkeley and Los Angeles: University of California Press, 1984); J. J. C. Smart, *Philosophy and Scientific Realism* (London: Routledge and Kegan Paul, 1963); Nicholas Rescher, *Scientific Realism: a critical reappraisal* (Dordrecht: D. Reidel, 1987); Peter J. Smith, *Realism and the Progress of Science* (Cambridge: Cambridge University Press, 1981).
33 See notes 4 and 6 above; also Hacking, *Representing and Intervening*.
34 Wesley C. Salmon, 'Epistemology of Natural Science', in Jonathan Dancy and Ernest Sosa (eds), *A Companion to Epistemology* (Oxford: Blackwell, 1992), pp. 292–6, esp. p. 295.
35 Ibid., p. 295.
36 Rescher, *Scientific Realism*, p. 61.
37 Ibid., p. 35.

38 Christopher Norris, *New Idols of the Cave: on the limits of anti-realism* (Manchester: Manchester University Press, 1997).
39 C. S. Peirce, *The Essential Writings* (New York: Dover, 1964); *Reasoning and the Logic of Things*, ed. K. L. Ketner (Cambridge, MA: Harvard University Press, 1992); *Essays in Philosophy of Science* (New York: Bobbs-Merrill, 1957).
40 Michael Devitt, *Realism and Truth*, p. 108.
41 David Hume, *A Natural History of Religion and Dialogues Concerning Religion* (Oxford: Clarendon Press, 1977).
42 Tooley, *Causation*.
43 Armstrong, *What Is a Law of Nature?*, p. 83.
44 Ibid., p. 83.
45 See Bloor, *Wittgenstein: a social theory of knowledge*; also Derek L. Phillips, *Wittgenstein and Scientific Knowledge: a sociological perspective* (London: Macmillan, 1977) and Peter Winch, *The Idea of a Social Science and its Relation to Philosophy* (London: Routledge and Kegan Paul, 1958).
46 Armstrong, *What Is a Law of Nature?*, p. 83.
47 Quine, 'Two Dogmas of Empiricism'.
48 Salmon, *Scientific Explanation and the Causal Structure of the World.*
49 Aristotle, *Posterior Analytics*, trans. Jonathan Barnes (Oxford: Clarendon Press, 1975), Book 1, chapter 2, 71b, 8ff.
50 See note 10 above; also Adolf Grunbaum and Wesley C. Salmon (eds), *The Limitations of Deductivism* (Berkeley and Los Angeles: University of California Press, 1988).
51 C. G. Hempel, *Aspects of Scientific Explanation* (New York: Macmillan, 1965) and *Fundamentals of Concept Formation in Empirical Science* (Chicago: University of Chicago Press, 1972).
52 See, for instance, Karl Popper, *The Logic of Scientific Discovery*, 2nd edn (New York: Harper and Row, 1959); *Conjectures and Refutations* (Harper and Row, 1963).
53 David-Hillel Ruben, *Explaining Explanation* (London: Routledge, 1992), p. 193.

8

Stuck in the Mangle: Sociology of Science and its Discontents

I

Wesley Salmon made a case (writing in the late 1980s) that the past four decades of scientific explanation had witnessed a crisis in philosophy of science very largely brought about by various shortcomings in the logical-empiricist model.[1] Chief among them was its failure to provide any adequate or convincing link between nomic regularity and those underlying causal mechanisms that explain why such regularities should ever be found or expected to occur. Indeed it had become the orthodox view – descending from Hume – that any such linkage must remain forever obscure, a product of our natural propensity to believe in the existence of knowable causes but still no part of any rigorous reflection on the scope and limits of scientific method. As this programme collapsed under pressures internal and external so philosophers – Salmon among them – began to explore the alternative resources held out by a return to causal-realist modes of explanatory reasoning.[2]

For there is clearly a sense in which philosophy of science can and should keep pace with scientific developments, especially those that exert a strong claim to have deepened and extended our knowledge of the world and hence (a fortiori) our knowledge of what counts as accurate, reliable or depth-explanatory knowledge of the world. Thus, for instance, we know a lot more than Hume about the sorts of inductive procedure – or inference to the best explanation – that have produced a whole range of scientific advances in various fields from astronomy and fluid mechanics to particle physics and molecular biology. By the same token we are much better informed than Locke about the underlying properties of manifold items – elements, compounds, natural kinds, sub-atomic and molecular structures, chemical affinities, etc. – whose 'real' (as opposed to their 'nominal') essences he famously despaired of defining within the limits imposed by the physical sciences of his age.[3] At any rate, there would seem small justification – sceptical prejudice aside – for maintaining a similar position nowadays with regard to the inscrutability of causes or for the recourse to nominalism (or anti-realism) in order to avoid unwanted ontological commitments. To adopt this line is to raise scepticism to a high point of a priori principle above all the surely undeniable evidence of scientific progress to date.

In any case, as Salmon argues, philosophy of science has itself moved on to a stage where there exist some well-developed alternatives to the kinds of problem encountered with the old logical-empiricist paradigm. These include causal realism in various forms, from philosophical semantics (Kripke and early Putnam) to the detailed scientific case-studies provided in his own work and by other contributions to the realist literature.[4] In addition – and perhaps more important – there is the fact that Hume can now be seen to have worked with a drastically restrictive idea of causal explanations, one that made no appeal to their counterfactual-supporting character. Thus, in David-Hillel Ruben's words, 'there is no possible world in which an object can have a dispositional feature and no structural basis whatever for that feature.'[5] In the same way there is no possible world – or none among the subset of physically possible (as distinct from abstractly conceivable) worlds – wherein such features would fail to support counterfactual arguments based on their hypothesized absence in this or that particular case. Thus, for instance: were it not for certain known features of the chemical-molecular composition of salt, then salt would not display the dispositional attribute of solubility in water. And again: we should have no adequate explanation for the fragility of glass if glass did not possess just that kind of molecular structure that accounts for its tendency to shatter under impact.

One could multiply examples to similar effect, all pointing up the reliance of scientific knowledge on modes of counterfactual (or subjunctive conditional) reasoning involved in the ascription of particular properties to particular objects or events. What causal explanations typically propose – or implicitly assume – is that a given event would not have occurred had the antecedent conditions differed in some crucial respect or had the needful cause (or conjunction of causes) not been operant. Although a version of this argument is present in some passages of Hume it remains undeveloped and – more crucially – as yet unprovided with the kinds of detailed depth-explanatory evidence that have emerged with the advancement of knowledge in fields like molecular biology and nuclear physics. At any rate, there would seem small warrant for maintaining the Humean (and logical-empiricist) sceptical position which makes it a permanent mystery how the logic of scientific enquiry could possible connect with matters of real-world (*de re*) causal explanation. Thus Armstrong distinguishes between Hume-type 'accidental' or 'single-case' uniformities – those based on local observations of regular recurrence – and other instances where we can properly appeal to laws of nature, since they meet all the above (counterfactual-supporting) requirements as well as exhibiting a genuine (explanatorily adequate) relation between universals.[6] On the same argument there is no real problem with the sorts of example contrived by philosophers like Goodman in order to revive the Humean 'puzzle' of induction by introducing a range of artificial, factitious or non-natural predicates.[7] For in such cases, according to Armstrong, 'it is plausible to say, on the basis of total science, that "grue" is a predicate to which no genuine, that is unitary, universal belongs. . . . Where there are no universals, there is no relation between universals.'[8]

This point becomes crucial in restricting the range of relevant (scientifically admissible) predicates, laws, regularities, inductive reasonings and so forth to those which can claim some genuine grounding in our knowledge of physical reality. The regularity theorist is clearly in need of some such restrictive clause if he or she is to claim even minimal (Humean) warrant for linking one observation with another on a basis of perceived similarity. But they will lack such a basis – and thus run into all the well-known puzzles about induction – if it is thought to consist in *nothing more* than regular concurrence, or to lack any further (depth-explanatory) means of distinguishing real from factitious predicates. What makes the difference, on Armstrong's account, is precisely the criterion of genuine 'sameness' which explains (for instance) why the attribute of greenness is a feature *both* of all emeralds sighted up to now *and* of all objects in the candidate class that could warrant that description in future sightings. Thus 'obviously he [the regularity theorist] ought in some way to restrict the principle to cases where natural predicates such as "green" are used, as opposed to unnatural predicates such as "grue".'[9] Of course the anti-realist – or the Humean sceptic – will raise all the usual objections to Armstrong's way of stating the case. They will regard his principle as by no means 'obvious'; enquire by what right (mere prejudice aside) he declares that the theorist *ought* to adopt this restrictive procedure; and remark that his distinction between 'natural' and 'unnatural' predicates is one that in effect prejudges the issue as to whether such differences hold *de re* (as a matter of physically instantiated 'laws of nature') or whether, conversely, they are not just the products of some chosen ontological scheme, method of inductive projection, or Goodmanian 'way of worldmaking'.[10]

Nor can it help to offer arguments from probability theory to the effect that, since greenness has been a property of all emeralds observed up to now, therefore we are justified – massively so – in basing any future identifications (or modes of inductive reasoning) on the presumed continuance of just that 'natural' property. For of course this is just the point of Goodman's sceptical argument: that if 'grue' = 'observed as green up to AD 2000 and as blue thereafter' then there is nothing to choose, in point of inductive warrant, between 'all emeralds are green' and 'all emeralds are grue.' Thus, in Armstrong's words, 'if it is just a formal argument from logical possibility being advanced, any differences between "green" and "grue" must be irrelevant. The observed sample of emeralds are green. The observed sample of emeralds are grue. The mathematics is the same in both cases.'[11] For Goodman – following Hume – the only way out of this dilemma is to bank on our relatively well-entrenched methods of inductive inference and use them as a basis for projecting future regularities of much the same kind. This mixture of 'entrenchment' and 'projection' is, he thinks, the best possible case for induction, given all the problems that must necessarily arise if one seeks to go beyond it to a stronger account based on notions of logical necessity or *de re* causal explanation. In which case Armstrong could have no grounds for his criticism of 'naïve' regularity theorists or his idea that there exist laws of nature (relations between universals) which provide the basis for just such an alternative account.

However one may doubt – with support from Armstrong – that Goodman's sceptical 'solution' is anything like an adequate answer to the problems thrown up by his own sceptical argument. Indeed that 'solution' is no more convincing than other attempts in a similar mode, among them Saul Kripke's oddly off-the-point suggestion that one can somehow resolve Wittgenstein's problem about rule-following (in arithmetic and elsewhere) by simply repeating Wittgenstein's argument that such practices derive what validity they have from our customs, conventions, language games, communally sanctioned 'forms of life', and so forth.[12] In neither case – Goodman or Kripke – does it help very much to be told that our worries about inductive inference or the status of arithmetical truths can be laid to rest simply by remarking that this is how we do things by general agreement within this or that cultural community. Moreover, it is by way of such conventionalist premises (often with reference to Wittgenstein) that a case can be made for other, more extreme forms of the current anti-realist or 'strong' sociological approach to issues in the history and philosophy of science.[13] Thus a good deal hangs on the attempt – by Armstrong and others – to demonstrate both the shortcomings of a Humean (regularity-based) theory of causal explanation and the case for a viable realist alternative.

Armstrong formulates that case as follows in a passage that will bear quoting at length since it offers some useful clarification.

> The task is to explicate the rationality of an inductive inference from P (say that all observed emeralds are green) to R (say that unobserved emeralds are green). I say that we can do this if we accept that P permits a non-deductive inference to a *law* Q, that all emeralds are green, from which it follows that R. But our concept of the law must be a suitable one. If our concept of law is simply the concept of P + R, the inference pattern becomes $P \rightarrow (P + R)$, $(P + R) \rightarrow R$, which reduces to $P \rightarrow R$. *That* inference pattern Hume was right to think non-rational. It remains to show that there is a concept of Q, which allows rational $P \rightarrow Q$. I believe that this demand can be satisfied if Q is conceived of as a relation between universals, and $P \rightarrow Q$ conceived of as a case of inference to the best explanation.[14]

That is to say, there is no answer to Hume's dilemma – to the problem of inductive inference – so long as one accepts any version of the logical-empiricist dichotomy between truths of observation (or instances of regular co-occurrence) and those higher-level theories, logical constructions or covering-law generalizations which seek to subsume (rather than explain) the causal processes involved. Such approaches must always at some point run into that dilemma – and also fall prey to Goodman-type puzzles of induction – since they can offer no adequate grounds for supposing that any regularities thus noted are instances of genuine (law-governed) causal connection rather than chance contiguities, random statistical bunchings, cases of the *post hoc, propter hoc* fallacy, or whatever. This failure has promoted various kinds of sceptical conclusion as regards laws of nature or our warrant for invoking *de re* as opposed to *dicto* orders of causal explanation. It is then a short step to Quinean/Kuhnian doctrines of ontological relativity, and – at the limit – to a wholesale (Rorty-type) anti-realism which counts 'reality' a world well lost for the sake of our freedom to reinvent it under whatever currently favoured or culturally salient description.[15]

Of course the anti-realist will not be much impressed by Armstrong's purely formal statement of the case as exemplified in the passage above. What is needed in addition is a wide range of corroborative instances – case-studies from the history of science – which can stand in for 'Q' and thus provide evidence that laws of nature are indeed the missing link (the strictly indispensable presupposition) required to make sense of inductive procedures and our knowledge of the growth of scientific knowledge. The same applies to those Kuhnian relativist or 'strong' sociological approaches which take it that everything is open to question – especially the truth-claims of science – *except* the appeal to history, cultural context, or socio-political interests as their baseline terms of explanation. Thus Kuhn's central doctrines (of paradigm shifts, framework relativism, incommensurability, radical meaning variance and so forth) are entirely dependent on his uncritical acceptance of the *history* of science as delivering factual truths whose effect is purportedly to relativize everything beyond their privileged domain. For if one asks – consistently with his own prescriptions elsewhere – what philosophy of history subtends this approach then the answer can only be: one incompatible with Kuhn's whole approach to science and philosophy of science. After all, as Peter Muntz aptly remarks, 'if no [scientific] observation has enough bite to establish any theory rationally, none of his historical examples have enough bite to establish any theory rationally.' In which case, logically, '[t]he most he can do is invite his readers to abandon all earlier philosophies of history and hope they will fade away.'[16]

However, this is not an option for Kuhn since he needs his historical examples – Aristotle and Galileo on swinging weights/pendulums, Priestley and Lavoisier on phlogiston/oxygen, etc. – in order to bear out his case for the paradigm-relative or shifting character of scientific theories and truth-claims. So it comes about that 'he allows [those examples] the absolute bite which he denies to all other facts of nature so called.' And again:

> history, so it seems, is for Kuhn the one exception. Here, for once in the whole and entire realm of human knowledge, we have a set of hard and brute facts which invite, indeed compel, us rationally to give up old paradigms for the philosophy of science advocated by Kuhn. . . . In short, if Kuhn is right in saying that paradigms are changed and accepted for irrational reasons, he must be wrong in saying so. For he offers us what he considers to be rational reasons for leaving behind all other philosophies of science – the very thing he maintains cannot be done.[17]

A similar argument can be made against those strong sociological approaches that take social facts (interests, values, priorities, etc.) as the one privileged or objectively given domain to which everything else must be relativized. Thus scientific theory and practice – along with philosophy of science – are viewed by contrast as culturally emergent activities whose truth-claims, methods and disciplinary values must always be subject to the undeceiving rigours of a sociological critique.

II

Andrew Pickering – himself no advocate of scientific realism – has noted this tendency as a major blind spot in the sorts of argument standardly adduced by David Bloor and other advocates of the 'strong' programme.[18] The social is treated as 'at most a *quasi-emergent* category', that is, an explanatory resource whose uses may require some reflexive allowance for its own culture-relative (or socially constructed) character but which none the less provides a stable reference point for demystifying other kinds of truth-claim.[19] Thus 'the SSK [Sociology of Scientific Knowledge] gaze only ever catches a fixed image of the social in the act of structuring the development of the technical and metaphysical strata of science. . . . SSK always seems to miss the movie in which the social is itself transformed' (Pickering, p. 452). As I say, Pickering is far from wishing to defend any variety of realism that would exclude or marginalize those social and cultural factors as they bear upon the history of scientific thought. Rather, his point is that they enter into a complex (overdetermined) relationship with other factors – some of them specific to particular scientific disciplines – which may in turn have a powerful transformative effect upon the very conditions of scientific knowledge-production. Thus where 'Bloor advances the Durkheimian argument that resistance to the strong programme arises from a sacred quality attributed to science in modern society,' Pickering counters with the moderate suggestion that 'perhaps in SSK the social has become the sacred' (p. 465n). 'Sacred', that is, in so far as it occupies a privileged (at most 'quasi-emergent') status in relation to which all other kinds of knowledge are treated as second-order or derivative.

Pickering is not the only commentator with broadly social-constructionist sympathies who has remarked on this odd reversal of priorities – or lack of reflexive self-awareness – among advocates of the strong programme. It is also a theme in the recent writings of Bruno Latour, himself a major inspiration for the movement and again, like Pickering, no friend of scientific realism.[20] Nevertheless, he has lately offered some pointed remarks about the non-symmetrical (self-privileging) role of arguments that routinely reduce almost everything – their own favoured methods excluded – to the level of socially constructed or culturally emergent 'knowledge'.[21] Such criticism is intended more by way of a corrective to current sociological excesses than as offering support for anything like a robust ontological and causal realism in philosophy of science. Thus Pickering talks about the 'mangle' of socio-cultural, metaphysical and techno-scientific practice conceived as a multi-factorial transformative process where no one determinant enjoys causal priority and where everything – from theories and obervation data to metaphysical world-views and social values – is subject to constant revision under pressure from the other factors involved. From this point of view 'the social should in general be seen as in the plane of practice, both feeding into technical practice and being emergently mangled there, rather than as a fixed origin of unidirectional, causal arrows' (Pickering, p. 452). Certainly he is far from wishing to reverse this latter (sociologically determined) order of priorities or from promoting the claims of causal realism in the history and philosophy of science. Nevertheless his argument does point to problems with the SSK programme even when construed in his own (fairly moderate) terms. That is to say, it raises the question as to what could possibly *count* as scientific knowledge were it not for the constraints upon 'technical practice' – along with various candidate metaphysical world-views – exerted by the nature of those manifold objects, processes and events which constitute the real-world physical domain.

'Technical practice' is the privileged term for Pickering since it follows from his strong-constructionist position that scientists (mathematicians and theoretical physicists included) are always in the business of devising such entities – from quaternions to quarks – through a process of what he calls 'resistance and negotiation'. In so far as reality is itself, on this account, a 'metaphysical' construct arrived at through the same process, it must also be seen as strictly negotiable and in no sense 'there' (ontologically given) as the object domain of scientific enquiry. Thus '[l]ike the technical culture of science and like the conceptual, like the social, like discipline – metaphysics is itself at stake in practice and just as liable to temporally emergent mangling there in interaction with all of those other elements' (Pickering, p. 456). However, one may doubt that this even-handed approach – this refusal to prioritize in matters of explanatory warrant – could possibly account for what is distinctive about science as a means of gaining more accurate knowledge of the world. For there is simply no conceiving of 'technical practice' as apart from those various specific 'disciplines' which themselves define what shall count as scientific knowledge at some given stage of development. Pickering's idea of the 'mangle' is suggestive enough as a handy metaphor for what goes on in the context of discovery, that is to say, at the stage where all sorts of interest – social, political, cultural, metaphysical and so forth – can be seen to have a strong (at times crucial) bearing on the conduct of scientific research. But it is much less apt – indeed highly misleading – when applied to the context of justification where scientific theories and truth-claims are subject to different, more rigorous criteria of evaluation.[22] For here 'the technical culture of science' is precisely what constitutes a genuine scientific 'discipline', i.e., a mode of knowledge acquisition which so far as possible excludes those (now extraneous) sociocultural factors and whose procedures are aimed solely towards providing the best, most perspicuous, theoretically refined, and explanatorily adequate account of the matter in hand. Pickering can scarcely avoid confusing these contexts, given his idea of the 'mangle of practice' as a machine that incessantly churns them up and creates a kind of sociocultural-metaphysico-technical-scientific mishmash. To this extent his quarrel with the 'strong' sociologists is a quarrel in name only, or one that exchanges their monocausal (sociologically reductive) account for an approach that, in effect, ends up by treating everything – the social included – as a product of short-term shifting interactions in the 'culture' of scientific practice.

Pickering's book *Constructing Quarks* is a full-scale application of this theory to the field of elementary particle physics and its various ontological visions and revisions.[23] His point is that such entitities are 'constructed' through a mixture of theory, mathematical modelling, oblique inference, and often quite extravagant conjecture which far outruns any currently available technology or means of observation. Where he differs from the strong sociologists, again, is in rejecting any simple (monocausal) appeal to supposedly determinate social factors that would take no account of their own constructedness in and through the all-transformative 'mangle of practice'. More specifically, what is lost to view on this theory is the extent to which 'technical practices' – including mathematical formalisms – may at times bring a strong counter-pressure to bear upon established habits of scientific thought and thus play a role in that perpetual process of 'resistance and negotiation' which keeps the mangle turning. The quotations above are drawn from a more recent essay ('Constructing Quaternions') where Pickering examines the work of the nineteenth-century Irish mathematician Sir William Hamilton.[24] What chiefly interests Pickering is Hamilton's shift from a realist metaphysics – the view that mathematical 'objects' have their correlates in time and space – to a formalist (indeed constructivist) approach where mathematical procedures are conceived as pertaining to a realm of *sui generis* operative rules without reference to anything outside their own domain.

Thus Hamilton started out convinced that arithmetic and algebra, like geometry, were modes of synthetic a priori knowledge, applicable to (but not derivable from) our experience of real-world temporal and spatial conditions. In this he followed Kant who had argued (in the 'Transcendental Aesthetic' of the first

Critique) that arithmetical operations such as counting were possible only on the basis of temporal succession – of time as the very form or condition of inner experience – while geometrical truths had to do with the form of outer (spatial) appearance as likewise given to the mind a priori through its capacity to bring intuitions under adequate concepts.[25] However, there were problems with this theory, as soon became apparent with the advent of more powerful mathematical formalisms and counter-intuitive (non-Euclidean) geometries. One such problem which concerned Hamilton was that of absurd quantities or irrational numbers – like the square root of minus one – that possessed well-defined mathematical uses but which could not be extended to the physical domain (or, in Kantian terms, the realm of phenomenal intuition) without creating all manner of paradoxes. So it was, on the standard account, that Hamilton abandoned his Kantian position and adopted a radically constructivist approach whereby to avoid these difficulties and prevent mathematics from getting into conflict with the dictates of commonsense realism.[26]

Pickering presents this conversion narrative as a typical instance of the 'mangling' that occurs when various conflicting priorities (technical, metaphysical, social, cultural, political and so forth) undergo the process of 'resistance and accommodation.' There is at least a prima facie case, he suggests, for 'understanding the transformation in Hamilton's metaphysics in the mid-1840s as an accommodation to resistances arising in technical-metaphysical practice' (Pickering, p. 456). More specifically: 'a tension emerged between Hamilton's Kantian a priorism and his technical practice, to which he responded by attenuating the former and adding to it an important dash of formalism' (ibid.). Here again his emphasis on 'technical practice' – on the role of specific (in this case mathematical) procedures and constraints – puts a distance between Pickering and advocates of the strong sociological or cultural-relativist approach. Thus he criticizes Bloor for assuming too readily that Hamilton's earlier philosophy of mathematics was a product of his deep-grained political conservatism, that is to say, his espousal of a Church and State creed which sought to uphold the values and interests of a dominant social class. On this account Hamilton 'was aligned with Coleridge and his circle and, more broadly, with "the interests served by idealism" – conservative, holistic, reactionary interests opposing the growing materialism, commercialism, and individualism of the early nineteenth century and the consequent breakdown of the traditional social order' (Pickering, p. 450). His philosophy would then fit into this picture as a kind of reflex metaphysics, a set of beliefs which enabled Hamilton to assimilate algebra and geometry to the Kantian realm of Understanding (i.e., as pertaining to our basic, a priori intuitions of space and time), while this was in turn subordinate to Reason as that which exercised legislative power in issues of religion and morality. To this extent algebra, as Hamilton conceived it, 'would always be a reminder of, and a support for, a particular conception of the social order . . . an "organic" social order of the kind which found its expression in Coleridge's work on Church and State' (p. 450). All of which placed him markedly at odds – so Bloor argues – with the Cambridge-based school of formalist mathematicians who rejected any such foundationalist (metaphysical) approach and who viewed mathematics as a strictly autonomous domain with its own special rules, methods, proof procedures and so forth. From a strong sociological viewpoint this attitude is best explained in terms of their position as 'reformers and radicals' keen to assert their professional independence of existing structures of power and authority. So it was that they adopted an anti-metaphysical stance whose chief attraction (on this reading) lay in its refusal to acknowledge such constraints upon the discipline – or the modes of technical practice – that defined mathematics as a specialized branch of knowledge.

Pickering very rightly criticizes Bloor for his reductive approach to these issues, that is to say, his 'strong' sociological idea that such differences of view can best be understood as reflex products of conflicting ideologies or opposed political interests. His objection takes the form – long familiar from debates within Western Marxist theory – of a principled insistence on the various 'mediations' (the knowledge- or discipline-specific values and priorities) that complicate the passage from socio-economic 'base' to cultural-ideological 'superstructure'.[27] To ignore these complexities is surely to risk just the kind of vulgar reductionist (monocausal) theory that thinkers like Bloor are quick to condemn when they find it at work in empiricist or positivist conceptions of scientific method. So there is much to be said for Pickering's emphasis on the extent to which mathematical formalisms and other such specialized modes of knowledge play a role in the process of 'resistance and accommodation' whereby scientists find themselves impelled to construct new theories or alternative means of getting around some particular obstacle to thought.

However, he is hard put to explain just *why* any such specific solution should commend itself as better – more adequate or convincing – than all the other possibilities that might have been canvassed in this or that context of debate. For as Pickering sees it, there is nothing *in the nature* of scientific truth, reason or progress which would allow us to distinguish some theories as representing a genuine advance in knowledge and others as affording no such advance through their failure to resolve outstanding problems or their lack of real-world explanatory power. Thus on his account everything goes into the 'mangle' – formalisms, theories, 'commonsense' intuitions, metaphysical commitments, social interests, the 'technical culture of science' – and what comes out is a compromise solution or a kind of patched-up hybrid theory, a product of the various competing claims that bear upon the context of discovery. In short, there is less difference than might at first appear between Pickering's radical constructivist view and Bloor's drive to reduce everything to the level of baseline sociopolitical interests, values, and motives. For in both cases it is the background 'culture' – no matter how complex or overdetermined – which finally sets the parameters of choice and which is taken to explain why scientists espoused one or another of the currently available options. That is to say, there is no appeal left open to the context of justification wherein certain theories prove to possess a high measure of predictive, conceptual or explanatory power while others prove wanting in just those crucial (scientifically decisive) respects.

The case of mathematics is of course the most difficult for anyone advancing a realist philosophy of scientific method and practice. Clearly there is a sense in which 'realism' as applied to mathematical entities (numbers, sets, classes, formalisms, axioms, proof procedures and so forth) cannot be construed in the same way as 'realism' applied to macro-physical objects, processes and events. Nor can it be treated as directly on a par with those quantum-theoretical formalisms – most notably Planck's Constant – whose peculiar virtue is to hold good across such a range of otherwise opaque referential contexts, as between wave and particle descriptions or values of particle location and momentum. Thus to be an anti-realist (or a van Fraassen-type constructive empiricist) with regard to scientific unobservables is not at all the same thing as to endorse anti-realism in respect of either mathematical entities or what J. L. Austin engagingly described as the everyday range of medium-sized dry goods.[28] On the other hand there is simply no accounting for scientific progress – for the extraordinary success of science in explaining such a range of physical phenomena – unless one acknowledges the degree of correlation between results arrived at on a purely mathematical basis and results arrived at through other (e.g., experimental or observational) procedures. This is not the place for a detailed account of the age-old debate among philosophers and scientists regarding the status of mathematics as a mode of knowledge strictly *sui generis* or one that affords knowledge of the world through some form of correspondence (if not pre-ordained harmony) between numbers and the structure of physical reality.[29] Sufficient to say that all the evidence – from Pythagoras to Galileo, Newton, Einstein and Planck – is such as to support a realist view at least in the sense that mathematics provides an indispensable means of constructing ever more powerful (rigorously formalized) natural-scientific theories and hypotheses. For these latter very often turn out to possess just the kind of predictive or explanatory power that can itself be explained only by reason of their bridging the otherwise insuperable gulf between pure and applied mathematics, or work in advanced theoretical physics and work under 'real-world' (at present attainable) laboratory conditions. At any rate, the anti-realist will have a much harder job in accounting for the manifold precise and proven correlations that exist across and between these realms of enquiry.

So there is reason to doubt Pickering's idea that any answer to problems like that of deciding between Kantian (intuition-based) and formalist philosophies of mathematics can only come about through a process of multi-factor 'mangling' where choices are dictated by pragmatic trade-offs. Certainly he is able to cite passages from Hamilton, early and late, which show him shifting from one to the other position and manifesting all the signs of 'resistance and accommodation'. Thus the Kantian Hamilton forthrightly declares his intent 'to connect . . . calculation with geometry, through some extension to a space of three dimensions'. Later on he has moved so far from this 'old metaphysics' as to view it as 'little supported by scientific authority' and as scarcely able to withstand the formalist challenge. 'Though not unused to calculation,' Hamilton writes,

> I may have habitually attended too little to the *symbolical* character of Algebra, as a Language, or organized system of signs: and too much (in proportion) to what I have been accustomed to consider its scientific character, as a doctrine analogous to Geometry, through the Kantian parallelism between the *intuitions* of time and space. (cited by Pickering, p. 454)

These passages might seem a perfect example of what Pickering has in mind, that is to say, an instance of 'pure' mathematical reasoning forced up against a strictly aporetic (non-negotiable) conflict of aims and priorities, and hence obliged to steer a compromise course through the 'mangle' of opposing views. Such is indeed Pickering's point: that in developing his theory of quaternions Hamilton was 'flying by the seat of his pants . . . struggling through dialectics of resistance and accommodation, reacting as best he could to the exigencies of technical practice, without much regard to or help from any *a priori* intuitions of the inner meanings of the symbols he was manipulating' (Pickering, p. 455).

No doubt this is an apt enough description of much that goes on in the original context of discovery, that is to say, in the often obscure (or heavily overdetermined) processes of thought whereby mathematicians, physicists and others seek a way beyond some particular problem thrown up in the course of enquiry. However, it can have no possible bearing on issues raised in the context of justification or with regard to the various pertinent criteria – of proof, consistency, explanatory power, predictive scope and so forth – which effectively decide what shall count as an advance (or a viable solution to outstanding problems) in the discipline concerned. For at this stage it cannot be simply a matter of the 'technical culture' (as Pickering describes it) operating as a kind of 'mangle' where everything – theories, hypotheses, material practices, metaphysical commitments, formalized procedures, etc. – is fed in to produce the kind of hybrid resultant that offers least resistance across the board. After all, the entire development of modern mathematics, theoretical physics and other allied disciplines (philosophical logic among them) can be seen as caught up in the same tension that characterized Hamilton's thought. It is expressed most clearly in Gaston Bachelard's account of the various 'epistemological breaks' that were necessary in order for science to relinquish certain deeply held commonsense assumptions.[30] Bachelard's interest is in just those periods of Kuhnian 'revolutionary' science when no consensus existed, and when thinking went beyond the presumed self-evidence of intuitive or epistemologically grounded knowledge. Such was, for instance, the phase that occurred between the development of non-Euclidean geometries by Riemann, Lobachevski and others and the emergence of relativity theory. These developments were – and still remain – strongly counter-intuitive in the sense that they belong to a realm of thought where (in Kantian terms) there is no question of obtaining synthetic a priori knowledge, or of bringing intuitions under adequate concepts.[31] They are none the less paradigmatic instances of that passage beyond the natural attitude – or the witness of 'naïve' sense certainty – which for Bachelard constitutes the hallmark of scientific progress.

What chiefly distinguishes Bachelard's from Kuhn's approach (as likewise from 'strong' sociologists like Bloor and cultural-constructivists like Pickering) is his care to delineate the specific constraints – the intra-disciplinary methods, procedures, conceptual frameworks and so forth – that alone make it possible to speak of 'progress' in this or that particular field. For Pickering, conversely, 'scientific culture . . . appears as a wild kind of machine built from radically heterogeneous parts, a supercyborg, harnessing material and disciplinary agency in material and human performances, some of which lead out into the world of representations, of facts and theories' (Pickering, p. 446). Once again, this may be true in the context of discovery where – allowing for some degree of rhetorical licence – Pickering's description can claim support from a good deal of recent, highly detailed research in the history of various scientific disciplines.[32] However, it is completely wide of the mark as regards the context of justification. For here one has the task of explaining – and not just explaining away – why some discoveries can now be seen as having brought about a genuine advance in scientific knowledge while others are best understood in terms of cultural bias, disguised social interest, metaphysical preconceptions, partisan interests or fixed 'commonsense' beliefs. The trouble with the strong sociological approach is that it leaves no room for these basic distinctions, and thus creates a present-day equivalent to Hegel's metaphysical 'night in which all cows are black'. That is to say, it makes a virtue of extending that approach to *all* kinds of theory, truth-claim, method, research programme and so forth, without prejudice as to their status or validity when judged by present-day scientific lights. Thus for Bloor and other advocates of the Strong Programme it is proper – only fair – that this principle of symmetry should apply in every case and hence that no distinction should be drawn between the contexts of discovery and justification. Nor is there any appeal left open to notions such as 'truth', 'reason' or 'progress', notions whose role in the discourse of mainstream science (and philosophy of science) is merely to reinforce existing structures of knowledge, authority and power. From which it must appear – if one follows this line to its Feyerarbendian conclusion – that astrology should be ranked with astronomy, voodoo magic with medical science, and phlogiston with oxygen as a chief component in the chemical process of combustion.[33]

III

As I have said, Pickering is keen to mark his distance from the more extreme forms of cultural-relativist or strong ('monocausal') social-determinist doctrine. However, his favoured metaphor of the 'mangle' is one that allows him little room for explaining just how – by what possible criteria – one could justify a choice (on scientifically valid grounds) between any of the above candidate items. In particular, his approach leaves it a mystery why mathematics should have played such a crucial role in the development of the modern physical sciences from Galileo and Newton to Planck, Einstein, Bohr and just about every current field of advanced research. Galileo's famous dictum – that 'the book of nature is written in the language of mathematics' – is unlikely to carry much weight nowadays with those of of a cultural-relativist or strong sociological persuasion. Nevertheless it is a truth borne out across such a range of fields and disciplines (from quantum theory to molecular biology, astrophysics and aerodynanmics) that the burden of proof must rest very squarely with those who espouse an anti-realist position of the sort that Pickering takes pretty much for granted.

The situation here is directly analogous to that which obtains in the case of scientific 'unobservables', or postulated objects – like the term *electron* when first introduced – which play a well-defined role in various strongly supported theories but whose existence (or ontological status) is a matter of dispute. One way out of this dilemma was to adopt the instrumentalist line according to which theoretical terms should be treated as just a convenient shorthand description for complexes of observational data.[34] Yet, as David Papineau remarks, '[i]t is scarcely credible to suppose that scientists who talk about "electrons" are in fact talking about the behaviour of oil drops, tracks in cloud chambers, and so on, and not about the small negatively charged objects which orbit the nuclei of atoms.'[35] Still, philosophers found it hard to credit the latter (realist) interpretation, given that 'observable circumstances' provided the only 'authorized ground' for applying such terms, and that nothing else could could possibly justify the scientists (or philosophers) in 'making some further insecure reference to invisible entities'.[36]

However, Papineau thinks this a false dilemma and one which found at least the outline of an adequate solution in Ramsey's (1931) theory of reference as applied to scientific unobservables.[37] On this account, in Papineau's words,

> 'electron' does not have its meaning fixed just by association with the observable symptoms of electron behaviour. It also gets its meaning from its role in a theory which postulates the existence of small particles which orbit atomic nuclei and are responsible for those observable symptoms. Ramsey showed how statements about electrons can be read as existentially quantified statements, which say that there exist particles which are small, negatively charged, orbit atomic nuclei, have certain observable symptoms, and so on.[38]

This theory is clearly capable of extrapolation to subsequent developments in fields – such as particle physics – where progress very often takes the form of positing ever more recondite 'objects' or 'entities' whose existence may at first be conjectured on purely theoretical grounds, and only later borne out by the advent of new, more powerful observational techniques.[39] At any rate, Ramsey's approach to this question is enough to cast doubt on those forms of (wholesale or qualified) anti-realist doctrine, van Fraassen's 'constructive empiricism' among them, which rest their case very largely on arguments from the crucial but problematic role within science of such unobservable entities.[40] And it also goes a long way towards explaining – contra Kuhn and the strong sociologists – why the 'theory-ladenness' of scientific terms (observation statements included) is no good reason to think such terms referentially opaque or lacking determinate content. For if the argument doesn't succeed with respect to unobservable entities then it (or some version of it) can scarcely be convincing with respect to those observation statements whose evidence – on van Fraassen's account – is our sole reason for supposing such entities to exist.

Papineau's chief point in all this is to defend a realist philosophy of mathematics which construes numbers, sets, classes, etc., as possessing the same ontological status as 'electron' in the discourse of early twentieth-century physics. Thus where Ramsey showed that 'talk about scientific unobservables derives from our ability to make existential claims about objects which are not immediately accessible,' so Papineau argues that 'this same ability makes it possible for mathematical claims to answer to proof-transcendent states of affairs.'[41] He thus takes a position squarely opposed to those varieties of anti-realist, instrumentalist or constructive-empiricist approach which deny that there could *ever in principle* exist any truth – whether of mathematics or the physical sciences – that transcended (i.e., held good quite apart from) our current best methods of proof or verification. This argument has received its best-known statement in the writings of Michael Dummett, where it is applied to a range of issues, from philosophical logic to the interpretation of natural languages, historical discourse and mathematical proof procedures.[42] The instance of mathematics is crucial for Dummett since it brings out most clearly the bearing of this debate on the status of truth-claims in general and scientific truth-claims in particular. That is to say, both thinkers (Dummett and Papineau) acknowledge the centrality of mathematics in every branch of the physical sciences where it has played an indispensable role in the process of theoretical, predictive and explanatory reasoning. For Dummett, this chiefly goes to show that since anti-realism is the best (indeed the only viable) philosophy of mathematics, therefore the same must necessarily apply to those various other fields and disciplines. For Papineau, conversely, it is the manifest implausibility of mathematical anti-realism that can both be brought out by comparison with arguments (such as Ramsey's) in philosophy of science, and read back into that wider context as a strong justification for scientific realism.

As I have said, the main point at issue between them is that of proof-transcendent (or verification-transcendent) truths. Dummett sees no way that such a claim could make sense given that truth *just is*, for all that we can possibly know, whatever counts as such according to our current best methods of proof or verification. This applies just as much to mathematical thorems as to truth-claims in the physical sciences, in the context of historical enquiry or in the business of assigning sense (and reference) to forms of natural-language utterance. Thus the most original aspect of Dummett's work is his way of combining two supposedly antagonistic lines of thought: a Fregean approach via the logical analysis of concepts or judgements and the Wittgensteinian context principle according to which those concepts or judgements are meaningful only by virtue of their role in some existing 'language game' or cultural 'form of life'. For if indeed it is the case – as Frege argued – that a word has significance only in the context of a sentence, then already this points towards Wittgenstein's cardinal precept of meaning as use.[43] And from here, once again, it is no great distance to the anti-realist conclusion that truth-values (like assignments of sense and reference) can be established only on the prior basis of some communal language game – or set of agreed verification procedures – which decide what shall count as a valid statement in this or that particular context.

For Dummett, therefore, the question whether numbers or mathematical entities actually 'exist' is an ill-formed question which merely betrays the influence of a lingering metaphysical prejudice. Thus we may not ordinarily want to say that 'there is such a number as 28' while still finding a use for such informative statements as 'there is a perfect number between 10 and 30 and that number is 28.' Even so it will seem odd – a pointless and somehow extravagant claim – to assert that the number 28 *exists* in some special (philosophically relevant) sense of the term. What Frege provides, on Dummett's account, is a means to discourage such empty metaphysical talk by simply translating the existence claim into a suitably contextualized redefinition which brings out the logical structure of judgements involved in mathematical reasoning. In which case any question of the type: do entities of this sort (e.g., unobservables, theoretical posits, numbers, sets or classes) *actually exist*? is a misconceived question that can best be recast as an enquiry into the conditions of meaningful expression – or of warranted assertability – which obtain for each instance of the kind. This theory is then extended by Dummett to include any truth-claim or any form of existential commitment whose criteria, on this view, can only be a matter of their functional dependence on the methods and procedures involved in our standard (communally sanctioned) linguistic practices or proof procedures. Thus anti-realism holds just as much for statements of historical fact or scientific warrant as for statements of mathematical truth. In each case there is nothing that could possibly count as a reason for supposing such truths to 'exist' – to possess some independent reality – outside or beyond our current best methods for ascertaining them.

For Papineau likewise this question of the status of mathematical 'entities' (numbers, theorems, proofs, constructions, etc.) is central to the whole issue between realists and anti-realists. That is to say, it offers a particularly striking instance of the way that anti-realist doctrines lead on to a Wittgensteinian conception of meaning as use which affords no account of validity or truth save that provided by the standard appeal to communal practices, language games, or forms of life. Thus, in Wittgenstein's view, there is ultimately no means of *justifying* the various mathematical procedures (addition, subtraction, multiplication and the like) beyond simply pointing to the rule that is followed in each case and remarking that this just *is* the way we do things when engaged in that particular practice.[44] Hence all the well-known difficulties about 'following a rule' that have lately received much attention from Wittgenstein's exegetes.[45] For if indeed it is the case that we can offer no more by way of justificatory argument – if mathematical judgements, as Papineau puts it, are 'fixed by the grounds which our practice authorizes as sufficient for their assertion' – then of course it is nonsensical to claim (with the realist) that there might be truths of mathematics for which as yet we possess no adequate proof or decision procedure. For to make this claim is necessarily to assume that there exist certain properties of numbers (additive relations, set-theoretical membership conditions and so forth) which determine their role in valid mathematical reasoning quite apart from what presently counts as such according to our current (maybe limited) grasp of their various powers and entailments.

On the contrary, the anti-realist argues: not only do we lack any grounds for maintaining the 'reality' (the proof-transcendent character) of mathematical truth but we are also quite devoid of conceptual resources for justifying those various accepted practices – such as asserting '2 + 2 = 4' – whose validity likewise has no better warrant than the following of communally sanctioned 'rules'. Moreover, this applies just as much to the kinds of basic logical premise (like excluded middle, non-contradiction, *modus ponens* or *modus tollens*) which might be taken – on the realist view – to define what it means to think logically quite aside from what we (or other communities) happen to do as a matter of customary practice. Here again the anti-realist flatly denies that there is or could be any such appeal beyond the fact of our doing things that way rather than in some other (to us 'illogical') way. What decides in such matters is the practice-based criterion of what counts for us as a valid instance of arithmetical addition, logical deduction, scientific reasoning 'to the best explanation' and so forth. For there is simply no further verification-transcendent 'reality' – whether of numbers, truth-preserving logical inference, or the way things stand in the world – that we could think of as somehow able to set us right if only we possessed a more adequate (maybe less parochial or culture-bound) set of criteria.

Papineau rejects this whole line of reasoning as inadequate to explain how advances come about – or how genuine discoveries are made – in mathematics and the physical sciences. On the Wittgensteinian (practice-based) view, as he puts it, 'the truth conditions of mathematical judgements cannot possibly transcend our grounds for asserting those judgements.'[46] On his own account, conversely, there is no making sense of an anti-realist doctrine which reduces 'truth' – in mathematics as elsewhere – to a matter of agreement with our current (presumptively best) ideas as to how such truth should be attained. For it is self-evident to the realist that there might always be truths which we are just not capable of grasping at present owing to some failure of logical grasp, some momentary or long-term lapse of judgement, some lack of information on the relevant topic, or again – as in cases of perceptual distortion – some failure to correct for the sorts of error which *could* be corrected through a causal-explanatory account of how those mistakes came about. In the case of mathematics, what makes this possible is the existence of numbers, theorems, set-theoretical constructions, etc., which possess an objective (*not* just a practice-based) validity and which cannot be reduced – so Papineau contends – to whatever we happen to accept at present as an adequate proof procedure or method of verification. In arithmetic, for instance,

> we have a theory which postulates the existence of objects with certain properties, namely just those properties which flow by logic from N =. We call these putative objects numbers. But the basis of our ability to make claims about numbers, namely our power of existential generalization, is independent of any further abilities we may have to prove such claims.[47]

Of course, he concedes, there is a certain sense – an uncontroversial and not very interesting sense – in which there must exist some 'discursive practice' (some Wittgensteinian language game or life-form) where 'N =' plays a communally recognized role. However,

> the *existence* of this practice does not *justify* N =, nor the arithmetical claims which follow from it. For, as we have seen, N = is not analytic, but a synthetic claim. which inflates our ontology by postulating entities we are not otherwise committed to. As such it cannot be justified just on the grounds that it is part of an established discursive practice. Analytic truths are justified by facts about linguistic usage. But synthetic claims require some other warrant.[48]

Such claims are justified, Papineau believes, partly on condition that mathematical entities (numbers, theorems, proof procedures, etc.) are subject to the same criteria that apply in the case of unobservable objects or postulates in a well-defined scientific theory. Thus 'mathematical terminology can be introduced, *à la* Ramsey, by existential quantification into theoretical contexts', which in turn gives grounds for supposing 'that mathematical discourse rests on no special vocabulary, but simply on the existential quantifier we use in general discourse'.[49] In other words: mathematical concepts and expressions have truth-values precisely (and only) in so far they refer to something beyond the internal 'rules' of this or that conventional language game.

Anti-realists may respond by claiming that numbers can be said to 'exist' merely in the sense that their existence is a matter of stipulative warrant, that is, an analytic truth guaranteed by such features as set-theoretical equivalence or the fact of equinumerosity. However, this leaves them with a twofold task of explanation. On the one hand there is the problem of accounting for advances in mathematical knowledge, cases – such as Hamilton's discovery of quaternions – which cannot be explained in anti-realist terms (i.e., on any theory that denies the existence of verification-transcendent truths) since they actually change what shall henceforth count as an adequate proof procedure. On the other there is the problem endemic to all forms of anti-realism, namely that of explaining how any such 'discourse' (mathematical, scientific, historical or whatever) could possibly gain purchase on a world – a discrete object domain – that was not just a construct out of its own language game or set of descriptive conventions. Thus the committed anti-realist in mathematics will have to offer some convincing account of 'what "existence" means for mathematical objects, and why it is different from fictional non-existence'.[50] That is, they will be placed in much the same position as those anti-realists with respect to scientific or historical truth-claims who must likewise have problems in distinguishing the various orders of factual, counterfactual (i.e., hypothetical), and fictive or other such non-referential discourse. For it is impossible to explain why mathematics should have played such a cardinal role in the advancement of scientific knowledge were it not the case that such 'entities' – and the methods and procedures for working with them – corresponded to something in the structure of physical reality.

Of course this argument will appear more obvious (or its denial more strongly counter-intuitive) when proposed in relation to applied rather than to pure mathematics. As concerns the latter there would seem at least a plausible case to be made for the idea that numbers, sets, classes, proof-theoretic constructions, etc., are abstract 'objects' derived by various kinds of purely formal operation and hence without reference outside the system wherein their functions or truth-values are defined. Nevertheless, as Papineau remarks, 'references to abstract mathematical objects, and in particular to the natural and real numbers, are also regularly made in the applied sciences, as when we say "the number of planets = 9" or "the distance-in-metres between two particles = 5.77".'[51] In such cases there is no clear distinction to be drawn between 'pure' and 'applied' mathematics since any realist commitment concerning the existence of planets or particles must entail a corresponding realist ontology as regards the pertinent numerical values or relations. Thus:

> An extension of this line of argument promises to vindicate not just mixed statements of applied mathematics . . . but also the axioms on which pure arithme-

> tic and analysis are based, such as that every number has a successor, or that every set of reals has a least upper bound. For these assumptions are presupposed in the mathematical calculations which we use to derive predictions from scientific theories, as when we add together the number of stars in different galaxies, say, or divide forces by masses, and so can be argued to be confirmed, along with the rest of such theories, when such predictions prove successful.[52]

Thus if 'pure' mathematics is reliable for truth across a wide range of 'practical' scientific fields then there can be no reason – anti-realist prejudice aside – for counting it a discipline whose axioms, theorems, or proof procedures are devoid of referential content. Or again, as Papineau puts it, '[i]f mathematical theories qualify as knowledge as part and parcel of scientific theories, then they will share this realist epistemological status with scientific theories.'[53]

IV

It is just this possibility that Pickering ignores in discussing Hamilton's mid-career conversion from a Kantian (intuition-based) philosophy of mathematics to a formalist approach supposedly devoid of all such realist commitments. Of course he might argue – as against this criticism – that the 'mangle of practice' is a metaphor designed precisely to emphasize the complex interplay of real-world practical constraints with the various concepts, mathematical formalisms, commonsense-intuitive or metaphysical commitments, etc., which all come to bear upon the process of scientific theory formation. These components can then be understood 'as positioned in fields of disciplinary agency much as machines are positioned in fields of material agency' (Pickering, p. 446). Clearly Pickering's 'much as' clause leaves some room for strategic manoeuvre with regard to the precise bearing of this simile and the extent to which the two kinds of 'agency' can really be viewed as analogous. However, the main problem with his metaphor is that it pushes the 'mangling' process to a point where all the relevant distinctions drop away – as between contexts of discovery and justification, human interests and material constraints, theoretical and other (observation-based or empirical) orders of enquiry – with the result that they all become mixed up in a pragmatist appeal to whatever emerges as good in the way of belief. In which case, at the end of this levelling process, they are all on a par with respect to their (notional) truth-content and must therefore be viewed as ultimately subject to the test of what counts – sociologically speaking – as a valid scientific 'practice' or a communally sanctioned techno-scientific 'form of life'. And from here (despite Pickering's express disclaimers) the way is clearly open to just those kinds of Wittgenstein-inspired cultural-relativist approach that he finds problematic when spelled out explicitly in the writings of theorists like Bloor.

All the same, there are passages in his essay where the metaphors take on a more precise content and a certain (albeit underdeveloped) potential for specific application. Thus: 'One can think of factual and theoretical knowledge in terms of representational chains passing through various levels of conceptual abstraction and multiplicity and terminating, in the world, on captures and framings of material agency' (Pickering, p. 446). The trouble with this, once again, is that Pickering's portmanteau phrase 'material agency' leaves it unclear just where or how any line could be drawn between causally efficacious real-world events or processes on the one hand, and on the other those practices of humanly contrived intervention and control that in some way affect what is there to be 'captured' and 'framed'. Roy Bhaskar makes this point more clearly when he distinguishes the two realms of 'intransitive' and 'transitive' phenomena, the one referring to an objective (mind-independent) ontological domain, the other to a field of complex – variously 'stratified' – interactions between knower and known or scientific practice and the objects/products of scientific practice.[54] This distinction is vital not only for avoiding deep-laid epistemological confusions of the kind that result when anti-realists of various colour (Wittgensteinians, cultural relativists or Rortian 'strong' redescriptivists) elect to view *everything* as a product of human agency, language games or 'forms of life'. It is also important in the context of advanced research – such as that carried out with high-energy particle accelerators or recombinant DNA techniques – where the object domain is not so much given as constructed (always within certain physically determined limits) by the application of new technologies. Any adequate account of such developments must therefore reckon with the detailed operations – and not just the general 'mangling' process – through which these novel entities emerge.

'Physics might be said to seek, among other things, somehow to describe the world, but what is mathematics for?'[55] Pickering's question is indeed hard to answer if one accepts his view of mathematical 'discovery' as a trade-off among various competing interests (disciplinary, metaphysical, applied-scientific, class-related or sociopolitical) whose outcome in any given case is overdetermined to the point of lacking any demonstrable reason or purpose. His own suggestion – taken up from Bruno Latour – is that the role of mathematical formalisms can best be captured by a range of metaphors that include 'joining, linking, association, and alignment . . . comparing mathematical structures to railway turntables, to crossroads and clover-leaf junctions, and to telephone exchanges' (Pickering, p. 422). Such metaphors are useful, he thinks, in weaning us away from any misplaced belief in the authority of scientific 'method' or any idea of somehow drawing a line – a sufficiently 'rigorous' line – between the contexts of discovery and justification. What they help us to see is the way that these linkages, like metaphors, 'serve as multipurpose translation devices, making connections among diverse cultural elements' (p. 422). In which case there can be no question of explaining Hamilton's mathematical 'advance' – or others like it

– in terms of a decisive contribution to knowledge both within and beyond its own specialized domain. On Pickering's account this distinction falls away to reveal how even the most abstract ('pure' or highly formalized) procedures of mathematical thought are subject to the kinds of motivating pressure – cultural, professional and so forth – which supposedly have no place in the context of justification. For if everything is a matter of cultural 'resistance and accommodation' then clearly there can exist no distinctive criteria (of truth, validity or progress) that could serve to demarcate the various domains of 'pure' and 'applied' knowledge. Thus it may well be, in Pickering's words, that 'the making of new associations – the construction of new telephone exchanges linking new kinds of subscribers – is nontrivial' (p. 422). But any 'discoveries' or conceptual 'advances' that result from this process will be viewed as such only in virtue of their offering some short-term pragmatic adjustment to the various pressures and counter-pressures that bear upon the context of discovery.

At this point the realist would surely counter by remarking that mathematics can indeed give us knowledge of the world (and hence provide a motive for pursuing such enquiries) since there exists a whole range of striking correlations between numbers, ratios, set-theoretic constructions, etc., and the structures of physical reality. Such, after all, is the chief claim for mathematics as a paradigm instance of 'pure' *and* 'applied' knowledge: that mathematical discoveries have so often provided a powerful means of describing, explaining and (crucially) predicting events in the macro- and micro-physical domains. Thus – to take perhaps the best-known example – the anti-realist will be hard put to explain how Newton's law of gravitation (that the force exerted by one body on another is proportional to their masses and varies inversely with the square of the distance between them) could possibly hold good were it not for the fact that this law captures a relevant truth about the world in exact mathematical terms. In Hilary Putnam's concise rendition,

> Newton's law . . . asserts that there is a force f_{ab} exerted by any body a on any other body b. The direction of the force f_{ab} is towards a, and its magnitude F is given by:
>
> $$F = \frac{gM_aM_b}{d^2}$$
>
> where g is a universal constant, M_a is the mass of a, M_b is the mass of b, and d is the distance which separates a and b.[56]

Putnam's chief purpose in citing this example is to argue the case for a realist construal of those various relatively 'abstract' mathematical entities (numbers, ratios, formalisms, etc.) which have played an important role in advancing our knowledge of the physical world. More specifically, he takes them as providing strong counter-evidence to the nominalist (or instrumentalist) claim that these are merely convenient fictions which correspond to nothing in the nature or the structure of that world. Thus '[t]he law presupposes, in the first place, the existence of forces, distances, and masses – not, perhaps, as real entities but as things that can somehow be measured by real numbers.'[57] Furthermore, it assumes a language rich enough to express a great number of derivative and corollary truths which likewise require a realist construal in order (quite simply) that their terms should have reference and not be reduced to just so many arbitrary or stipulative products of definition.

Putnam goes on to extend this argument – the necessity of quantification over abstract entities – to a range of other items including mathematical sets. His point in each case is that the nominalist can have no adequate grounds for adopting a position so contrary to the best methods of the physical sciences. In short: 'quantification over mathematical entities is indispensable for science, both formal and physical; therefore we should accept such quantification; but this commits us to accepting the existence of the mathematical entities in question.'[58] As should scarcely need saying, this book appeared in 1972, at a time when Putnam was still committed to defending a strong realist stance in various philosophical fields of enquiry. However, its arguments are by no means discredited by his subsequent retreat to a doctrine of so-called 'internal realism' which meets the nominalist (or the Rorty-style pragmatist) on just about every major point.[59] For it is a truth borne out by the entire history of modern (post-Galilean) science that mathematical advances have always gone along – not of course *pari passu* but in complex patterns of mutual reinforcement and exchange – with advances in our knowledge of the physical world.

This point is not lost upon Pickering despite his avowed commitment to a theory of paradigm change which disallows any notion of objective truth, any claim that there exists a distinctive logic of scientific enquiry or any reference to a real-world (non-culturally-emergent) object domain. Thus, if 'Hamilton's achievement in constructing quaternions is of considerable historical interest,' then this has to do with its decisive impact on later scientific developments. More specifically: 'It marked a turning point in the development of mathematics . . . the introduction of noncommuting quantities into the subject matter of the field as well as an exemplary set of new entities and operations, the quaternion system, that mutated over time into the vector analysis central to modern physics' (Pickering, p. 423). In that case, however, there is a question that cannot be answered on his own theory of how changes come about through the 'mangle' of practice. For this metaphor does nothing to explain why it is that certain changes give rise to all manner of profound or far-reaching consequences, while others exert a more limited influence in some specialized field of enquiry, and others again prove largely or wholly devoid of substantive implications. That Hamilton's discovery possesses such great 'historical interest' is a matter of its having opened the way for developments in theoretical physics – in quantum mechanics especially – which would otherwise lack an adequate conceptual grounding.[60] These in turn have produced a wide range of 'applied' scientific advances – from transistors to lasers and nanotechnology – as well as explaining various phenomena (such as black-body radiation) that had hitherto defied the best efforts of physical theory.

Of course there are deep-laid metaphysical problems about the interpretation of quantum mechanics which continue to exercise philosophers of science (along with theoretical physicists) and which can scarcely be ignored by anyone defending a realist approach to these issues.[61] Still there is no reason – foregone prejudice aside – to view quantum mechanics as a clinching counter-argument to any form of ontological realism with respect to entities in the sub-atomic or micro-physical domain. One response – the so-called 'hidden variables' theory espoused by physicists from Einstein and de Broglie to David Bohm – seeks to avoid the conceptual anomalies of the standard (Copenhagen) interpretation by adopting a causal-realist ontology which can yet find room for such puzzling phenomena as that of superluminal (faster than light) communication between widely separated particles in a state of anti-polarized doublet spin.[62] Others again – J. S. Bell among them – have stated these problems in a highly paradoxical or counter-intuitive form while not denying that there might conceivably exist some (so far elusive) realist solution that respects the commitment to localized causal influence and the impossibility, as Einstein saw it, of 'spooky' simultaneous action at a distance.[63] However, the realist can still point out that these difficulties are epistemological (rather than ontological) since they have to do with the present limits of scientific knowledge and theory, rather than with some ultimate mystery in the nature of quantum phenomena. Thus the case is much the same as with those other once or still unobservable posits – from 'electrons' to 'mesons' and 'quarks' – whose ontological status has been (or is) bound up with the limits of available technology or continuing refinements in the means of observation. So it is that nineteenth-century scientists moved from an instrumentalist to a realist view of those various putative 'elementary' particles whose existence was at first a matter of conjecture – of inference to the best explanation – and then borne out by the advent of later, more advanced investigative techniques.[64]

In any case the realist will do much better to follow Michael Devitt's sensible advice and adopt a range of positions corresponding to the level or degree of confirmatory evidence for this or that candidate item. Thus 'Scientific Realism does take the posits of science pretty much at face value. However, it is committed to most of those posited by the theories we have good reason to believe: it is commited to most of science's "confident" posits.'[65] From this point of view it is only reasonable – given the conceptual anomalies of quantum mechanics – to maintain an attitude of qualified acceptance with regard to the sorts of 'posit' required by the currently more speculative branches of the physical sciences. On the other hand, the realist is perfectly entitled to argue that this attitude respects the great achievements of quantum mechanics – not least its enormous success in explaining such a range of otherwise inexplicable data – while keeping such scruples firmly on the side of our present limited understanding. Moreover, she can point with some justice to the kinds of manifest conceptual confusion – the various extravagant conjectures, hypotheses and notions of a wholly mind-dependent (as distinct from partially observer-relative) physical domain – that have figured prominently in recent debate.[66] What the realist seeks to maintain, in short, is a due sense of the crucial distinction between ontological and epistemological issues, that is to say, between questions of the general form: 'How can we be justified in positing the existence of a physical world whose properties may transcend our current best powers of description or explanation?' and questions of the general form: 'What can we presently claim to know respecting that world and its various constituent features?' In the case of quantum mechanics – given the ongoing controversy as regards its most basic metaphysical claims – the realist should perhaps 'see this as evidence that quantum theory is not to be trusted at this stage as a guide to reality'.[67] Here perhaps, Devitt ventures, we should follow Feyerabend and adopt a *pro tempore* instrumentalist outlook pending further developments.

At any rate realism is by no means compromised or forced to abandon its central thesis (i.e., the objective, verification-transcendent, mind-independent status of physical reality) by its sometimes adopting a qualified or agnostic stance *vis-à-vis* certain entities in the scientific object domain. For to take this line is simply to acknowledge the non-finality of scientific knowledge as we have it and the fact that future developments may well turn out to resolve current problems through some decisive technological or theoretical advance. Indeed it is just this standing possibility – evidenced by the entire record of scientific progress to date – which confirms the realist in his or her claim that truth cannot be internal (or relative) to this or that paradigm, discourse, conceptual scheme, verification procedure, etc. If this were the case then there could be no explaining how genuine discoveries come about, whether (as with Hamilton) in relatively abstract realms such as that of pure mathematics or again, through the kinds of 'applied' scientific research that deal more directly with the often recalcitrant nature of physical reality. Where Pickering's theory breaks down is in its failure to provide any adequate account of those *specific* 'resistances' that Hamilton met with in his passage from a Kantian (intuition-based) to a formalist conception of mathematical reasoning. Thus 'Hamilton continually arrived at technical results and then had to scratch around for interpretations of them – starting with the search for a three-dimensional geometric interpretation of his initial four-dimensional formulation of quaternions and ending up with biquaternions' (Pickering, p. 456). This may be an apt enough description when applied to the context of discovery, i.e., to those various journal entries and passages of anecdotal comment where Hamilton reflects on the processes of thought that led up to his discovery.

However, it conspicuously fails to explain (1) why these particular forms of resistance arose at specific points in Hamilton's thinking; (2) why he (and others) encountered the limits of an intuition-based reasoning *more geometrico* at this precise stage in the history of mathematics and physics; and (3) why Hamilton's conceptual advance – along with kindred developments such as non-Euclidean geometry – laid the ground for such a range of later scientific achievements.

Certainly there is not much guidance to be had from Pickering's idea of a 'temporally emergent mangling' process wherein all the various elements – conceptual, metaphysical, disciplinary, cultural, sociopolitical and so forth – undergo changes that can only be described by deploying so vague and all-purpose a metaphor. Nor is this idea much clarified by his recourse to phrases like 'the technical culture of science', designed to make allowance for the kinds of specificity – the intra-disciplinary problems and constraints – that bear upon mathematicians or physicists in particular fields of research. For here again the word 'culture' acts (as so often) to dissolve any real sense of knowledge-constitutive interests, disciplines, procedures, etc., and to carry the suggestion that science *just is* whatever counts as such by the lights of some given (past or present) cultural practice. Thus in the end there is not so much distance between Pickering and the advocates of a 'strong' sociological approach, despite his express reservations in that regard. What unites them is a habit of collapsing the distinction between context of discovery and context of justification, along with a kindred failure to explain how particular disciplinary interests – whether in mathematics or various branches of the pure and applied sciences – retain their specificity while also (at times) providing the basis for some new interdisciplinary advance. Imre Lakatos put the case well when he remarked (alluding to Kant's famous dictum) that 'philosophy of science without history of science is empty' while 'history of science without philosophy of science is blind.'[68] To which one might add that there is nothing more blinkered than sociology of science when overly fixated on its own methodological aims and concerns.

References

1 Wesley Salmon, *Four Decades of Scientific Explanation* (Minneapolis: University of Minnesota Press, 1989); also Salmon, *Scientific Explanation and the Causal Structure of the World* (Princeton, NJ: Princeton University Press, 1984).

2 See, for instance, D. M. Armstrong, *Universals and Scientific Realism*, 2 vols (Cambridge: Cambridge University Press, 1978) and *What Is a Law of Nature?* (Cambridge University Press, 1983); J. Aronson, R. Harré and E. Way, *Realism Rescued: how scientific progress is possible* (London: Duckworth, 1994); John W. Carroll, *Laws of Nature* (Cambridge: Cambridge University Press, 1994); Michael Devitt, *Realism and Truth* (Oxford: Blackwell, 1984); Jarrett Leplin (ed.), *Scientific Realism* (Berkeley and Los Angeles: University of California Press, 1984); Nicholas Rescher, *Scientific Real-*

ism: a critical reappraisal (Dordrecht: D. Reidel, 1987); Sean Sayers. *Reality and Reason: dialectic and the theory of knowledge* (Oxford: Blackwell, 1985); Peter J. Smith, *Realism and the Progress of Science* (Cambridge: Cambridge University Press, 1981); Michael Tooley, *Causation: a realist approach* (Oxford: Blackwell, 1988).

3 See Hilary Kornblith, *Inductive Inference and its Natural Ground: an essay in naturalistic epistemology* (Boston, MA: MIT Press, 1993) and Kornblith (ed.), *Naturalizing Epistemology* (MIT, 1985); also Rom Harré and E. H. Madden, *Causal Powers* (Oxford: Blackwell, 1975); Brian Skyrms, *Causal Necessity* (New Haven: Yale University Press, 1980); Ernest Sosa (ed.), *Causation and Conditionals* (London: Oxford University Press, 1975).

4 See entries under notes 1, 2 and 3 above; also Saul Kripke, *Naming and Necessity* (Oxford: Blackwell, 1980); Alvin Plantinga, *The Nature of Necessity* (Oxford: Clarendon Press, 1976); Hilary Putnam, *Philosophical Papers*, vols 1 and 2 (Cambridge: Cambridge University Press, 1975); Stephen Schwartz (ed.), *Naming, Necessity, and Natural Kinds* (Ithaca, NY: Cornell University Press, 1977); David Wiggins, *Sameness and Substance* (Oxford: Blackwell, 1980).

5 David-Hillel Ruben, *Explaining Explanation* (London: Routledge, 1990), p. 227.

6 D. M. Armstrong, *What is a Law of Nature?*

7 Nelson Goodman, *Fact, Fiction, and Forecast* (Cambridge, MA: Harvard University Press, 1955).

8 Armstrong, *What is a Law of Nature*, p. 57.

9 Ibid., p. 58.

10 Nelson Goodman, *Ways of Worldmaking* (Indianapolis: Bobbs-Merrill, 1978).

11 Armstrong, *What is a Law of Nature*, p. 58.

12 Saul Kripke, *Wittgenstein on Rules and Private Language* (Oxford: Blackwell, 1982).

13 See, for instance, David Bloor, *Wittgenstein: a social theory of knowledge* (New York: Columbia University Press, 1983) and Derek L. Phillips, *Wittgenstein and Scientific Knowledge: a sociological perspective* (London: Macmillan, 1977).

14 Armstrong, *What is a Law of Nature?*, pp. 58–9.

15 See W. V. Quine, *Ontological Relativity and Other Essays* (New York: Columbia University Press, 1969); Thomas S. Kuhn, *The Structure of Scientific Revolutions*, 2nd edn (Chicago: University of Chicago Press, 1970); Richard Rorty, *Objectivity, Relativism, and Truth* (Cambridge: Cambridge University Press, 1991).

16 Peter Muntz, *Our Knowledge of the Growth of Knowledge: Popper or Wittgenstein?* (London: Routledge and Kegan Paul, 1985), p. 119.

17 Ibid., p. 119.

18 David Bloor, *Knowledge and Social Imagery* (London: Routledge and Kegan Paul, 1976).

19 Andrew Pickering, 'Concepts and the Mangle of Practice: constructing quaternions', in Barbara Herrnstein Smith and Arkady Plotnitsky (eds), *Mathematics, Science, and Postclassical Theory* (*South Atlantic Quarterly*, 94/2, Spring 1994), pp. 417–65. All further references indicated by 'Pickering' and page number in the text. See also Pickering, *The Mangle of Practice: time, agency, and science* (Chicago: University of Chicago Press, 1995).

20 Bruno Latour, *We Have Never Been Modern* (Cambridge, MA: Harvard University Press, 1993).

21 See, for instance, Barry Barnes, *About Science* (Oxford: Blackwell, 1985); Peter L. Berger and Thomas Luckmann, *The Social Construction of Reality: a treatise on the sociology of knowledge* (Harmondsworth: Penguin, 1967); David Bloor, *Wittgenstein: a social theory of knowledge*; Harry Collins, *Changing Order: replication and induction in scientific practice* (Chicago: University of Chicago Press, 1985); Steve Fuller, *Social Epistemology* (Bloomington: Indiana University Press, 1988); Bruno Latour and Steve Woolgar, *Laboratory Life: the social construction of scientific facts* (London: Sage, 1979); Andrew Pickering, *Constructing Quarks: a sociological history of particle physics* (Edinburgh: Edinburgh University Press, 1984); Steven Shapin, 'History of science and its sociological reconstructions', *History of Science*, 20 (1982), pp. 157–210; Steve Woolgar (ed.) *Knowledge and Reflexivity: new frontiers in the sociology of knowledge* (London: Sage, 1988).

22 See Hans Reichenbach, *Experience and Prediction* (Chicago: University of Chicago Press, 1938).

23 Pickering, *Constructing Quarks*.

24 Pickering, 'Constructing Quaternions'. See also Hamilton, 'Quaternions' and 'Letter to Graves on Quaternions', in *The Mathematical Papers of Sir William Rowan Hamilton*, vol. 3, *Algebra* (Cambridge: Cambridge University Press, 1967), pp. 103–5 and 106–10.

25 Immanuel Kant, 'Transcendental Aesthetic', in *Critique of Pure Reason*, trans. Norman Kemp Smith (London: Macmillan, 1964), pp. 65–91.

26 See Thomas L. Hankins, *Sir William Rowan Hamilton* (Baltimore: Johns Hopkins University Press, 1980).

27 See, for instance, Louis Althusser, *'Philosophy and the Spontaneous Philosophy of the Scientists' and Other Essays*, ed. Gregory Elliott (London: Verso, 1990); Dominique Lecourt, *Marxism and Epistemology: Bachelard, Canguilhem and Foucault* (London: New Left Books, 1975); David-Hillel Ruben, *Marxism and Materialism: a study in Marxist theory of knowledge* (Brighton: Harvester, 1977); John Mepham and David-Hillel Ruben (eds), *Epistemology, Science, Ideology* (Atlantic Highlands: Humanities Press, 1979).

28 Bas C. van Fraassen, *The Scientific Image* (Oxford: Clarendon Press, 1980).

29 See, for instance, Michael Detlefson, *Proof and Knowledge in Mathematics* (London: Routledge, 1992); Philip Kitcher, *The Nature of Mathematical Knowledge* (Oxford: Clarendon Press, 1983); Hilary Putnam, *Mathematics, Matter and Method* (Cambridge: Cambridge University Press, 1979); William Asprey and Philip Kitcher (eds), *History and Philosophy of Modern Mathematics* (Minneapolis: University of Minnesota Press, 1988).

30 See, for instance, Gaston Bachelard, *La Formation de l'esprit scientifique* (Paris: Corti, 1938); *The Philosophy of No: a philosophy of the new scientific mind* (New York: Orion Press, 1968); *The New Scientific Spirit* (Boston: Beacon Press, 1984); also Lecourt, *Marxism and Epistemology* and Mary Tiles, *Bachelard: science and objectivity* (Cambridge: Cambridge University Press, 1984).

31 See Gordon G. Brittan, *Kant's Theory of Science* (Princeton, NJ: Princeton University Press, 1978) and Michael Friedman, *Kant and the Exact Sciences* (London: Harvester, 1992).

32 See entries under note 21 above.

33 See Paul K. Feyerabend, *Against Method: outline of an anarchistic theory of knowledge* (London: New Left Books, 1975).
34 See, for instance, Bas C. van Fraassen, *The Scientific Image*; Michael Gardner, 'Realism and instrumentalism in nineteenth-century atomism', *Philosophy of Science*, 46/1 (1978), pp. 1–34; Ian Hacking, *Representing and Intervening* (Cambridge: Cambridge University Press, 1983); Mary Jo Nye, *Molecular Reality* (London: MacDonald, 1972).
35 David Papineau, *Philosophical Naturalism* (Oxford: Blackwell, 1993), p. 190.
36 Ibid., p. 190.
37 Frank P. Ramsey, *Philosophical Papers*, ed. D. H. Mellor (Cambridge: Cambridge University Press, 1990).
38 Papineau, *Philosophical Naturalism*, p. 190.
39 See Hacking, *Representing and Intervening*.
40 Van Fraassen, *The Scientific Image*.
41 Papineau, *Philosophical Naturalism*, p. 190.
42 Michael Dummett, *Truth and Other Enigmas* (London: Duckworth, 1978); also *Frege: philosophy of language*, 2nd edn (Duckworth, 1981).
43 Gottlob Frege, 'On Sense and Reference', in *Translations from the Philosophical Writings of Gottlob Frege*, ed. Max Black and P. T. Geach (Oxford: Blackwell, 1952), pp. 56–78; Ludwig Wittgenstein, *Philosophical Investigations*, trans. G. E. M. Anscombe (Oxford: Blackwell, 1958).
44 Wittgenstein, *Philosophical Investigations*; also *On Certainty*, ed. G. E. M. Anscombe and G. H. von Wright, trans. D. Paul and G. E. M. Anscombe (Oxford: Blackwell, 1969); *Remarks on the Foundations of Mathematics*, trans. Anscombe (Oxford: Blackwell, 1956).
45 See, for instance, Saul Kripke, *Wittgenstein on Rules and Private Language*.
46 Papineau, *Philosophical Naturalism*, p. 190.
47 Ibid., pp. 190–1.
48 Ibid., p. 191.
49 Ibid.
50 Ibid.
51 Ibid., p. 192.
52 Ibid.
53 Ibid.
54 Roy Bhaskar, *Scientific Realism and Human Emancipation* (London: Verso, 1986); *Reclaiming Reality: a critical introduction to contemporary philosophy* (Verso, 1989).
55 Pickering, 'Constructing Quaternions', p. 422.
56 Hilary Putnam, *Philosophy of Logic* (London: George Allen and Unwin, 1972), p. 36.
57 Ibid., p. 37.
58 Ibid., p. 57.
59 See, for instance, Hilary Putnam, *The Many Faces of Realism* (La Salle: Open Court, 1987); *Realism With a Human Face* (Cambridge, MA: Harvard University Press, 1990); *Renewing Philosophy* (Harvard University Press, 1992).
60 For a lucid account of these developments, see Mark Kac and Stanislaw M. Ulam, *Mathematics and Logic* (Harmondsworth: Penguin, 1968).
61 See, for instance, Arthur Fine, *The Shaky Game: Einstein, realism, and quantum theory*

(Chicago: University of Chicago Press, 1986); Peter Gibbins, *Particles and Paradoxes: the limits of quantum logic* (Cambridge: Cambridge University Press, 1987); John Honner, *The Description of Nature: Niels Bohr and the philosophy of quantum physics* (Oxford: Clarendon Press, 1987); Joseph M. Jauch, *Are Quanta Real? a Galilean dialogue* (Bloomington: Indiana University Press, 1973); Henry Krips, *The Metaphysics of Quantum Theory* (Oxford: Clarendon Press, 1987); Michael Redhead, *Incompleteness, Nonlocality and Realism: a prolegomenon to the philosophy of quantum mechanics* (Clarendon Press, 1987).

62 A. Einstein, B. Podolsky and M. Rosen, 'Can quantum-mechanical description of reality be considered Complete?', *Physical Review*, ser. 2, 47 (1935), pp. 777–80. See also J. S. Bell, 'On the Einstein–Podolsky–Rosen Paradox', *Physics*, 1 (1964), pp. 195–200; S. V. Bhave, 'Separable hidden variables theory to explain the Einstein–Podolsky–Rosen Paradox', *British Journal for the Philosophy of Science*, 36 (1986), pp. 467–75; James T. Cushing, *Quantum Mechanics: historical contingency and the Copenhagen hegemony* (Chicago: University of Chicago Press, 1994); D. Bohm and B. J. Hiley, *The Undivided Universe: an ontological interpretation of quantum theory* (London: Routledge, 1993); J. R. Lucas and P. E. Hodgson, *Spacetime and Electromagnetism* (Oxford: Clarendon Press, 1990).

63 J. S. Bell, *Speakable and Unspeakable in Quantum Mechanics: collected papers on quantum philosophy* (Cambridge: Cambridge University Press, 1987).

64 See Gardner, 'Realism and Instrumentalism' and Hacking, *Representing and Intervening*.

65 Devitt, *Realism and Truth*, p. 122.

66 See, for instance, P. C. W. Davies, *Other Worlds* (London: Dent, 1980) and some of the interviews collected in Davies (ed.), *The Ghost in the Atom* (Cambridge: Cambridge University Press, 1986).

67 Devitt, *Realism and Truth*, p. 122.

68 Imre Lakatos, *Philosophical Papers*. Vol. 1: *The Methodology of Scientific Research Programmes*, ed. J. Worrall and G. Corrie (Cambridge: Cambridge University Press, 1978), p. 102.

9

But Will It Fly?
Aerodynamics as a Test Case for Anti-Realism

I

With a few notable exceptions (one of which I take as my point of departure here) aerodynamics has not figured prominently in the debate between realist and anti-realist philosophers of science.[1] It is a challenging case for several reasons, among them the fact that theory and practice were for a long period – some two millennia – so far out of step that thinkers (from Aristotle to Leonardo and Newton) came up with theories which, if true, would have ruled out the very possibility of heavier-than-air flight.[2] However, it is also of interest with regard to the claims put forward by cultural relativists and adherents to the 'strong' programme in sociology of knowledge. For on their account aerodynamics would present a particularly apt and compelling example of the way in which scientific 'facts', truth-claims, explanations, theoretical conjectures and so forth, are always produced under specific sociohistorical conditions and are thus subject to various cultural pressures that make for acceptance or rejection.[3] To the realist, conversely, such evidence may bear in all sorts of interesting ways upon the original context of discovery but can have no purchase on issues concerning the context of justification. That is to say, there is no argument – at least no scientifically or philosophically respectable argument – from the range of sociocultural factors that may be shown to have influenced the course of enquiry in this or that field to the idea that scientific truth *just is* the resultant of those various factors.[4]

I should perhaps lay my cards on the table right now and say that I shall here be defending the latter (i.e., the realist and anti-conventionalist or non-cultural-relativist) view. However, it is not hard to see how the history of thinking about aerodynamics might be treated to a full-scale applied 'deconstruction' in the strong sociological mode. Such an argument would aim at revealing its reliance on various metaphysical preconceptions at different stages in its history; also (more recently) its responsiveness to pressures and incentives – social, political, cultural, commmercial, military – which have acted upon it in ways ignored by any strictly 'internalist' approach. Hence the anti-realist's claim: that knowledge is socially constructed in the sense of permitting no appeal to criteria of truth (of method, validity, experimental warrant, theoretical consistency, etc.) beyond those sanctioned by some authorized research community or some given range of knowledge-constitutive interests and values. For the realist, conversely, it is nothing short of self-evident that aerodynamics has exhibited a growth of knowledge – albeit by no means a steady or smoothly progressive growth – that is manifest in its various accomplishments to date, from airfoil theory to the design and construction of flying machines with ever more advanced performance characteristics. We should therefore apply the basic principle of inference to the best explanation and conclude that there exist certain laws of nature – those investigated by the science of aerodynamics – whose valid formulation and putting into practice is precisely what explains why aircraft should fly in accordance with just those laws. To which the anti-realist (or strong sociologist) will predictably respond that this begs all the pertinent questions concerning what has counted as genuine or worthwhile knowledge from one context of enquiry to the next. For if indeed it is the case, following Kuhn, that scientific observations are always 'theory-laden' and theories always 'underdetermined' by the best evidence to hand, then there is just no room for any talk of trans-paradigm progress or for ideas of a real-world (belief-independent) object domain where laws of nature would somehow hold whatever our current state of knowledge concerning them.[5]

Such at any rate is the argument advanced by those who would reject a 'weaker' version of the sociological claim, that is to say, an approach that conceded the relevance of cultural and sociohistorical factors but which still found adequate reason to distinguish between context of discovery and context of justification. I shall argue that this latter approach gets it right – 'right' by something other (and more) than just our own present-day techno-scientifico-cultural lights – and that its truth is borne out in striking fashion by the history of pure and applied aerodynamic research. That history has been detailed with wonderful elegance and insight by N. R. Hanson, himself a keen pilot and student of aerodynamics as well as a philosopher-historian of science.[6] His main point – in brief – is that the early theorists were working on a range of conjoint premises and assumptions that were justified in their own terms, i.e., within the limits of currently available knowledge, but which consistently blocked the way to any further advance. Thus to Newton (in Book II of the *Principia*) it seemed self-evident that the only way in which a plane surface could generate lift was through the pressure exerted on its underside by the oncoming rush of air particles. In that case any lift thus produced would be subject to the well-established law whereby fluid resistance was calculated as the sine-square product of its angle of incidence. And indeed, as Hanson remarks, '[t]his particulate model of such an ideal fluid comports well with such a sine square law of resistance' (Hanson, p. 34). However, wh en applied to practical cases, air behaves *not* as an 'ideal fluid' (frictionless, inviscid, incompressible, irrotational, etc.) but in quite different ways which the early theorists – Newton among them – failed to take into account. It was just these non-ideal characteristics that enabled air to generate lift when flowing over a surface (optimally a well-shaped airfoil section) which exploited the negative differential pressure on its topside to greatly increase the available upward force. And yet, as Hanson well shows, once this effect was recognized then the mathematics fell into place as a powerful means of both grasping the relevant principles and modelling new, more efficient airfoils. The theory of propellers was likewise transformed when their blades were thought of not as simply twisting their way forward through the air on a corkscrew principle but rather as rotating wings whose performance could be greatly improved by developing a suitable airfoil section.

It is worth quoting Hanson at length to bring out the extent to which 'pure' theory (or one well-entrenched version of it) could get out of touch with the practical possibilities of heavier-than-air flight. 'Suppose,' he suggests,

> *à la* Leonardo and Newton, that the only thing that holds birds aloft is the pressure upon the underside of the wing; then a bird such as an albatross (moving at an airspeed of 15 miles an hour) – in order to achieve the value for F [i.e., for effective lift] compatible with these other known parameters – must angle its wings' incidence up to something like 60 degrees! A Boeing 707, to move as it does at take off or cruising speed, would also have to tilt its wings up to about 65 degrees – given the sine squared law! This would force calculations for the associated drag which would be totally out of the question. Birds and planes could not soar and glide as they do with wings tilted up like snow plough blades! The sine squared law requires that we'd have to minimize the angle of incidence in order to keep the drag factor within the bounds of conceivability. This would necessitate enlarging the value for the wing area such that the actual area of a 707's wing would be about the size of two football fields. Either that, or boost the value of V^2, by fantastic increases in propulsive power – the result of which will be fantastic increases in gross weight. Reflections such as these forced many to conclude that, by mechanical means alone, birds *couldn't* fly. (Hanson, p. 34)

On the other hand, Hanson is far from maintaining that mathematical formulae (or considerations of 'theory') have acted solely as a brake upon progress in the field of aerodynamics. Indeed, he has some eloquent pages on the signal contribution of theorists – from Daniel Bernoulli to F. W. Lanchester – whose ideas about the basic principles of flight were well ahead of anything currently achievable in practical terms. Bernoulli, in his 1738 treatise, was the first to offer a formula for the precise relation between pressure and kinetic energy when a fluid flows over a solid object. ('$H = p + {}^1/_2qv^2$ where H stands for the total pressure, p for the external pressure exerted on the fluid, q for the fluid's density, and v for the velocity of the fluid'; cited by Hanson, p. 365.) Although its implications were of course not fully understood at the time, this theorem none the less became the basis for all subsequent progress in fluid mechanics and aerodynamic research. More than that: it achieved the perspicuous formulation of a principle – a transfactually efficacious law of nature – that held good for all times and places and whose truth depended not at all on its having been discovered under certain (historically or culturally specifiable) conditions.

Hanson's point – argued in detail through this and other examples – is that aerodynamics could never (so to speak) have got off the ground as a genuine science were it not for the close if uneven pattern of reciprocal exchange between theory and practice. Thus Bernoulli's theorem pointed a way beyond the dead-end of Newton's sine-squared law and established the parameters for every significant advance in modern airfoil design. It explained, in short, 'how wings that are curved so that the air moves faster across the top than the bottom might provide lift. It would accomplish this, not by increasing the pressure below the shape, as Newton's analysis required, nor by compressing the air beneath the shape, as Leonardo conjectured, but by *decreasing* the pressure above the wing' (Hanson, p. 365). This not to say – far from it – that the history of aerodynamics can be viewed as a smooth progression where theoretical advances go hand-in-hand with the accumulation of practical experience on the part of engineers and designers. Indeed, Hanson makes a point of remarking how often theory has run ahead of practice (as with Bernoulli's ground-breaking formula) and also, conversely, how a theory like Newton's can constitute a long-term block to the achievement of practical goals. Thus it was not until 1809 that Sir George Cayley first managed to articulate the basic principle of heavier-than-air flight, observing that 'the whole problem is confined within these limits, viz., – To make a surface support a given weight by the application of power to the resistance of air' (cited by Hanson, p. 367). In this simple statement – as Hanson sees it – was potentially contained the entire future history not only of aerodynamic theory but of aviation as a practical science whose achievements could scarcely be guessed at by its early theorists. In other words, there is a certain unignorable messiness about the context of discovery – a lack of precise or punctual coordination between theory and practice – which cannot but strike the historian of science when he or she looks back over the record of achievement to date.

Hanson has some hair-raising anecdotes from his own experience as a pilot of high-speed aircraft which emphasize this simple fact: that discoveries very often come about through a process of trial and (potentially disastrous) error which cannot be avoided by any amount of pre-applied abstract theory. One such incident occurred when he fitted some snap-fasteners to the wing of his racing plane in order to secure close-fitting covers as protection against the elements. When he next took off he encountered severe vibration problems and very nearly lost control as the plane accelerated or decelerated past certain critical airspeeds. On reflection these velocities (240 and 300 knots) turned out to be related in a simple arithmetical series which suggested something other than a chance occurrence or a merely mechanical fault. He then began applying Mil's method of 'concomitant variation', i.e., the attempt to explain this anomaly by asking what additional factors were now present whose effect might be such as to disrupt or complicate the otherwise known regularities of aerodymanic behaviour.[7] The answer, of course, was the snap-fastenings which Hanson had attached to the forward section of the wing, thinking them small enough not to cause any significant change in the airflow pattern or the aircraft's handling characteristics.

However, he reasoned, there must be a more adequate (theoretical) explanation for just why the fastenings had created an effect so far out of proportion to their physical size. And the solution dawned when he consulted two authorities (Prandtl and Lanchester) who had produced pioneering work in the field of airfoil theory and design. Thus:

> The snaps I had affixed to the wings extended into the boundary layer. What I had done in effect was to spread the boundary layer and introduce turbulence in the interface. Prandtl's analysis showed that the thickness of the fluid layer affected by viscosity is inversely proportional to the square root of the speed. . . . At the right speed – or in my case the wrong speed – the boundary layer was lowered until the clips poked through it. The resulting turbulence caused the vibrations I felt. (Hanson, p. 387)

One could wish for no better, more vivid illustration of the difference between conditions obtaining in the context of discovery and those obtaining in the context of justification. The one kind of research operates mainly through ad hoc experiment, intuitive guesswork and hands-on practical experience. The other has to do with conceptual-explanatory hypotheses – in this case drawn from mathematics and theoretical physics – that are more abstract in character but which none the less possess a genuine power to explain some particular anomaly. Which is also to say that philosophy of science goes badly wrong if it *either* loses sight of that distinction *or* attempts to raise it to a high point of dogma, thus preventing any two-way traffic or reciprocal influence between those contexts. On the one hand, theories such as Newton's can constitute an obstacle to the kinds of advance that might otherwise have occurred through practical research unencumbered by the weight of received theoretical ideas. On the other – as with Bernoulli's theorem or Hanson's Prandtl-derived explanation of the snap-fastener effect – there is simply no dispensing with theory if one wishes to progress beyond the basic (pre-reflective) stage of taking each event as it comes and seeking no further reason for this or that novel occurrence.

Indeed the main thrust of Hanson's argument, here and elsewhere, is to cast doubt on that standard (logical-empiricist) distinction between 'context of discovery' and 'context of justification'.[8] His point is that major discoveries can – and very often do – come about through a process of thought which has its own distinctive kind of logic even though it fails (necessarily fails) to meet all the justificatory criteria laid down by exponents of the standard view. This 'logic of discovery' was an idea perfectly acceptable to many philosophers from Aristotle to Mill and Peirce, though ignored – or rejected – by those who assume that the only respectable manner of proceeding in philosophy of science is one that finds no place for matters of historical background, cultural context or motivating interest.[9] To some extent this attitude is understandable, given the sorts of claim advanced by 'strong' sociologists of knowledge, or even by those – Andrew Pickering among them – who adopt a more moderate cultural-relativist stance.[10] All the same, as Hanson remarks, 'Aristotle and Peirce thought they were doing something other than psychology, sociology, or history of discovery; they purported to be concerned with a *logic* of discovery; theirs was a philosophical inquiry about the formal structure of reasoning which constitutes scientific innovation' (p. 289). Peirce called this the *retroductive* method and saw it as resulting from the sense of some perceived anomaly – some as yet unexplained phenomenon – which set the scientist off in pursuit of a likewise novel but logically arrived-at solution.[11] His point (and Hanson's) is that this sort of case cannot be adequately dealt with on the standard hypothetico-deductive model that takes no account of such 'extraneous' factors or consigns them to a context of discovery assumed to be of interest only to historians (not philosophers) of science.

Thus, according to Hanson, 'there is far more scope for the exercise of reason and analysis within the "context of discovery" than most philosophers of science have been willing to grant' (p. 299). A good example is Newton's discovery of the inverse-square law of gravitation whose general form – if not its precise details – revealed itself to him through the absence at that time of any theory that would adequately explain a whole range of otherwise unrelated phenomena. While the HD (hypothetico-deductive) account 'pictures the scientist with a readymade theory and a store of initial conditions in hand, generating from these testable observation statements', the RD (retroductive) theory by contrast 'pictures him as possessing only the initial conditions and an upsetting anomaly, by reflections upon which he seeks an hypothesis . . . to explain the anomaly and to found a new theory' (p. 293). This latter approach has the great virtue of fitting many cases where the logic of enquiry has involved some perceived aberration – such as planetary motions at variance with the model of Newtonian celestial mechanics – which can be explained only by inferring the existence of other (as yet unobserved or unobservable) bodies.

In this particular case the astronomers came up with two such candidate bodies – Neptune and Vulcan – in order to close the worrisome gap between theory and observation opened up by the anomaly (the precessional aberration) of the planet Uranus. These arguments were 'formally one and the same', Hanson writes. That is, they both took the form:

> (1) Newtonian mechanics is true; (2) Newtonian mechanics requires planet *P* to move in exactly this manner, *x*, *y*, *z*, . . . ; (3) but *P* does not move *à la x*, *y*, *z*; (4) so either (a) there exists some as-yet-unobserved object, *o*, or (b) Newtonian mechanics is false. (5) (4b) contradicts (1), so (4a) is true – there exists some as-yet undetected body which will put everything right again between observation and theory. The variable '*o*' took the value 'Neptune' in the former case; it took the value 'Vulcan' in the latter. And these insertions constituted the zenith and the nadir of classical celestial mechanics – for Neptune *does* exist, while Vulcan does not. (Hanson, p. 31)

In other words, such retroductive reasoning involves just the kinds of procedure – and the same possibilities of error – that have characterized many signal episodes in the history of scientific thought. That they sometimes produce the wrong answer (or the wrong value for some given variable) is a risk exemplified in numerous cases such as Priestley's 'phlogiston', hypothesized to explain the process of combustion, or the 'aether' as a putative means of accounting for the propagation of light-waves. However, the theory of retroduction has a good claim to capture what is most important about such episodes and also what distinguishes the false starts from the genuinely productive (truth-yielding) lines of enquiry. That is, it views the context of discovery as properly subject to rational assessment, and *not* as just a kind of messy preliminary stage where reasons, interests and motives are all mixed up and where philosophers of science should wisely fear to tread. For it is then a short step – and one quickly taken by the strong sociologists of knowledge – to the argument that since HD accounts have little to tell about science as it is actually practised, therefore we had better ignore all talk of the 'context of justification' and focus our attention solely and exclusively on what goes on at that preliminary stage. Moreover, it is assumed that there is no place here for assessments of scientific truth or warrant. After all, such assessment would itself presuppose the existence of justificatory standards – context-transcendent criteria – which of course it is their main purpose to deny.

II

A great virtue of Hanson's arguments is that he holds out against this reductive drive to equate scientific thought with the context of its cultural emergence while readily conceding the frequent lack of fit between theory and practice, pure and applied research, or mathematical rigour and the adjustments required to make practical use of such findings. Aerodynamics again provides a useful test case since its history exhibits these tensions in particularly striking form. Thus theoreticians such as Euler and Bernoulli manged to bring great precision and refinement even to a 'highly fluid and unstructured' object domain. In their work, as Hanson describes it, '[t]he relationships between velocity and pressure, between boundary layers and turbulence, between flow direction and "lift", are beautifully mapped within the elegant algebra' (p. 282). However, this was achieved at the cost of rendering their results largely irrelevant to the needs of civil engineers, ship-designers, bridge-builders and (later) aeronautical pioneers who found them unusable in practice. For the theorists worked on the assumption of an ideal fluid – one 'utterly lacking in resistance and viscosity' – while in fact this condition is nowhere to be met with in the realm of applied fluid mechanics. As a result there developed an alternative science – hydraulics – which eschewed such theoretical refinement in favour of what can only appear 'a chaotic collection of recipes, hints, descriptions, and techniques – a plumber's tool box' (p. 283). Morever, it is still the case (as Hanson readily acknowledges) that there remain 'standard problems' of aerodynamics which cannot be resolved through the kinds of approach that take mathematics – or the limit-point ideal of a purely formalized theory – as the paradigm of adequate science. Thus:

> Many perplexities arise through the required use of partial, nonlinear differential equations of the second order in time – for which no *general* mathematical description can grind out past or future 'state-descriptions' of phenomena, comparable to what is encountered with the linear differential equations of the first order as encountered within Newtonian mechanics. (p. 283)

However – and this is the crucial point – there is simply no explaining the basic principles of flight (or the very possibility of constructing a heavier-than-air flying machine) except on the assumption that Bernoulli's theorem and other such mathematical formalisms must correspond to something in the nature of aerodynamic reality.

Of course this way of putting it will raise all the usual questions as to just what *kind* of 'correspondence' exists between theories (or statements) and the various items – from objects and events to laws of nature – that supposedly constitute their object domain and hence determine their truth-value in any given case. For the anti-realist such talk merely manifests a failure to grasp the elementary point: that we possess no criteria of 'truth' or 'reality' save those which play a role in our current theories or present best methods of verification.[12] Thus Richard Rorty would reject this whole argument out of hand as just another unfortunate example of what goes wrong when philosophers cling to bad old foundationalist (or epistemological) habits of thought.[13] And it would scarcely impress a constructive empiricist such as Bas van Fraassen, one who could admit (for all practical purposes) the existence of objects 'corresponding' to the data of empirical observation, but who would certainly draw the line at 'laws of nature' and the sorts of unobservable entity very often required for those laws to have explanatory content.[14] Then again, Hanson's example might seem to support the anti-realist case with regard to mathematics, suggesting as it does that the applied science of aerodynamics can have no use for such 'ideal' or 'elegant' formalisms as those worked out by theorists like Bernoulli. Thus the kinds of counter-factual hypothesis involved in such reasoning (*'if there were* an inviscid, nonresisting, and irrotational fluid, it would be observed to do the following things . . .') were such as to render wholly irrelevant 'the *facts* concerning what kinds of fluid there actually are' (Hanson, p. 283). From which the anti-realist can happily conclude that in aerodynamics – as in other scientific fields – all the evidence points to a wholesale collapse of the correspondence theory in whatever shape or form.

However, this reading would entirely mistake the purpose and the import of Hanson's argument. For in fact, as emerges on a closer reading, he is committed not only to the crucial relevance of theoretical work in aerodynamics but also to the claim that there *does* in fact exist a precise correspondence between theory and practice without which that science would indeed never have 'got off the ground'. This claim is developed most fully in his essay 'A Picture Theory of Meaning', where Hanson starts out from Wittgenstein's early (*Tractatus*) theory of the relation between logic, language and the world, but develops that theory in a novel direction. The argument is complex and I shall offer no more than a sketch of its main outlines. It involves the idea that various kinds of formal representation – e.g., maps, graphs, diagrams, flow-charts, linear and non-linear equations, multi-dimensional or multi-parametric formulas – are 'pictures' that relate to their respective object domains through a process of transformative modelling which, no matter how abstract, still preserves some aspect of verisimilitude. Thus, for instance, 'a curve of any slope on a data graph is representable (in principle) algebraically, by way of a Cartesian transformation' (p. 20). More specifically, in the case of aerodynamics, one can plot the various physical coordinates for a given wing configuration (span, chord, aspect ratio, upper- and lower-surface airfoil camber, angle of incidence, etc.), along with its performance data (airspeed, stalling speed, lift/drag ratio, point of transition from laminar to turbulent flow) by means of a parametric chart or graph which captures precisely those salient features.

To this extent, as Hanson puts it, '[t]he charts *are* the wings; they are everything aerodynamically significant about the wings' (p. 22). Perhaps they are not 'pictures' in anything like the literal sense of that term, the sense that has given Wittgenstein's commentators (not to mention the later Wittgenstein) so much trouble.[15] But they *are* 'structural representations' which can be both veridical (i.e., preserve the relevant data through various successively more abstract stages of formal or algebraic treatment) and capable of testing under physical (e.g., wind-tunnel) conditions when translated back into airfoil shape. Such charts, diagrams and graphs are informative, Hanson writes,

> because they share structures with the actual wings in question – dynamical structures, not geometrical structures. They provide a pattern, through which the

> multiform and chaotic manifestations of the original appear as correlated parameters. . . . Thus from knowledge of what is in fact the numerical value of x (the respective angle of attack) supposed to obtain with [airfoil section] NACA 2412, plus a knowledge of the airspeed, one can infer to a value for 2412's coefficient of lift, coefficient of drag, and its trailing edge turbulence. . . . To know the shape of the general parameter-configuration, and the value of one parameter, is to be positioned for inference to all the other parameters which describe 2412 . . . 2412 is completely delineated *via* the drawn parameter interactions. (Hanson, pp. 22-3)

So it would clearly be wrong to put Hanson down as an anti-realist either with respect to the laws of aerodynamics or in regard to those various 'abstract' (algebraic) theories whose formal nature – or remoteness from the practical context of discovery – very often resulted in a long delay before their implications were fully understood. For it is still the case, as he remarks of Lanchester's pioneerings efforts in the field, that 'this transformational technique has become one of the algorithmic glories of contemporary aerodynamic theory' (p. 43). What is more, such successes could not have been achieved – and aviation history could scarcely have made a start – except on the basis of productive exchange (at times a mutual irritation) between theory and practice.

No doubt Hanson's work could be quarried by a strong sociologist of knowledge with a view to demolishing any such realist claim. This counter-argument would seize on his many striking examples of the way that the science of aerodynamics evolved through a process of largely uncoordinated fits and starts according to the pressures and incentives brought to bear by diverse competing interests. Nor it is hard to imagine how a more moderate anti-realist such as Pickering might put those examples to work in support of his cultural-emergence theory, that is, his idea of scientific paradigm shifts as somehow brought about through a multi-factorial 'mangling' process.[16] However – as I have argued – there is still the large problem of explaining just why some of these developments turned out to revolutionize their field, or to constitute a real theoretical or practical advance, while others have exerted no such claim on the attention of later scientists or historians and philosophers of science. For the strong sociologists this problem simply doesn't arise since any talk of scientific 'truth', 'discovery', 'progress', etc., must always be construed as relative to this or that socially legitimized discourse of knowledge. Pickering quite rightly queries this approach on account of its narrow (monocausal) insistence on the priority of social over other factors and – coupled with that – its regular deployment of a self-exempting clause whereby the social sciences enjoy unique explanatory power. Yet in his case also there is no real attempt to hold the line against cultural relativism by clearly distinguishing the various components (scientific, technological, social, professional, career-driven and so forth) which make up the context of discovery.

After all, as Pickering views it, '[e]verything within the multiple and heterogeneous culture of science is, in principle, at stake in practice'. And again:

> Trajectories of cultural transformation are determined in dialectics of resistance and accommodation played out in real-time encounters with temporally emergent agency, dialectics which occasionally arrive at temporary oases of rest in the achievement of captures and framings of agency and associations among multiple cultural extensions.[17]

I have cited this passage – I think a fair sample – as indicative of the problems that Pickering confronts when attempting to explain his theory of the 'mangle' in other than circular or redundant terms. After all, what could be the source of such 'resistance' if not that scientific theories on occasion come up against recalcitrant evidence which forces a more or less drastic revision of hitherto accepted ideas? Or again: what need for this 'accommodating' process if the only constraints upon paradigm change are those internal to a given techno-scientific 'culture' with endless possibilities for restoring equillibrium between its various component parts? For if the claim means anything, in scientific terms, then it must have to do with science's capacity for constructive self-criticism in response to such problems, rather than its power to make them vanish by adopting some compromise solution. Here, as so often in cultural-relativist writings, one detects the influence of Quine's 'Two Dogmas' and his claim that any theory can always be saved, no matter how strong the counter-evidence, by a process of redistributing predicates or making suitable adjustments elsewhere in the overall fabric of accredited belief.[18]

There is a similar question about Pickering's idea of 'temporally emergent agency' as that which somehow negotiates the passage between successive paradigms or across the various levels (theoretical, metaphysical, technological, socio-political, etc.) that constitute 'scientific culture'. For it is hard to make out just what might be involved in a 'real-time encounter' – a 'dialectical' encounter at that – where all these dimensions (and others besides) were simultaneously at stake, and where all were subject to a state of flux that allowed only for 'oases of rest' amid the trackless and shifting sands. This may be to press rather heavily on Pickering's local choice of illustrative metaphors. However, those metaphors bear the main burden of argument, here and elsewhere in his essay, very often (one suspects) for lack of any alternative conceptual formulation that would bring the problems more clearly into view. Thus what can be meant by those 'captures and framings of agency' (subjective or objective genitive?) that somehow apply to both the practice of scientific enquiry and the object domain towards which that practice is directed? Or how can it be thought that advances in knowledge come about through a process – a 'trajectory of cultural transformation' – that so completely blurs the line between that object domain and the realm of communal values, interests, research priorities and so forth? No doubt it is the case, as Pickering says, that 'one needs to think about the intentional structure of human agency in order to understand this process.'[19] But there is not much hope of improved understanding if this agency is relativized to cultural context (itself vaguely defined), and if 'plans and goals' cannot be specified except as 'emergent from existing culture and at stake in scientific practice', itself once again 'liable to mangling in dialectics of resistance and accommodation'.[20] What is left at the end of all this is a painful sense of tail-chasing argument and a total failure, on Pickering's part, to explain how science could ever have progressed beyond the stage of merely reflecting or confirming the accepted wisdom of its time.

These are not just isolated problems of style or phraseology. Rather they indicate a deep confusion – one rife among contemporary cultural theorists – with regard to the issue of scientific realism and its bearing on their own more favoured methods of enquiry. This confusion is most pronounced in the strong sociological doctrine which on principle suspends all questions of truth, validity, or scientific warrant and which places every theory – past and present – on an equal footing from its own investigative viewpoint.[21] One argument often put forward in this context is the standard anti-realist meta-induction to the effect that, since most of science hitherto has turned out false or been rendered obsolescent, therefore it is extremely probable that the same will apply to most (if not all) of our current best theories, explanatory 'laws', ontological posits and so forth. In its semantic form – as summarized by Putnam – the argument holds: 'just as no terms in the science of more than fifty (or whatever) years ago referred, so it will turn out [that] no term used now . . . refers.'[22] However, this clearly raises further issues with respect to the character and scope of referring expressions, the extent to which scientific terms (or theories) do in fact conserve some stability of reference across paradigm changes, and – most crucially – the realist case for drawing a firm (categorical) line between ontological and epistemological matters.[23]

Thus, taking the last point first, there is no good reason to accept the sceptical meta-induction if it argues fallaciously from the evidence of theory change concerning various scientific posits to the claim that there cannot exist any objects (any real-world, theory-independent, non-culturally-emergent objects) quite aside from our beliefs about them. That there might indeed be truths beyond human reach – truths that we were simply not capable of grasping on account of some inbuilt limit to our cognitive or intellectual powers – is thus no argument in support of anti-realism but, on the contrary, the strongest case for adopting a realist position. For clearly such truths can be verification-transcendent in the sense of not depending on anything that we (or some future community of enquirers) might happen to believe regarding them. To think otherwise – like Dummett and his fellow anti-realists – is simply to confuse ontology with epistemology.[24] Nor is this case much strengthened when they raise the question as to how, given the presumed non-finality of science as we have it, there could ever be grounds for according our best current theories a truth-value greater than that now attributed to previous (now obsolete) theories. After all, as Michael Devitt commonsensically remarks,

[o]ur observable physical ontology has expanded through the ages with the discovery of new species, new planets, etc., but very little has been abandoned; cases like those of witches and the planet Vulcan are rare. There is nothing in the history of theory-change to shake the view that we have, in a steady accumulative way, give or take a mistake or two, discovered more and more of the entities that exist ('absolutely') in the observable world.[25]

If this point has been lost upon many recent theorists, then the reason is not far to seek. First, it results from their view of semantics (questions of meaning and reference) as prior to epistemological issues, and therefore – on the standard descriptivist construal – as blocking the way to any possible realist ontology. Hence the idea that each and every theory change brings about a shift of referential scope (or ontological commitment) in any terms that apparently survive unchanged across and between theories. Thus the argument runs, in Devitt's formulation: '"*a*" refers only if the (weighted most of) descriptions commonly associated with it apply to one and only one object. But they don't. So "*a*" does not refer. So, by the schema, *x* does not exist.'[26] To which the realist can perhaps best respond – despite some of the wilder speculations of quantum theory – that the universe existed for aeons before there were humans around to observe it and will more than likely continue to do so long after they have vanished from the scene.

This is also to say that philosophers of science get things backwards (come up with preposterous arguments in the literal sense of that term) if they suppose that the ontological issue – that of realism versus anti-realism – cannot be settled except by appealing to semantic criteria. For it is surely the case, as Devitt remarks, that the physical sciences have advanced much further and are currently in much better shape than philosophical semantics or theories of meaning and reference. Moreover, it is far from clear that there exists a necessary (logical) connection between the thesis of semantic priority and the claims of anti-realism, ontological relativity, radical meaning variance and the like. Thus, for instance, Hartry Field has argued persuasively that a term such as 'mass' can undergo quite drastic shifts of theoretical usage – as between the discourses of Newtonian and Einsteinian physics – while retaining sufficient continuity of reference to enable meaningful comparisons (and judgements of scientific progress) from one such 'paradigm' to the next.[27] Field's most useful innovation is the idea of 'partial reference', construed as the capacity of terms like 'mass' (or 'atom', 'electron', 'gene', etc.) to pick out certain features of the physical domain whose properties are still subject to dispute but whose existence (as objects of referring expressions) is not for that reason open to doubt.

As Devitt puts it: 'Newton's term "mass" "partially refers" to each of two physical quantities posited by contemporary physics – relativistic mass (= total energy/c^2) and proper mass (= non-kinetic energy/c^2) – but does not determinately refer to either'.[28] However, this gives no reason to suppose – with

Kuhn, Quine or the strong sociologists – that such terms must therefore be exposed to all the vagaries (the 'theory-laden' shifts of meaning) that preclude any valid comparison betweeen alternate or successive paradigms. On the contrary, it is quite possible to explain – on this model of 'partial reference' – how Newton's account was superseded by Einstein's while retaining a certain (restricted but well-defined) range of applications and values. In which case 'we should accept that there is such a physical quantity as Newtonian mass . . . , for that quantity is either relativistic mass or proper mass (although there is no fact of the matter *which* it is).'[29] Any 'indeterminacy' that obtains in present-day (post-Einsteinian) allusions to Newton's mechanics is best accounted for precisely by means of the later theory's more extensive scope, its redefinition of inherited terms, and its power to reveal a significant duality of reference in the phrase 'Newtonian mass'. Thus 'Newton was not so much wrong in his ontology but he . . . did not make all the distinctions that ought to be made. Einstein's ontology differs from Newton's but it does not (always) eliminate Newton's.'[30] From which it follows – contra the anti-realists – first that meaning variance is no obstacle to explaining scientific progress, and second that ontological issues (questions of the order: what entities exist?) are logically prior to issues in the realm of epistemology, semantics or interpretation theory.

So the strong sociologists are by no means backing a safe horse when they take it for granted that realist arguments lack any adequate philosophical defence and base their approach on the standard (post-Kuhnian) topoi of theory change, radical meaning variance or paradigm incommensurability. For these doctrines are not only open to doubt for the kinds of technical reason advanced by philosophers like Field and Devitt. They are also, as I have argued, responsible for much confusion in the broader sphere of current debates about science and its presumed role as a legitimizing force *vis-à-vis* the structures of social authority and power. Thus it can fairly be claimed that the entire programme of Strong Sociology of Knowledge rests on the validity of certain very dubious premises, among them the semantic thesis (i.e., that issues of meaning and reference are prior to questions of ontology) and – following from this – the idea of scientific knowledge as a cultural or social construct whose truth-claims can always be relativized to the discourse of this or that interpretive community. What is so remarkable about this current relativist orthodoxy is the fact that it questions just about everything in the methods, procedures and truth-claims of the physical sciences while exempting its own methods, procedures and truth-claims from anything like so rigorous a process of critical self-examination. Thus it is taken as a simply unchallengeable truth – self-evident to all but benighted old-style realists – that science (or the dominant self-image of science) now enjoys such enormous cultural prestige on account of its success in having captured the high ground of socially authorized 'knowledge', rather than its having developed through stages of ever more refined and exacting physical enquiry. For this latter claim simply cannot make sense if indeed it is the case – as these commentators argue – that scientific terms (theories, hypotheses, observation statements) must always be construed relative to the norms and conventions of some existing 'language game' or cultural 'form of life'.

However, one can turn the argument around and ask what follows for the strong sociological programme if it can be shown to rest on less than self-evident (indeed very shaky) theoretical grounds. All the more so since they – the advocates of this programme – are left with the massive burden of proof which comes of their adopting a sceptical stance (or an attitude of principled nescience) with regard to the very idea that science might just occasionally get something right. It seems to me that an attentive reading of Hanson on aerodynamics should go a long way towards correcting this currently widespread overestimate of the social or conventional factors involved in judgements of scientific truth.[31]

References

1 Norwood Russell Hanson, 'A Picture Theory of Theory-Meaning' and 'The Theory of Flight', in *What I Do Not Believe, and Other Essays*, eds. Stephen Toulmin and Harry Woolf (Dordrecht: D. Reidel, 1971), pp. 4–49 and 333–90.

2 See Hanson, 'The Theory of Flight'; also Clive Hart, *The Prehistory of Flight* (Berkeley and Los Angeles: University of California Press, 1985); Richard von Mises, *Theory of Flight* (New York: McGraw-Hill, 1945).

3 See, for instance, Barry Barnes (ed.), *Sociology of Science* (Harmondsworth: Penguin, 1972); Barry Barnes and Steven Shapin (eds), *Natural Order* (London: Sage, 1979); David Bloor, *Knowledge and Social Imagery* (London: Routledge and Kegan Paul, 1976); Augustine Brannigan, *The Social Basis of Scientific Discoveries* (Cambridge: Cambridge University Press, 1981); Steve Fuller, *Social Epistemology* (Bloomington: Indiana University Press, 1988); Bruno Latour and Steve Woolgar, *Laboratory Life: the social construction of scientific facts* (London: Sage, 1979); Andrew Pickering, *Constructing Quarks: a sociological history of particle physics* (Edinburgh: Edinburgh University Press, 1984); Pickering, *The Mangle of Practice: time, agency, and science* (Chicago: University of Chicago Press, 1995); Steven Shapin, 'History of science and its sociological reconstructions', *History of Science*, 20 (1982), pp. 157–211; Shapin, *A Social History of Truth: civility and science in seventeenth-century England* (Chicago: University Chicago Press, 1994); Steve Woolgar (ed.), *Knowledge and Reflexivity: new frontiers in the sociology of knowledge* (London: Sage, 1988).

4 See, for instance, W. H. Newton-Smith, *The Rationality of Science* (London: Routledge and Kegan Paul, 1981).

5 Thomas S. Kuhn, *The Structure of Scientific Revolutions*, 2nd edn (Chicago: University of Chicago Press, 1970). For further discussion see also Gary Gutting (ed.), *Paradigms and Revolutions* (Notre Dame, IN: Notre Dame University Press, 1980); Ian Hacking (ed.), *Scientific Revolutions* (London: Oxford University Press, 1981); and John Krige, *Science, Revolution and Discontinuity* (Brighton: Harvester, 1980).

6 N. R. Hanson, *What I Do Not Believe*. All further references indicated by 'Hanson' and page-number in the text.

7 John Stuart Mill, *A System of Logic* (London: John W. Parker, 1843).

8 See Hans Reichenbach, *Experience and Prediction* (Chicago: University of Chicago Press, 1938); C. G. Hempel, *Aspects of Scientific Explanation* (New York: Macmillan, 1965) and *Fundamentals of Concept Formation in Empirical Science* (Chicago: University of Chicago Press, 1972); R. B. Braithwaite, *Scientific Explanation* (Cambridge: Cambridge University Press, 1953).

9 See also Peter Lipton, *Inference to the Best Explanation* (London: Routledge, 1993) and G. Harman, 'Inference to the best explanation', *Philosophical Review*, 74 (1965), pp. 88–95.

10 See Pickering, *The Mangle of Practice*.

11 Charles Sanders Peirce, *Reasoning and the Logic of Things*, ed. K. L. Ketner (Cambridge, MA: Harvard University Press, 1992); *Essays in Philosophy of Science* (New York: Bobbs-Merrill, 1957); and Peirce, *The Essential Writings* (New York: Dover, 1964).

12 See, for instance, Michael Dummett, *Truth and Other Enigmas* (London: Duckworth, 1978).

13 See especially Richard Rorty, *Consequences of Pragmatism* (Hemel Hempstead: Harvester, 1982) and *Objectivity, Relativism, and Truth* (Cambridge: Cambridge University Press, 1991).

14 Bas C. van Fraassen, *The Scientific Image* (Oxford: Clarendon Press, 1980) and *Laws and Symmetry* (Clarendon Press, 1989).

15 Ludwig Wittgenstein, *Tractatus Logico-Philosophicus*, trans. D. F. Pears and B. F. McGuiness (London: Routledge and Kegan Paul, 1961).

16 Pickering, *The Mangle of Practice*.

17 Andrew Pickering, 'Concepts and the Mangle of Practice: constructing quaternions', in Barbara Herrnstein Smith and Arkady Plotnitsky (eds.), *Mathematics, Science, and Postclassical Theory* (*South Atlantic Quarterly*, 94/2, Spring 1995), pp. 417–65; p. 447.

18 W. V. Quine, 'Two Dogmas of Empiricism', in *From a Logical Point of View*, 2nd edn (Cambridge, MA: Harvard University Press, 1961), pp. 20–46.

19 Pickering, 'Constructing Quaternions', p. 447.

20 Ibid., p. 447.

21 See entries under note 3 above.

22 Hilary Putnam, *Meaning and the Moral Sciences* (London: Routledge and Kegan Paul, 1978), p. 25.

23 See especially Hartry Field, 'Theory change and the indeterminacy of reference', *Journal of Philosophy*, 70 (1973), pp. 462–81; 'Conventionalism and instrumentalism in semantics', *Nous*, 9 (1975), pp. 375–405; 'Logic, meaning, and conceptual role', *Journal of Philosophy*, 74 (1977), pp. 379–409; also Arthur Fine, 'How to compare theories: reference and change', *Nous*, 9 (1975), pp. 17–32; Michael E. Levin, 'On theory-change and meaning-change', *Philosophy of Science*, 46 (1979), pp. 407–24; David Papineau, *Theory and Meaning* (London: Oxford University Press, 1979); Crispin Wright, *Realism, Meaning and Truth* (Oxford: Blackwell, 1987).

24 See Dummett, *Truth and Other Enigmas*; also Michael Luntley, *Language, Logic and Experience: the case for anti-realism* (London: Duckworth, 1988); N. Tennant, *Anti-Realism and Logic* (Oxford: Clarendon Press, 1987); Crispin Wright, *Realism, Meaning and Truth* (Oxford: Blackwell, 1987).

25 Michael Devitt, *Realism and Truth* (Oxford: Blackwell, 1984), p. 142.

26 Ibid.

27 See entries under note 23 above.

28 Devitt, *Realism and Truth*, p. 148.

29 Ibid., p. 148.

30 Ibid.

31 I should note that Hanson's work is itself very often – though in my view mistakenly – cited in support of anti-realist, social-constructivist, or cultural-relativist approaches to philosophy of science. At any rate there is a marked tension between those passages that may plausibly be thought to sustain such a reading and the kinds of argument I have drawn upon here. In this connection see, for instance, Hanson, *Patterns of Discovery: an inquiry into the conceptual foundations of science* (Cambridge: Cambridge University Press, 1958); *The Concept of the Positron: a philosophical analysis* (Cambridge: Cambridge University Press, 1963); *Observation and Explanation: a guide to the philosophy of science* (New York: Harper, 1971).

10

Why Strong Sociologists Abhor a Vacuum: Shapin and Schaffer on the Boyle/Hobbes Controversy

I

In what follows I propose to address some problems concerning the 'strong' sociology of knowledge and its bearing on the history and philosophy of science. I shall refer mainly to Shapin and Schaffer's classic study *Leviathan and the Air-Pump*, a text that is very often cited in support of the strong sociological programme.[1] For this is, one may readily agree, a work of great scholarly erudition and detailed historical research. Moreover it ducks none of the relevant philosophic and methodological issues, and argues its cultural-relativist case with impressive consistency and vigour. Nevertheless it is a deeply problematical work in ways that shed a useful (if negative) light on the entire research programme currently pursued by members of this burgeoning academic movement.[2] At any rate, its arguments deserve close attention partly on their merits as historical scholarship and partly on the grounds of their widespread influence among cultural theorists who have tended to endorse them with less than adequate critical scrutiny.

Their book focuses on the issue between Boyle and Hobbes concerning the status of experimental evidence – of 'facts', 'observations' and reliable 'witness' – in the natural sciences. For Boyle and the new experimentalists this was the only sure route to genuine scientific knowledge, that is to say, the kind of knowledge best obtained by careful observation under controlled laboratory conditions, and ideally through the repeated practice of such experiments by members of a like-minded expert community.[3] This also provided a model, so Boyle believed, for the proper conduct of social and political affairs, a model of rational consensus arrived at through free and open participant debate among those duly qualified to hold some opinion in the matter. Such questions were especially urgent at this time – the mid to late seventeenth century – since the English Civil War stood out in memory as a cautionary instance of the evils that resulted from a breakdown of consensus values and the splitting of society into various rival groups and factions. Thus to Boyle it seemed that the experimental method offered the best and safest (that is, the least dogmatic and divisive) procedure for attaining a fair measure of agreement in matters that were open to dispute.[4]

One such matter was the long-running quarrel between 'plenists' and 'vacuists', or those (like Hobbes) who held to the idea of nature as everywhere containing some element or portion of material substance, and those (like Boyle) who argued that a vacuum could exist not only in theory but as a condition physically achievable by devising a suitable apparatus. Hence the controversies that swirled around Boyle's famous air-pump, constructed as it was – with immense labour – to bear out the vacuist hypothesis and a number of related proposals concerning atmospheric pressure, the composition of air, the presence (or absence) of immaterial 'essences', and so forth. For Hobbes, conversely, this experimental programme failed in every aspect of its self-professed scientific, social and moral endeavour.[5] As science – or natural philosophy – it remained at the superficial level of mere observation, abjuring any interest in those underlying causes which a scientific theory ought properly to explain. Besides, according to Hobbes, the air-pump was at best an imperfect mechanism whose construction was prone to all manner of leakage or seepage, and which thus did nothing to support the vacuist theory. Moreover, there were other explanations – some of which Hobbes rehearsed at great length in point-for-point rebuttal of Boyle's claims – which could just as well account for the observed phenomena, even granting (*concesso non dato*) the latter's validity as 'scientific' evidence. In social and moral terms, likewise, the method seemed to Hobbes more a symptom of looming catastrophe than a procedure devised to avert such catastrophe through the use of disciplined experiment as a substitute for speculative theory, private judgement or the clash of metaphysical systems. Indeed, he argued, there was nothing so divisive – so likely to foment civil discord – as a method that confined competence in these matters to a small group of self-authorized expert individuals with the power (exercised through bodies like the nascent Royal Society) to determine what should count as valid scientific practice.

Behind all these various objections – ontological, scientific, social and moral – lay Hobbes's commitment to a vision of nature and society which absolutely required the plenum hypothesis in order to make good its claim. For of course it was Hobbes's radically materialist ontology – the great scandal of *Leviathan* – which led him to associate vacuist talk with confused (superstitious or scholastic) talk of spirits, immaterial essences, Aristotelian 'forms' and suchlike metaphysical arcana. In his view the only basis for a stable political order was the mechanist philosophy which barred all appeal to occult (non-physical) entities and which treated nature and society alike as subject to a uniform, exceptionless order of causal determination.[6] Thus his quarrel with Boyle was not so much about the programme of mechanism – to which both parties subscribed, albeit for different reasons – but more about the scope or explanatory power that any such theory should properly claim. For Boyle it was the greatest virtue of his method that it remained content with conducting experiments, producing observations, attempting to replicate those same experiments on as many occasions (and before as many witnesses) as possible, and thus securing widespread assent among those qualified to judge. Any further speculation – as to causes or laws – was a needless metaphysical extravagance that could only threaten the consensus achieved through this wise self-regulating ordinance. For Hobbes, on the other hand, natural philosophy was a useless (and worse than useless) endeavour if it gave up the quest for law-governed causal explanations that went beyond the manifest or merely observational domain. That is to say, its chief purpose was to provide a full and systematic account of those various mechanisms whose workings pervaded the physical, psychological and civic-institutional realms. This in turn required that nature and society alike be treated as as a plenum of material bodies interacting with each other at every level and leaving no space – no explanatory gap – for the intrusion of theories such as Boyle's vacuum hypothesis.

Thus the issue between them had as much to do with politics – with the problems of achieving a viable Restoration settlement – as with matters of ontology and epistemology. Where Hobbes pinned his faith to an absolutist conception of political power Boyle saw the way to renewed civil peace through a process of enquiry where differences of view could best be reconciled by experiment, observation and the will to abide by the rules of civilized discourse. Where Boyle so far as possible avoided polemics – dissenting only mildly from the gloss placed upon his work by religious thinkers such as the Cambridge Neoplatonist Henry More – Hobbes took on a whole range of opponents, among them the scholastic philosophers and religious sectaries of various persuasions. And where Boyle admitted some margin of doubt with regard to his experimental findings – even conceding Hobbes's point that a perfect vacuum might not be physically attainable owing to design problems or material defects – Hobbes could be satisfied with nothing less than absolute certainty respecting those claims that were to count as genuine scientific knowledge. In his view the vacuist hypothesis amounted to just the kind of empty speculative thinking – the conjuring up of metaphysical quiddities and essences – which had so disfigured philosophy and had lately wrought such murderous strife between civil and religious factions. Nor was this threat in any way diminished by Boyle's reasonableness in debate, his appeal to shared observation or his refusal to adopt a dogmatic line on the vacuist versus plenist controversy. (Thus Boyle: 'by "vacuum", I understand not a space, wherein there is no body at all, but such as is either altogether, or almost totally devoid of air'; cited in *LAP*, p. 46.) For to Hobbes the very idea of an *experimentum crucis* – a decisive test carried out with special apparatus under laboratory conditions in the presence of qualified observers – was just another sign of that sectarian spirit for which the only truths were those vouchsafed to some elect group of self-authorized expert witnesses.

Thus their two conceptions of science (or natural philosophy) were opposed point for point on every issue from ontology to epistemology and thence to the ethics and politics implied by this nascent experimental culture. In Hobbes's view its supposed great virtues – of tolerance, careful observation, cooperative enquiry, avoidance of dogmatic 'metaphysical' commitments – served merely to disguise its partisan and prejudicial character. Worst of all was its appeal to the *experience* of qualified observers, that is to say, their *beliefs* as tested and refined through *perceptual acquaintance* with the kinds of phenomena that Boyle and his associates sought to produce by their experiments with the air-pump. For Hobbes such knowledge was wholly unreliable, resting as it did on the dubious testimony of the senses, and subject to all kinds of error and distortion brought about by preconceived ideas. Moreover, it failed to reckon with the frequent misunderstandings and communicative breakdowns which resulted from the usage of technical terms – such as 'plenum', 'vacuum' or 'aether' – that assumed diverse (incommensurable) meanings in diverse contexts of debate. Hence the great evil, as Hobbes saw it, of a method which abandoned the sure path to knowledge – that of a priori demonstrative reasoning – and which thus opened the way to all manner of civil and intellectual discord. For what could be the value of experimental 'witness', no matter how often repeated, if there always existed a strong possibility that various observers would see different things or construe the 'same' things according to different identifying marks or criteria? And what could be the grounds of a shared commitment to the good of experimental practice if that practice turned out to be observer-relative, and those grounds nothing other than a private (or, at most, a factional group-based) motion of assent to the 'evidence' thus produced? Furthermore, why credit the experimenters' claim to avoid mere quibbles of scholastic definition – or to be simply concerned with natural phenomena – if indeed it could be shown (as Hobbes maintained) that their observation language was itself shot through with metaphysical beliefs and commitments? In short, there was nothing to commend their programme save its power to conjure assent among those (relatively few) individuals – experts, patrons, Fellows of the Royal Society, supporters of the new political settlement in this pseudo-scientific guise – who had most to gain from its acceptance as a matter of disinterested seeking after truth.

It is therefore not surprising that Shapin and Schaffer incline very markedly to Hobbes's side of the argument despite all their strong sociological professions of strict even-handedness as between the two parties. This bias is evident in various ways throughout the book but directly stated only in their closing peroration. Thus: 'As we come to recognize the conventional and artifactual status of our forms of knowing, we put ourselves in a position to realize that it is ourselves and not reality that is responsible for what we know. Knowledge, as much as the state, is the product of human actions. Hobbes was right' (*LAP*, p. 344). This passage bears the mark of various present-day doctrines – from the strong sociological programme to Foucault's Nietzschean (but also strikingly Hobbesian) genealogies of power/knowledge – which likewise adopt an attitude of extreme scepticism towards any 'discourse' that professes an attachment to values of disinterested truth or justice.[7] Thus Hobbes stands to Boyle and the Royal Society as Foucault (or indeed Shapin and Schaffer) stand to the received 'Whiggish' tradition of thinking in history and philosophy of science. That is to say, in each case the former parties reject the idea of scientific 'facts' as established through experiment and observation, or of progress – in science and society alike – as best achieved by a liberal (consensus-based) culture whose values are reflected in a wider liberal polity. For each of them, moreover, this argument consorts with a deep mistrust of such naïve 'Enlightenment' talk and a Hobbesian desire to expose its origins in the omnipresent struggle for power and recognition.

There are other, more specific reasons why Shapin and Schaffer should have adopted so sceptical a view of Boyle's programme, and should therefore have set out to rescue Hobbes from the massive condescension of posterity. Thus Hobbes's main objections to that programme – as summarized above – are couched in terms that cannot but remind us of various contemporary *idées recues* in epistemology, philosophy of science and the sociology of knowledge. These include (1) the theory-laden character of observation statements; (2) the consequent problem of meaning-change (or shifts in the reference of crucial terms) across or between paradigms; (3) the thesis of ontological relativity, arrived at by combining (1) and (2); and (4) the idea – most congenial to strong sociologists – that for all these reasons there can be no criterion of scientific truth or warrant that is not 'internal' or 'relative' to some given language game, research programme, cultural 'form of life', etc.[8] From which they deduce (item 5) the need for a strictly symmetrical – non-partisan – approach as between truth-claims that still have a place in the discourse of accredited knowledge and those that now figure, on a 'Whiggish' account, as mere episodes in the history of failed or obsolete scientific endeavour. In other words, they will attempt so far as possible to suspend all judgements of validity and truth, thus challenging the received idea of Boyle as among the great founders of modern scientific method, and of Hobbes as an old-fashioned dogmatist who wrongly and stubbornly set his face against the cause of scientific progress.

This is not to say that Shapin and Schaffer have no use whatever for questions of 'truth', 'objectivity' or 'proper method'. On the contrary, they write, 'we will be confronting such matters centrally.' However, they will be treated 'in a manner slightly different from that which characterizes some history and much philosophy of science' (*LAP*, pp. 13–14). That is, they will be subject to the strong sociological imperative which requires that they should always be placed, so to speak, under erasure and allowed none of the substantive meaning – or 'proper' attributive force – which typifies their usage in the discourse of normal (pre-Kuhnian) scientific historiography. Thus: '"truth", "adequacy", and "objectivity" will be dealt with as accomplishments, as historical products, as actors' judgements and categories. They will be topics for our enquiry, not resources unreflectively to be used in that enquiry' (p. 14). In short, it is the social construction of knowledge – of facts, theories, experimental procedures, canons of legitimate method and so forth – which Shapin and Schaffer take as their topic, and which therefore (they argue) precludes any claim to assess the issue between Boyle and Hobbes in point of scientific truth. Hence their frequent appeal to Wittgenstein on 'language games' and 'forms of life' as the furthest one can go towards establishing criteria for what counts as truth within this or that cultural context.[9] Just as, for Wittgenstein, 'the term "language-*game*" is meant to bring into prominence the fact that the *speaking* of language is part of an activity or of a form of life,' so Shapin and Schaffer will 'treat controversies over scientific method as disputes over different patterns of doing things and of organizing men to practical ends. . . . *That* is what the Hobbes–Boyle controversies were about' (p. 15).

That they might also have been about something else – about issues of scientific truth and method with decisive implications for our knowledge of physical reality – is a notion that the authors quickly rule out as just another product of the mainstream approach. Thus:

> here we see the germs of a standard historiographic strategy for dealing with the Hobbes-Boyle controversy, and, arguably, for handling rejected knowledge in general. We have a dismissal, the rudiments of a causal explanation of the rejected knowledge (which implicitly acts to justify the dismissal), and an asymmetrical handling of rejected and accepted knowledge. First, it is established that the rejected knowledge is not knowledge at all, but error. This the historian accomplishes by taking the side of accepted knowledge and using the victorious party's causal explanation of their adversaries' position as the historian's own. Since the victors have thus disposed of error, so the historian's dismissal is justified. (p. 11)

Of course there will be no such presumptive asymmetries – no siding with the victor or the verdict of received scientific opinion – in Schapin and Schaffer's account. On the contrary, they will approach the issue between Boyle and Hobbes in a non-partisan spirit that on principle eschews all judgements of truth and falsehood, and which thus avoids what the authors regard as a circular, self-confirming or downright collusive enterprise. Although not exactly defending Hobbes against the weight of orthodox opinion, they will present his case as strictly on a par with that of Boyle in point of this various motivating interests and sociopolitical commitments. For on this account 'the problem of generating and protecting knowledge is a problem in politics, and, conversely, the problem of political order always involves solutions to the problem of knowledge' (p. 21).

In which case – following the strong sociological principle of symmetry – there is no distinction to be drawn between context of discovery and context of justification, or those aspects of science that are now of interest mainly from a cultural-historical viewpoint and those that can still lay claim to some measure of genuine 'scientific' warrant.[10]

The authors' chief warrant for this is that most accounts of the Hobbes/Boyle controversy and other such well-known disputes – like that between Priestley and Lavoisier concerning the process of combustion – have been written by philosophers or historians of science who 'naturally' (that is to say, unthinkingly) endorse the standard scientific verdict. Their own aim, conversely, is to adopt a distancing 'stranger's perspective' that allows them to suspend all the values and assumptions – the items of 'self-evident' knowledge – which define our present-day scientific culture. Among the models for this approach is that of recent anthropologists and cultural historians who seek not so much (in the old 'Whiggish' mode) to interpret other cultures by the lights of our own, more 'advanced' or 'rational' beliefs, but rather to turn this process around and offer an alternative 'view from afar' on various aspects of present-day Western society. 'As part of [this] exercise,' the authors write,

> we shall be adopting something close to a 'member's account' of Hobbes's anti-experimentalism. That is to say, we want to put ourselves in a position where objections to the experimental programme seem plausible, sensible, and rational. . . . Our goal is to break down the aura of self-evidence surrounding the experimental way of producing knowledge, and 'charitable interpretation' of the opposition to experimentalism is a valuable means of accomplishing this. . . . We want to show that there was nothing self-evident or inevitable about the series of historical judgements in that context which yielded a natural philosophical consensus in favour of the experimental programme. Given other circumstances bearing upon that philosophical community, Hobbes's views might well have found a different reception. They were not widely credited or believed – but they were *believable*; they were not counted to be correct – but there was nothing inherent in them that prevented a different evaluation. (*LAP*, p. 13)

I have cited this passage at length because it offers a clear and forthright statement of just about every item in the strong sociological programme. What counts as scientific 'truth' – so the authors maintain – is determined *neither* by the way things stand in reality, *nor* by any special merit – any 'inherent' truth-related virtue – in those theories or procedures that happen to gain widespread communal assent. Rather, it is a product of the reception history (or the cultural pressures making for acceptance or rejection) to which all truth-claims are constantly exposed and which thus provide the ultimate court of appeal in matters of scientific 'fact'. If we now think that Boyle scored some decisive points in his quarrel with Hobbes, then this is best explained by his greater success in passing off the experimental programme – along with its congruent ideological values – as an effective guarantor of civil stability and peace.

All the same (so Shapin and Schaffer maintain) things might have gone otherwise with the 'progress' of science had history taken a different sociopolitical course. Thus it *could* have turned out that the Restoration settlement inclined not so much towards the construction of a viable – albeit fragile and selective – consensus but more towards the Hobbesian absolutist idea of an all-powerful state apparatus for suppressing any form of civic or religious dissent. And in that case – so the authors maintain – we might now take a very different view of the Hobbbes/Boyle controversy. For if absolutist values had prevailed in the contest for social legitimacy, then there is a strong likelihood that natural philosophy would have followed suit and adapted its procedures – its canons of scientific truth and method – to a Hobbesian plenist ontology. That is to say, this alternative 'settlement' would have left no room either for such speculative notions as the vacuum hypothesis or for those potential schisms and rifts which Hobbes viewed as the inevitable upshot of a culture given over to the private vagaries of mere 'experimental' witness. Thus on his account, quite simply, 'the elimination of vacuum was a contribution to the avoidance of civil war', and the rebuttal (or suppression) of vacuist theories 'the elimination of a space within which dissension could take place' (*LAP*, pp. 108–9).

II

So it is – by combining their 'stranger's perspective' with a degree of explicit counter-factual licence – that Shapin and Schaffer set out to undermine the received view of Boyle's experimental method as a paradigm of scientific truth, discipline and (not least) moral integrity. Their book is without doubt a classic of its genre: exhaustively researched, vigorously argued, and replete with just the kinds of telling detail that must give pause to any 'Whiggish' believer in scientific progress as a steady accumulation of agreed-upon facts and ever more adequate explanatory theories. It also brings out very clearly – as with other recent studies of this kind – the extent to which even 'exemplary' science (such as Boyle's experiments with the air-pump) may involve various complicating factors of mechanical imperfection, perceptual error, latitude for differing (observer-relative) interpretations and so forth.[11] However, these virtues should not obscure the very real problems with Shapin and Schaffer's approach when the focus shifts – as at some point it must – from the context of discovery to the context of justification. That is to say, the sociology of knowledge may offer good reasons to revise our estimate of how and why certain conflicts developed over the conduct of 'proper' scientific enquiry in a given sociocultural context. It may also – on occasion – bring about some change in our judgement of an individual scientist's claim to have followed the best (most reputable) methods in pursuing such enquiry. What it *cannot* justifiably claim is to undermine the very grounds of belief in science as a truth-seeking enterprise and – beyond that – in the existence of a real-world (belief-independent) object domain whose nature, properties or causal dispositions may be discovered to the best of our ability. The first is an epistemological point about the difference between matters of *belief* – no matter how widely shared – and items of *knowledge* whose claim to that title comes of their having so far withstood the most rigorous test procedures. (That there are various degrees or weightings of justified belief commitment, shading off into legitimate knowledge claims at the highest end of the scale, is no argument against this basic distinction.)[12] The second is an ontological point about the difference between those sorts of issue and such questions as: what objects really exist? what are their constitutive features? and: how could there be truths concerning those objects that transcend our current (perhaps our best possible) powers of observation or conceptual grasp? There is widespread confusion on both points in the strong sociological literature. And this confusion is particularly evident – or so it seems to me – in the marked anti-realist or conventionalist bias that shows up at numerous points in *Leviathan and the Air-Pump*.

Some aspects of that bias I have mentioned already. They include the thesis of radical meaning variance as between different usages of terms such as 'vacuum', 'plenum', 'pressure', 'density', 'atmospheric weight', 'elastic power', 'corporeal substance', 'aether' and so forth. On the orthodox post-Kuhnian view, these terms would function altogether differently in the theories of Boyle and Hobbes, with no common ground – no shared ontology or range of 'commensurable' referents – whereby to adjudicate the issue between them. Shapin and Schaffer standardly italicize these terms (or place them within quotation marks) so as to signal their own non-commitment with regard to the rival theories. In the same way – by strict application of the Principle of Symmetry – one might wish to adopt a nescient stance with regard to the conflicting explanatory claims of Priestley's 'phlogiston' and Lavoisier's 'oxygen' as agents in the process of combustion. However, this would demand a quite extraordinary effort of willed counter-factual imagining, or – more bluntly – of feigned ignorance concerning large areas of established scientific knowledge.

Shapin and Schaffer find justification in Ernest Gellner's 'Charitable Principle' according to which it is best to proceed by imputing a maximum of truth, rationality, or plain good sense to any theory or hypothesis under review.[13] This approach works out – roughly speaking – as a social-science equivalent of Donald Davidson's generalized 'Principle of Charity', that is to say, his idea that in construing other people's meanings, intentions or beliefs we are obliged to count them presumptively 'right in most matters' since otherwise we should lack any basis for mutual understanding.[14] No doubt these claims have an element of truth when applied to such abstractly bothersome issues as the very possibility of translating across languages, cultures, scientific paradigms, etc. Indeed they are quite useful as a basic resource in countering some of the more extreme varieties of cultural-relativist doctrine. Still, this approach gives rise to the obvious problem – whether in Gellner's or in Davidson's version – that it makes all beliefs come out true (more likely than not) irrespective of whatever more specific grounds we might have for granting or withholding assent. Thus it runs clean up against the awkward fact that *some* beliefs have been – and are – plain wrong no matter how firmly adhered to by the members of this or that cultural community.

This problem is perhaps less pressing for Davidson, concerned as he is with philosophical issues in the nature of linguistic understanding, and not so much with matters of applied methodology in the social or natural sciences. All the same it is hard to see – as I have argued elsewhere in greater detail – how Davidson's generalized Principle of Charity could offer much help in the practical business of figuring out a speaker's meanings, beliefs or intentions.[15] For this also involves the standing possibility that they might be mistaken (apt to say wrong or off-the-point things) for various reasons such as cultural bias, false information, partial understanding, perceptual distortion, doctrinal adherence and so forth. Or again, they might have all the right beliefs but simply not manage to express them adequately owing to some occasional slip of the tongue or some other, more persistent habit of aberrant usage. To a very large extent it is our grasp of such distinctions that enables us (in keeping with Davidson's rule) to optimize the truth-content of what people say while also, on occasion, suspending that rule – when they say something patently false – or casting around for some alternative explanation (circumstantial, psychological, linguistic, etc.) where that seems more appropriate.[16]

Thus the trouble with Charity in its wholesale form is that it leaves no room for such fine discriminations on a scale ranging from downright falsehood – or systematically distorted belief – to perceptual error, gaps of knowledge, fallacious inference, or mere peculiarities of linguistic expression. That is to say, by counting other people (or cultures) necessarily 'right on most matters' it distracts attention from the various ways in which they might in fact be wrong – misinformed or deceived – in respect of some particular belief. By the same token it tends to underrate our capacity for interpreting such false beliefs in (for instance) sociological, psychological or causal-explanatory terms. For there is no lack of 'charity' in seeking to explain just how and why – under what sorts of circumstance or owing to what preconceived habits of thought – they were led to some erroneous conclusion. After all, we ourselves are capable of learning from past mistakes by reflecting on the various contributory factors (from perceptual distortion to ideological bias) whose influence on our judgement we had hitherto failed to recognize. Thus it is better – more 'charitable' in a genuine (non-condescending) sense of that term – to apply a principle of symmetry here and assume that others are likewise placed with regard to both the possibility of error and the prospects of improved understanding.

However, this is *not* the same principle of symmetry adopted by strong sociologists of knowledge such as Shapin and Schaffer. On the contrary, their approach comes out more in line with Davidson's methodological injunction to count other people (other language communities or cultures) necessarily 'right on most matters', since otherwise the parties would share so little in the way of operative truth-conditions that any mutual understanding would constitute a well nigh miraculous event. So it is, according to Shapin and Schaffer, that the good historian/sociologist of science should properly proceed when dealing with arguments – such as those of Hobbes contra Boyle – which possess little or no credibility from a present-day scientific standpoint. That is, they should adopt a 'stranger's perspective' *vis-à-vis* the experimental culture bequeathed to modern science by Boyle and his Royal Society colleagues, and a more sympathetic (or member's) view with regard to just those aspects of Hobbesian natural philosophy which nowadays figure as so many items in the history of obsolete ideas. This is, they acknowledge, a difficult procedure, but one that is of particular value in the context of a cultural 'life-form' so massively self-evident as that of modern experimental science. Not that this approach should (or could) be undertaken from a position of genuine ignorance concerning the relevant scientific history. Thus: 'we need to *play* the stranger, not to *be* the stranger. A genuine stranger is simply ignorant. We wish to adopt a calculated and an informed suspension of our taken-for-granted perceptions of experimental practice and its products' (*LAP*, p. 6). Nevertheless this strategy clearly involves quite a large-scale exercise in the redistribution of persuasive (or at any rate 'believable') scientific claims. For it is, we recall, a major plank in their argument that Hobbes's views 'might well have found a different reception'; that there was 'nothing self-evident or inevitable' about the process which led to their marginalization in text-book history of science; and moreover, that the quarrel between Hobbes and Boyle would most likely have been viewed in a different light then and now had political history taken a different direction.

What this amounts to – in effect – is the wholesale Davidsonian Principle of Charity transposed from the philosophico-semantic to the sociocultural or history of science domain. Thus it requires that we count Hobbes not (perhaps) 'right in most matters' but at any rate justified according to the lights of his own philosophic and political life-form, a life-form grounded in principles and values diametrically opposed to those of Boyle and the new experimentalists. Only thus – so Shapin and Schaffer argue – can we attain the kind of critical distancing effect that allows for a genuine 'stranger's perspective' on Boyle's and our own scientific culture. And this in turn requires thinking of Boyle's programme as a set of purely *conventional* practices with no stronger claim to scientific truth than the fact of their having prevailed at a certain time and for certain historically and culturally contingent reasons. After all, '[c]ausal talk is grounded in conventions which Boyle's reports exemplify, just as the construction of the matter of fact is conventional in nature' (*LAP*, p. 52). We are simply not placed to adjudicate the issue between Hobbes and Boyle since the two of them proposed quite different (incommensurable) claims as to what should properly count as scientific truth. Thus for Hobbes such truth was a matter of absolute (demonstrative) warrant and not to be achieved by any mere establishment of facts, least of all the sorts of error-prone, observer-relative, experimentally produced 'fact' proposed by Boyle and his colleagues. For the latter, conversely, it was just those facts that carried genuine scientific weight since they were arrived at through a process of fair and open debate among suitably qualified observers. This in turn seemed to Hobbes a grosss dereliction of the scientist's (or the natural philosopher's) proper responsibility. For on his view that duty could only be discharged through the kind of rigorously consequent reasoning from first principles that admitted no doubt – no leeway for dissent – with respect to their ultimate conclusions.

Thus the task of natural philosophy was to offer nothing less than a *full and rationally compelling* account of the law-governed (causal) mechanisms that were operative throughout the physical domain. To relinquish this task – as Boyle did when he confined himself to observation and experiment, refusing to speculate on ultimate causes – was for Hobbes a disgraceful admission of defeat not only in natural-scientific but also in civic or sociopolitical terms. For it threatened once again to plunge society into just those kinds of fractious dispute that came of promoting private opinion over absolute truth, or the vagaries of mere subjective belief over the virtue of unqualified rational assent. In short,

> [o]ne solution (Boyle's) was to set the house of natural philosophy in order by remedying its divisions and by withdrawing it from contentious links with civic philosophy. Thus repaired, the community of natural philosophers could establish its legitimacy in Restoration culture and contribute more effectively to guaranteeing order and right religion in society. Another solution (Hobbes's) demanded that order was only to be ensured by erecting a demonstrative philosophy that allowed no boundaries between the natural, the human, and the social, and which allowed for no dissent within it. (*LAP*, p. 21)

Hence the need, as Shapin and Schaffer see it, to suspend our normal (acculturated) habits of judgement with regard to the Hobbes/Boyle controversy and accept that Hobbes had his own reasons – no less 'valid' than Boyle's – for taking a principled stance on these issues.

Their argument possesses great power and persuasiveness up to a certain point, that is to say, when sociology of knowledge passes over into issues of ontology, epistemology and philosophy of science. Beyond that point – or so I would contend – it evinces a number of deep confusions that are often to be found in the current sociological literature. Of course this objection would cut no ice with

Shapin and Schaffer, since in their view such criticisms are all of a piece with the mainstream 'asymmetric' approach to the history of science. That is to say, it assumes that very different criteria apply when dealing on the one hand with successful (currently accepted) scientific theories, and on the other with theories which are nowdays counted false or inadequate, and are thus thought to stand in need of explanation from some alternative (e.g., sociological, cultural, or psycho-biographical) standpoint. To reject this partisan approach, so its adversaries claim, is the first step towards a more even-handed procedure that would not take for granted the massive self-evidence of present-day scientific 'fact'. For it is precisely the nature of such putative facts – their warrant, status, modes of production, conditions of experimental testing and so forth – that was the main point at issue between Boyle and Hobbes. To the latter's way of thinking,

> factual knowledge, based on sensory impressions, did not have an epistemologically privileged position. It did not matter how one proposed socially to process such factual knowledge, the limitations remained. . . . If the aim was certain knowledge and irrevocable assent, then the way towards it could not traverse anything as private and unreachable as individual states of belief. Knowledge, science, and philosophy were set on one side; belief and opinion on the other. (*LAP*, p. 102)

For Boyle, conversely, such beliefs were one (intermediate but necessary) stage in a process that led from trained observation to widespread agreement on 'matters of fact' among members of the relevant knowledge community. Hobbes was therefore wrong in equating this new experimental procedure with the kinds of 'private' and 'unreachable' opinion – mere phantasms of the individual mind or brain – whose effect was supposedly to foment civil discord by removing science from the realm of absolute, demonstrative truth. What he failed to acknowledge was the self-corrective mechanism that came into play when matters of belief were subject to various forms of critical assessment. Among these latter – as Shapin and Schaffer point out – were the replication of Boyle's experiments under controlled laboratory conditions and the possibility of 'virtual witnessing', that is, of appealing to an ever wider community through the description of those same experiments in carefully written-up form. So it was that 'facts' could be sifted out from mere 'opinions', or beliefs possessing a high degree of established probability from opinions grounded in nothing more reliable than private sense certainty. Granted all this – so Boyle maintained – there was simply no reason to insist, like Hobbes, that natural philosophy was failing in its proper (scientific and sociopolitical) function if it gave up the claim to establish truths as a mattter of absolute demonstrative warrant. Indeed, such claims would more probably open the way to just that kind of dogmatic a priori thinking which had hitherto exerted so baneful an influence in science and politics alike.

This was why Boyle preferred to take no position – at least no fixed doctrinal position – with respect to the largely metaphysical dispute between hard-line plenists and vacuists. Of course, his own researches disposed him to accept the vacuist case in so far as it allowed in principle for just those sorts of observational finding which he and his colleagues sought to produce by means of the air-pump experiments. However, it was no part of this programme to settle the issue as regards the possibility of achieving a *perfect* vacuum, or to demonstrate – contra Hobbes and the plenists – the non-existence of that *horror vacui* in nature propounded by many thinkers from Aristotle down. For the new experimentalists,

> all that could be expected of physical knowledge was 'probability', thus breaking down the radical [Hobbesian] distinction between 'knowledge' and 'opinion'. Physical hypotheses were provisional and revisable; assent to them was not obligatory, as it was to mathematical demonstrations; and physical science was, to varying degrees, removed from the realm of the demonstrative. The probabilistic conception of physical knowledge was not regarded by its proponents as a regrettable retreat from more ambitious goals; it was celebrated as a wise rejection of a failed project. By the adoption of a probabilistic view of knowledge one could attain to an *appropriate* certainty and aim to secure *legitimate* assent to knowledge-claims. (*LAP*, p. 24)

Thus Hobbes was simply missing the point when he thought to discredit the entire enterprise by arguing against it from first principles, i.e., from the presumed *impossibility* of creating a perfect vacuum to the conclusion that there must be some fault in the apparatus – some constant leakage despite all the efforts to create a perfect seal – which rendered the air-pump experiments utterly worthless. For Boyle such reasonings were powerless to establish any item of scientific fact, committed as they were to an article of faith which allowed no scope for its own refutation through careful and repeated testing.

This is why he felt able to concede various points to Hobbes, among them the possibility that *some* degree of leakage might always have occurred; that a *perfect* vacuum might not have been achieved (or might perhaps be unachievable under terrestrial conditions); and moreover, that no experimental evidence could count against the plenist hypothesis of a subtle, undetectable, all-pervasive 'aether' which would *always necessarily* fill any space – any notional 'vacuum' – created by the best efforts of the new science. From Boyle's viewpoint there was nothing at stake in such issues except, on Hobbes's side, a dogmatic commitment to metaphysical postulates which led to his construing the whole debate in typecast ('plenist' versus 'vacuist') terms. Thus, again, Boyle could see no reason to object when Hobbes argued that the experimental programme rested on mere probability – or the amassing of evidence that tended in favour of certain scientific conclusions – rather than claiming absolute demonstrative warrant. For it was only (he believed) such dogmatists as Hobbes who continued to insist on this obsolete idea of natural philosophy as a mode of knowledge ideally exempt from all the problems of establishing scientific truth under less than ideal experimental conditions.

In short, Boyle's concern was to establish and maintain the most reliable methods of enquiry, given both the limits of currently existing technology and the limits placed upon scientific knowledge by the non-availability of ultimate (i.e., a priori or metaphysical) explanations. 'He would create a new discourse,' as Shapin and Schaffer describe it,

> in which the language of vacuism and plenism was ruled out of order, or at least managed so as to minimize the scandalous disputes that, in his view, it had engendered. The receiver [i.e., the supposedly evacuated air-pump chamber] was a space into which one could move this paradigmatic experiment. And the discourse and social practices in which talk about this experiment was to be embedded constituted a space in which disputes might be neutralized. (*LAP*, p. 42)

That this programme fell far short of Hobbes's requirement – that it amounted, as he thought, to a mere accumulation of private 'opinions' with no better claim to method or truth than the ravings of sectarian zealots – was for Boyle just a sign that he had misunderstood the very nature of the new science. All the more so in that Hobbes was violently opposed to just those aspects of experimental method which enabled Boyle and his colleagues to adopt a noncommittal (hence non-adversarial) stance on such issues as the plenist/vacuist controversy and the existence or non-existence of an all-pervasive aether. Where Hobbes saw this as a plain admission of defeat – worse than that, as an opening to renewed civil strife through the lack of any firm ontological commitment – Boyle considered it the best way forward to a rational consensus within the scientific community and a *modus vivendi* with other beliefs (such as those of the deists and Cambridge Platonists) which sought some accommodation with the new science. Thus Hobbes's approach had nothing to commend it, whether as a purported contribution to science or as a means of securing that common peace which is mankind's chief concern. Such has indeed been the verdict of most commentators since, especially those in the philosophy of science mainstream whose effect – as Shapin and Schaffer argue – is simply to endorse the dominant self-image of scientific method as bequeathed by (among others) Boyle and his new experimentalist colleagues.

Still one may doubt that it is possible to follow their prescription and apply a principle of strict symmetry as between Boyle's and Hobbes's arguments. More precisely: though of great interest and value as an exercise in sociohistorical method, their approach goes seriously wrong when extended to issues in philosophy of science. For there is plentiful evidence – much of it provided by Shapin and Schaffer – that Hobbes not only misunderstood various aspects of the experimental programme but was also in error about numerous points of (by now) well-established scientific fact. Of course, it may be said that this begs all the questions raised by a study like *Leviathan and the Air-Pump*, questions that are simply pushed out of sight by any such confident appeal to science (or mainstream philosophy of science) as the sole arbiter of its own methods, truth-claims, canons of inductive warrant and so forth. From this perspective there is nothing – orthodox prejudice aside – that could justify talk of Hobbes's having 'misunderstood' Boyle on certain crucial points, or his having been simply wrong with regard to vacuum-related phenomena. For to take such a line is also to adopt a whole series of assumptions, values, beliefs and methodological criteria which effectively prejudge the issue between them by endorsing the received historical view and siding with Boyle against Hobbes. Thus a bias comes in with the very idea of 'natural *phenomena*' (things manifest to the senses) which lend themselves to empirical observation and eventually yield matters of fact – or agreed-upon scientific knowledge – through a process of repeated experiment and a wise avoidance of theories and hypotheses not borne out by the evidence. Such was the 'language game', the experimental 'form of life', that Boyle and his colleagues were the first to devise and which still – so it is argued – constitutes a paradigm for research in the history and philosophy of science.

However, once again, this attitude of scrupulous even-handedness turns out to have some awkward (not to say absurd) consequences when Shapin and Schaffer discuss the content of Hobbes's scientific theories. For their method requires that they present those theories from a strictly non-adjudicative standpoint which suspends all issues of scientific truth or falsehood, and which thus treats it as an open question – at least for the purposes of sociological enquiry – whether Hobbes or Boyle had the better explanation with regard to the air-pump experiments. From which it follows (by the principle of symmetry) that Boyle's researches should be no less subject to investigative treatment in this manner. After all, '[t]he practices involved in the generation and justification of proper knowledge were part of the settlement and protection of a certain kind of social order,' while conversely '[o]ther intellectual practices were condemned and rejected because they were judged inappropriate or dangerous to the polity that emerged in the Restoration' (*LAP*, p. 342). So it was, according to Shapin and Schaffer, that Boyle achieved fame for his signal contributions to scientific knowledge while Hobbes's excursions into natural philosophy now merit only a footnote in the history of pseudo-science. For if indeed the experimental method was just one 'language game' or 'life-form' among others – if the reasons for its success were socio-politico-cultural through and through – then clearly there could be no deciding the issue in point of scientific method, validity or truth.

Of course, one might construe them as making an alternative, more moderate or 'weak' sociological claim. On this view they would not so much be challenging the authority of science – or mainstream philosophy of science – to adjudicate in matters that properly lie within its own explanatory scope. Rather, they would be arguing (less controversially) that social factors have a great deal to do with the particular projects that scientists pursue, with the interests or values that motivate their work, and also – in some cases more than others – with their research methods, experimental techniques, evaluative criteria and so forth. However, this construal is firmly ruled out by the authors' various statements of intent, in particular those concerning the principle of symmetry. For it is clear that they mean to extend that principle far beyond the token recognition that science might conceivably have taken a different course in some alternative possible world. Thus: 'we intend to display scientific method as crystallizing forms of social organization and as a means of regulating social interaction within the scientific community' (*LAP*, p. 14). On this view there is simply no room for any notion – any argument (say) in the critical-realist mode – that scientific theories, truth-claims or hypotheses may sometimes actually get into conflict with the established belief-systems or the dominant social ideologies of their time. Still less can it be thought that such conflicts might develop through the commitment to certain distinctive values – such as careful observation, shared endeavour, the freedom to question and criticize – which are not (or not always) just a reflex product of vested power-knowledge interests. For if scientific method must be seen as 'crystallizing forms of social organization', and if the method that proves most successful is *for that very reason* necessarily ranged on the side of social authority and power, then of course those values can amount to nothing more than a species of self-serving rhetoric.

III

What is particularly striking about all this is the extent to which it manifests a Hobbesian bias in Shapin's and Schaffer's approach, despite their professions of strict impartiality (or parity of esteem) as between the rival parties. Indeed one could argue that a good many recent developments in social and cultural theory have their closest analogue – if not perhaps their direct source – in Hobbes's conception of political power as an all-pervasive mechanism which, rightly understood, should leave no space for dissenting views or failure to comply with the mandates of a sovereign authority. This elective bias can be seen most clearly in Foucault's ubiquitous deployment of the phrase 'power/knowledge' as a means of exposing the coercive drive that purportedly motivates all professions of disinterested, truth-seeking enquiry.[17] Moreover, in Foucault as in Hobbes, it leads to an indiscriminate levelling of the various 'discourses' or 'regimes of truth' – scientific, philosophic, ethical, political, economic, sociocultural and so forth – which all tend to figure as minor variations on the standard power/knowledge theme. Hence Foucault's markedly Hobbesian attitude towards arguments that invoke such typecast 'liberal' values as truth, reason, enlightenment, progress or emancipatory critique. This in turn goes along – as various commentators have noted – with his bleak view of the prospects for social change and his idea that various supposedly distinct sociopolitical orders (from the totalitarian to the liberal-democratic) in fact differ only in the extent to which effects of power are diffused and internalized by subjects whose compliance is thereby assured without the need for coercive or 'spectacular' punitive sanctions.[18] In other words, we are deluded – dupes of a typical 'Enlightenment' narrative of progress and truth – if we suppose that knowledge could ever be detached from the workings of an omnipresent disciplinary regime all the more efficient for its having dispensed with such overtly repressive mechanisms.

Shapin and Schaffer are a deal more circumspect (or less prone to dramatic overstatement) in making their case against received ideas of scientific method and practice. All the same, their programme bears a strong resemblance to Foucault's, not least in its espousal of a Hobbesian view concerning the relation between politics and natural philosophy. Thus they agree in stressing (among other things) the ubiquity of power/knowledge effects, the production of 'truth' through agencies of social control, and the circular nature of any appeal to those emergent cultural values – such as method, observation, evidence, rational consensus and so forth – that are treated as products of a certain phase in the history of shifting discursive formations. Moreover, there are aspects of Shapin and Schaffer's approach that align them even more closely with Hobbes on the issue as to whether the new experimental philosophy was in any way capable of settling the argument between plenists and vacuists. For very often, when citing Hobbes against Boyle, their commentary modulates from direct quotation to a hybrid mode of *oratio obliqua* (or indirect style) whose effect is to suspend any judgement concerning the validity or otherwise of Hobbes's scientific claims. The following passage is a fair sample of this strategy at work.

> How, then, did Hobbes describe the air? In the *Dialogus*, he said that his 'hypotheses' about the air's constitution were twofold: 'first, that many earthy particles are interspersed in the air, to whose nature simple circular motion is congenital; second, the quantity of these particles is greater in the air near the Earth than in the air further from the Earth'. Hobbes mobilized the tripartite typology of visible matter, invisible matter, and a fluid space-filling aether which he had outlined in *De corpore*. . . . In each of the phenomena Hobbes now examined, he used the contrasting fluidity of pure air and earthy effluvia to explain the observed effects. In so doing, Hobbes showed how it was always possible to generate such explanations from his two hypotheses of fluidity and firmness: he also showed that the invocation of absolute vacuity was both unnecessary and unphilosophical. (*LAP*, pp. 120–1)

This passage would bear a great deal of rhetorical unpacking but I shall here focus only on the features most relevant in the present context of discussion. One is the way that it presents a whole range of interlinked Hobbesian hypotheses, some of which (like the greater density of air particles in regions closer to the Earth's surface) possess a good claim to genuine scientific warrant, while others – like that of an endemic 'circular motion' governing such particles or the existence of an all-pervasive 'aether' that occupies regions inaccessible to even the subtlest of 'earthy' particles – cannot be accorded any measure of credence consonant with our best scientific knowledge.

It was the two latter claims that Hobbes most often deployed against Boyle, the one (circular motion) as a means of explaining how air would always necessarily re-enter any space apparently emptied by the air-pump experiments, and the other (his aether hypothesis) as a fall-back argument adopted by default when there seemed no way that the apparatus could be infiltrated by 'invisible matter' or refined 'earthy particles'. In both cases there is a great mass of evidence – much of it adduced by Boyle – which demonstrates the lack of scientific warrant for Hobbes's claims and their incompatibility with various results produced under carefully controlled laboratory conditions. That is to say, there is no good reason to accept hypotheses that start out from some dogmatic article of faith – like the sheer *impossibility* of a vacuum – and proceed to 'explain' discrepant phenomena by adopting so unlikely a theory as that which would have every region of space (including the air-pump chamber) promptly refilled by the constant circulation of particles as required by the Hobbesian plenist ontology. Still less can there be good warrant for a theory – that of the omnipresent aether – which serves no explanatory purpose save to rescue Hobbes's position in the face of mounting experimental proof that air could indeed be exhausted from the chamber to a point far below atmospheric pressure if not (as Hobbes unrealistically demanded) to the point of an absolute vacuum.

These issues are obscured, in Shapin and Schaffer's rendition, by their attributing to *Hobbes* what was in fact (as they concede elsewhere) a main plank in Boyle's argument, namely that 'the invocation of absolute vacuity was both unncessary and unphilosophical' (*LAP*, p. 121). For of course it was Hobbes – the acknowledged absolutist in politics as in matters of natural philosophy – who refused to accept anything less than a demonstration from first principles both that a perfect vacuum could exist, and also that the air-pump exhibited such absolute 'integrity' (i.e., perfect freedom from leakage, whether by 'subtle' particles or – *per impossibile* – by that all-pervasive 'aether') as to rule out any possible alternative theory. The trouble with Shapin and Schaffer's account is that it strives so hard to do justice to Hobbes on the strong sociological Principle of Symmetry that it ends up with a curious reversal of roles whereby Boyle comes out as a thinker in the grip of his own preconceived ('vacuist') beliefs and Hobbes – improbably enough – as one who rejected such dogmatic claims. And this despite other passages in *Leviathan and the Air-Pump* where the authors provide copious documentary evidence that it was Hobbes who declared his arguments proof against mere 'experimental' findings, and Boyle who declined to adopt any stance – any absolute position either way – on the plenist versus vacuist controversy.

For Boyle, therefore, '[s]ettling the question of a vacuum was not what this experiment was about, nor were questions like this any part of the experimental programme' (*LAP*, p. 45). Rather, they concerned the more modest claim – of no interest to dogmatists like Hobbes – that the air-pump exhibited certain experimentally induced phenomena which, through constant improvements in its design and construction, could offer strong probabilistic grounds for rejecting the plenist hypothesis. Thus:

> Boyle's 'vacuum' was a space 'almost totally devoid of air': the incomplete fall of the mercury indicated to him that the pump leaked to a certain extent. The finite leakage of the pump was not, in his view, a fatal flaw but a valuable resource in accounting for experimental findings and in exemplifying the proper usage of terms like 'vacuum'. (p. 46)

And again: 'by "vacuum", I understand not a space, wherein there is no body at all, but such as is either altogether, or almost totally devoid of air' (ibid.). It was precisely by adopting this agnostic stance – this refusal to meet his plenist critics on their own intransigent terms – that Boyle sought to avoid what he saw as a fruitless (because merely metaphysical) dispute between rival dogmatic creeds. He was even willing to suspend judgement with regard to the existence or non-existence of an 'aether' since this issue could neither be settled by experiment nor subjected to testing by any method available to the physical sciences.

For Shapin and Schaffer – again adopting a Hobbesian standpoint – this all goes to show that Boyle's programme was by no means free of dogmatic commitments, working as it did to equate genuine science with the order of experimentally produced 'matters of fact', and thus to exclude any theories (such as the plenist or aether hypotheses) that were simply not amenable to observation or experiment. Another passage from *Leviathan and the Air-Pump* may help to bring out the subtle bias that enters their discussion whenever it is a matter of Hobbes versus Boyle on the issue of scientific method. 'Once again,' they write,

> Hobbes picked out a central problem of the air-pump research programme. Boyle laboured to establish the air's spring as a matter of fact, eschewing any systematic attempt to explain the spring or to prove the vacuum. Hobbes's polemic disproved the vacuum and offered a physical explanation of the apparent spring. Hobbes's interlocutor [in the *Dialogus physicus*] agreed that 'Your hypothesis pleases me more than that of the elastic force of the air. For I see that the truth of the vacuum or of the plenum depends on the former's truth, whereas from the truth of the latter, nothing follows for either part of the question.' It was Boyle's nescience, and his *recommendation* of nescience as an appropriate philosophical stance, that Hobbes found objectionable: hence his efforts in the *Dialogus* to show how *easily* his two hypotheses could explain all the phenomena whose cause Boyle said he was unable to find. (*LAP*, p. 121)

However, it is an odd line of argument that requires us to suspend all judgements of truth or falsehood as between Boyle's and Hobbes's scientific theories while strongly implying that Hobbes got the better of Boyle – or at least uncovered a weakness in the latter's position – since Boyle himself adopted just such a 'nescient' attitude with respect to certain matters of causal explanation. Indeed, it might be said that Shapin and Schaffer get the issue completely upside-down by pressing so far with their Principle of Symmetry. For it is simply not the case that Hobbes in any sense 'disproved the vacuum', that he 'offered a physical explanation of the apparent spring', or again, that he managed to show 'how *easily* his two hypotheses could explain all the phenomena whose cause Boyle said he was unable to find'. On the contrary: Hobbes's hypotheses (especially that concerning the inherent circular motion of particles) were borne out neither by experiments at the time nor by the long-run progress of research in pneumatics, fluid mechanics and other related fields. Moreover, Boyle had adequate grounds – given the existing state of scientific knowledge – for adopting an agnostic stance on some matters and refusing to advance causal explanations where the evidence so far was insufficient to bear them out. For it would scarcely recommend Boyle's scientific practice had he jumped to such conclusions merely on the strength of some foregone metaphysical prejudice.

All of which suggests that Shapin and Schaffer are more predisposed towards Hobbes's side of the issue than might appear from their professions of strict neutrality as between the rival parties. This results from their methodological premises (1) that 'truth' is a product of discursive definition; (2) that each discourse has its own criteria of method, validity, experimental warrant, etc.; and (3) – with reference to Wittgenstein – that those criteria cannot be further justified except by pointing to the role they play in some particular language game, cultural life-form or scientific research programme. The resulting approach is laid out very clearly in the book's introductory chapter. Thus the authors will examine the 'form of life that Boyle proposed for experimental philosophy', and will do so (they claim) without bias or prejudice as to its scientific value. And again:

> We identify the technical, literary, and social practices whereby experimental matters of fact were to be generated, validated, and formed into bases for consensus. We pay special attention to the operation of the air-pump and the means by which experiments employing this device could be made to yield what counted as unassailable knowledge. We discuss the social and linguistic practices Boyle recommended to experimentalists; we show how these were important constitutive elements in the making of matters of fact and in protecting such facts from items of knowledge that were thought to generate conflict and discord. Our task here is to identify the conventions by which experimental knowledge was to be produced. (*LAP*, pp. 18–19)

At first glance there is nothing in this passage that leans either way on the issues at stake between Boyle and the anti-experimentalists. However, that appearance turns out to be deceptive if one looks more closely at the authors' way of framing those issues. First there is the view of Boyle's programme as a set of 'conventions' – of 'social and linguistic practices' – whose role in the production of 'experimental knowledge' is best explained in sociopolitical or cultural-historical terms. Along with this goes the notion that Boyle's apparatus (the air-pump) was just one item in a range of allied technologies that extended all the way from his methods of laboratory observation, via expert bodies such as Gresham College and the Royal Society, to the various 'literary' or descriptive techniques whereby Boyle and his colleagues sought to recruit the widest possible support among members of an informed lay public.

This aspect of the programme – 'virtual witnessing', as Shapin and Schaffer term it – was all the more vital (and all the more suspect to critics like Hobbes) since its findings laid no claim to absolute demonstrative warrant and had therefore to be tested in the public domain on the basis of good observational practice and consensus over any results achieved. Thus:

> Hobbes attacked Boyle as a vacuist despite the latter's professions of nescience on the vacuist-plenist debates of the past. Of greater epistemological significance was Hobbes's attack on the generation of matters of fact, the constitution of such facts into the consensual foundations of knowledge, and Boyle's segregation of facts from the physical causes that might account for them. (*LAP*, p. 19)

Hence the new experimentalists' adoption of a distinctive, markedly empiricist style in the writing-up of their research: a style characterized chiefly by detailed description, rhetorical self-effacement, an avoidance of large metaphysical claims and an appeal to shared ('commonsense') standards of probity and truth. Hence also – it is argued – the success of their project when measured against the conspicuous failure of Hobbes's rival claims. For the experimental 'life-form' and its cognate style were much better attuned to the social and political climate that emerged during this period.

What is chiefly problematic about all this is the idea – pretty much taken for granted by strong sociologists of knowledge – that scientific truth can be reduced *without remainder* to the currency of in-place consensus beliefs or prevailing social values. There is a further irony in the fact that Boyle and his associates adopted an approach which, on the face of it, differed very little from that of the strong sociologists. That is to say, their conception of good method in the natural sciences was one that emphasized the communal dimension of enquiry, that played down issues of absolute (context-transcendent) validity or truth, and which accordingly placed a high value on agreement arrived at through open participant debate. However, this resemblance turns out to have sharp limits if one asks how it is – by what curious trick of perspective – that Shapin and Schaffer can portray Boyle (in his quarrel with Hobbes) as spokesman for a rising hegemonic class whose interests shaped his most basic ideas of scientific method and enquiry. The answer, I would suggest, is that they take no account of the single most important distinctive feature of the new science: namely its commitment to methods of research – like the *experimentum crucis* – that could always challenge existing beliefs through the production of anomalous or discrepant findings. However, this argument will carry no weight with those of an anti-realist, a cultural-relativist, or a strong sociological persuasion. For to their way of thinking it is always the case that any results thus produced will turn out to confirm the various motivating interests and priorities that first gave rise to that particular programme of research.

Of course, such ideas go clean against the more traditional post-Galilean view of scientific progress. On this view, significant discoveries occur at the point where thinkers manage to break with some existing (often deep-laid) set of beliefs and thus arrive at a better, more adequate theory on the basis of improved – technologically enhanced – observation plus a readiness to criticize much of what counts as the established scientific wisdom. William Empson puts the case well when he remarks – in a review of E. A. Burtt's *The Metaphysical Foundations of Modern Science* – that 'it is always unsafe to explain discovery in terms of a man's intellectual preconceptions, because the act of discovery is precisely that of stepping outside preconceptions.'[19] But such ideas are nowadays widely written off as just a relic of that old, self-aggrandizing image of the scientist as a lone defender of reason and truth against the forces of doctrinal adherence, religious obscurantism, social conformity and the like. Thus there exists a large literature – especially in the case of Galileo – dedicated to the purpose of debunking such heroic myths by purporting to show how mixed were the motives (and very often how dubious the experimental methods) that produced some apparent great leap forward in the history of scientific thought.[20] As I have said, Shapin and Schaffer are at some pains to distance their approach from this particular brand of ill-disguised wholesale anti-scientism. Nevertheless, they share some of its main tenets, among them the social-constructionist view that 'matters of fact' (along with ideas of experimental method, observational warrant, valid reasoning on the evidence, etc.) must always be relativized to some pre-existing currency of socially accepted beliefs. From which it follows that nothing could count as a genuine *experimentum crucis* in the sense of a decisive test carried out under controlled laboratory conditions and involving the use of a given apparatus (like Boyle's air-pump) whose properties, behaviour and observed characteristics were potentially such as to falsify certain claims and to provide strong evidence – if not absolute proof – for the validity of certain others.

Hence the importance that Shapin and Schaffer attach to their idea of Boyle's controversial device as just one item or nodal point in a whole range of cognate 'technologies' – descriptive, mimetic, discursive, social, political, juridical and so forth – which played an equally important role in gaining support for his scientific claims. On this account the air-pump receiver functioned as 'a space into which one could move [a] paradigmatic experiment' in just the same way that 'the discursive and social practices in which talk about this experiment was to be embedded constituted a space in which disputes might be neutralized' (*LAP*, p. 42). Thus there is no real difference, in Shapin and Schaffer's view, between a physical apparatus designed to produce and to replicate experimental 'matters of fact' and the various discourses that circulated around and about that apparatus. After all, 'matters of fact in Boyle's new pneumatics were machine-made' and his mechanical philosophy in turn 'used the machine not merely as an ontological metaphor but also, crucially, as a means of intellectual production' (p. 26). Moreover, 'the "vacuum" Boyle referred to in his *New Experiments* was a new item in the vocabulary of natural philosophy; it was an operationally defined entity, reference to which was dependent upon the working of a new artificial device' (p. 80). Thus the standard accounts get it wrong – reveal their attachment to a naïve and discredited realist ontology – when they take for granted the difference in status between 'objective' facts about Boyle's apparatus, its physical construction, refinements of detail, operative conditions, etc. and disputes concerning the precise interpretation to be placed on Boyle's experiments. For this is to suppose that the air-pump worked – and produced verifiable results – in consequence, on the one hand, of its having been constructed with a view to achieving the best, most objective experimental data, and on the other of its having enabled the discovery of certain hitherto unknown facts about atmospheric pressure, the constitution of air and related phenomena. Such was no doubt Boyle's conviction and that of his fellow experimenters. But it cannot hold up – so Shapin and Schaffer contend – against the kinds of evidence their book brings to bear concerning the interest-relative status of 'matters of fact' or the socially constructed character of scientific knowledge. More specifically, it ignores their cardinal point: that the very apparatus in which (or through which) these experiments were conducted was itself an artificial construct, a disputed social or discursive 'space' whose attributes varied with the range of construals placed upon it by various interested parties.

IV

Thus the air-pump figures as a test case not only for the argument between Hobbes and Boyle but also for the present-day dispute between realists and anti-realists in philosophy of science. On Shapin and Schaffer's anti-realist view, it offers a richly documented field for their claim that what counts as scientific truth – or as good scientific method – is decided very largely by social, cultural and other such 'extraneous' factors. Just as the air-pump deconstructs (so to speak) into the various discourses or social practices that vied for possession of it, so the whole edifice of natural science becomes something more like a force field traversed by multiple contending theories, methods and agencies of power/knowledge. This is why the authors go into great detail on the physical construction of the air-pump and on Boyle's continual efforts to improve it in response to his opponents. There are three main points to be noted, they suggest:

> (1) that both the engine's integrity and its limited leakage were important resources for Boyle in validating his pneumatic findings and their proper interpretation; (2) that the physical integrity of the machine was vital to the perceived integrity of the knowledge the machine helped to produce; and (3) that the lack of its physical integrity was a strategy used by his critics, particularly Hobbes, to deconstruct Boyle's claims and to substitute alternative accounts. (*LAP*, p. 30)

There is no single statement in this passage to which one could take exception from a realist standpoint. That is to say, each point is fully compatible with an account of the air-pump experiments that holds the line between truth and falsehood, knowledge and belief, matters of (genuine) scientific fact and 'matters of fact' accepted as such on a less than adequate observational or theoretic basis. Nor could there be any quarrel – even by defenders of Boyle's programme – with the claim that his apparatus was by no means perfect, that most likely it suffered from a certain (albeit 'limited') degree of atmospheric leakage, and moreover that an absolute vacuum might be beyond reach of terrestrial attainment. For these concessions – though seized upon avidly by his critics then and since – in fact did nothing to impugn the validity of Boyle's experimental methods. Rather, they served to make his point, as against dogmatists like Hobbes, that such methods were the best available for reaching an informed (observation-based and rationally warranted) consensus on matters that allowed of no absolute demonstrative proof.

Also they preserved a due sense of the distinction – often blurred in recent debate – between ontological and epistemological issues. In this context the line was drawn between questions of the order: 'What is a vacuum?' or 'Can a vacuum exist in nature?' and questions of the order: 'How can we know – on what evidential basis – whether such a state has been (or can be) achieved by the best experimental methods?' Boyle's intent was to keep these issues apart by conceding the fallible nature of his research programme – the possibility of leakage, inaccurate measurement, interpretive anomalies and so forth – while none the less holding to the realist (ontological) premise that there existed an independent physical object domain that was in no sense a construct of the various beliefs that he or others might hold concerning it. For Shapin and Schaffer, conversely, the experimental 'discourse' or 'form of life' must be seen as a set of dominant conventions which effectively *decided in advance* what should count as a valid observation or a justified inference. Thus in Boyle's view, 'assent was to be secured through the production of experimental findings, mobilized into matters of fact through collective witness' (*LAP*, p. 152). But here, as so often in their book, we must allow for the invisible quotation marks around phrases such as 'experimental findings' and 'matters of fact'. For any findings thus arrived at would be deemed acceptable – or promoted into matters of fact – only on condition that they met the requirements laid down for the conduct of scientific discourse within that particular community. And those requirements, in turn, would exclude any finding that failed to promote the group interest in attaining consensus on just such amenable 'matters of fact' as suited its collective purposes. In short, '[t]he establishment of a set of accepted matters of fact about pneumatics required the establishment and definition of a community of experimenters who worked with shared social conventions: that is to say, the effective solution to the problem of knowledge was predicated upon a solution to the problem of social order' (p. 282). In which case – so the logic of this argument goes – scientific 'discoveries' by the experimental method could only confirm whatever seemed currently good in the way of belief.

The problems with this strong sociological approach are basically the same problems that arise with Kuhnian talk of paradigm shifts and radical meaning variance or Quinean talk of ontological relativity and the theory-laden character of observation statements.[21] That is to say, such ideas are very often pushed to the point where it becomes quite impossible to explain how science could ever have made progress by discarding certain theories under pressure from recalcitrant evidence and retaining others – always subject to criticism and possible future disconfirmation – if the evidence bore them out. Thus for Kuhn (as likewise for the strong sociologists) 'there is no standard higher than the assent of the relevant community.'[22] And again: any such standard 'must, in the final analysis, be psychological or sociological. . . . It must, that is, be a description of a value system, an ideology, together with an analysis of the institutions through which that system is transmitted and enforced.'[23] This last phrase is of particular interest with regard to what I have noted as the Hobbesian bias – the language of enforcement or coercive social authority – that typifies Shapin and Schaffer's approach to the Boyle/Hobbes controversy. For in their view there is nothing that could settle such issues except the appeal to *force majeure*, a socialized nexus of power/knowledge interests which dictate the very terms of any 'settlement' arrived at among those authorized to decree what shall count as good scientific method. Least of all can it be thought – as the realist would surely maintain – that despite the sundry pressures exerted by this or that interest group, cultural life-form, sociopolitical consensus, etc., still there is always an appeal open to discovery procedures (or experimental checks) which may sometimes produce a decisive break with conventional habits of belief.

It is just this idea that Shapin and Schaffer set out to undermine by every means at their methodological disposal. Thus they proceed, first, to treat Boyle's 'findings' as an artefact of his chosen apparatus (the air-pump); second, to deconstruct that apparatus into the various 'discourses' surrounding it; and third, to represent the vacuum itself as a notional topos – or discursive 'space' – whose attributes varied in accordance with the interests (the conflicting ideological investments) of parties to the air-pump debate. 'Scientific knowledge, like language, is intrinsically the common property of a group or else nothing at all. To understand it we shall need to know the special characteristics of the groups that create and use it.'[24] These sentences are Kuhn's – from *The Structure of Scientific Revolutions* – but they might just as well have been cited from *Leviathan and the Air-Pump* or other recent works of a kindred (anti-realist and strong sociological) persuasion. They are perhaps the more remarkable for appearing at the close of Kuhn's 1969 postscript, where he is chiefly concerned to answer those critics who had raised problems with his talk of incommensurable paradigms, of radical meaning-variance between theories, and of scientific knowledge as framework-relative and socially constructed. That his defence relies heavily on Quine's 'Two Dogmas' – along with late Wittgenstein on language games and forms of life – is a clear indication that these criticisms were not misplaced and that Kuhn was indeed a cultural relativist, as gladly proclaimed by his many followers in the human and social sciences.

Shapin and Schaffer are more guarded than Kuhn in their explicit formulations of the relativist case or their statements concerning the socially constructed character of scientific 'facts'. They also provide much more in the way of detailed historical and sociocultural background, thus (from their point of view) avoiding the charge – often brought against Kuhn – of failing to explain why changes in the currency of scientific knowledge should ever come about. But their explanation is entirely in terms of those social pressures or motivating interests that are taken to decide what passes for fact (or for good scientific method) among members of a given community. That is to say, it starts out from a strong sociological position and then proceeds – in purely circular style – to find that position everywhere confirmed by just the sorts of evidence that carry most weight with strong sociologists of knowledge. Of course it might be argued, in defence of their approach, that this objection simply misses the point, since Shapin and Schaffer are under no illusions regarding the interest-relative character of their own claims or the need for sociologists to be fully 'self-reflexive' in applying the Principle of Symmetry. Such is the stance they maintain, for fairly obvious reasons, in their opening methodological remarks. However, that principle tends to be dropped – or applied in highly selective fashion – when they move on to the main business of discussing Boyle's air-pump experiments. For it then becomes clear that *some* 'matters of fact' (those pertaining to the natural sciences) are to get the full relativist treatment, while others (those in the social or cultural domain) are for some reason granted exemption.

Here, as with Kuhn, there is a curious blind spot that can only be accounted for in terms of the authors' very marked disciplinary bias. After all, one would think that the natural sciences had a much better claim to such privileged status in virtue of their methodological refinement, their well-defined discovery procedures, their strong record of explanatory and predictive success, and – not least – their openness to criticism on rational and constructive grounds. The same can scarcely be said of the social and human sciences where issues of method are deeply divisive, where 'facts' or 'discoveries' are always contested, where explanatory claims have at best a limited purchase, and where interpretation goes a long way (if not perhaps all the way) down. So there is, on the face of it, something highly suspect about the transfer of epistemic confidence by which Shapin and Schaffer treat social facts as a source of reliable evidence while adopting an attitude of a priori scepticism towards Boyle's experimental programme. To put it bluntly, we have progressed much further in our knowledge of the physical world through the methods of the natural sciences than we have in our knowledge of the human world (of history, society, cultural practices, motives, intentions, values, etc.) through sociology and allied disciplines.

A similar point could be made about the turn towards anti-realism in recent philosophy of science, based as it is – very often – on the idea that scientific truth-claims are 'constructed in' (or relative to) some particular language game, paradigm, discourse or conceptual-semantic scheme. For the realist can then quite reasonably ask why on earth we should adopt such a range of ill-defined ('essentially contested') theories when at least one alternative lies ready to hand. This alternative is an inference to the best explanation which takes it that science has made real progress towards the better understanding of a physical world that none the less exists – and exerts its various causal powers – independently of our various beliefs concerning it.

References

1 Steven Shapin and Simon Schaffer, *Leviathan and the Air-Pump: Hobbes, Boyle, and the experimental life* (Princeton, NJ: Princeton University Press, 1985).

2 See, for instance, Barry Barnes, *About Science* (Oxford: Blackwell, 1985); David Bloor, *Knowledge and Social Imagery* (London: Routledge and Kegan Paul, 1976) and *Wittgenstein: a social theory of knowledge* (New York: Columbia University Press, 1983); Harry Collins, *Changing Order: replication and induction in scientific practice* (Chicago: University of Chicago Press, 1985); Steve Fuller, *Social Epistemology* (Bloomington: Indiana University Press, 1988) and *Philosophy of Science and its Discontents* (Boulder, CO: Westview Press, 1989); K. Knorr-Cetina and M. Mulkay (eds), *Science Observed* (London: Sage, 1983); Bruno Latour and Steve Woolgar, *Laboratory Life: the social construction of scientific facts* (London: Sage, 1979); Andrew Pickering, *Constructing Quarks: a sociological history of particle physics* (Edinburgh: Edinburgh University Press, 1984); Andrew Pickering (ed.), *Science as Practice and*

Culture (Chicago: University of Chicago Press, 1992); Andrew Ross, *Strange Weather: culture, science and technology in the age of limits* (London: Verso, 1991); Steve Woolgar, *Science: the very idea* (London: Tavistock, 1988).

3 See Shapin and Schaffer, *Leviathan and the Air-Pump* for a detailed bibliography of Boyle's works and other relevant sources.

4 See Robert Boyle, *The Sceptical Chymist* [1661] (London: Dent, 1911).

5 For a detailed bibliography of relevant works by and on Hobbes, see Shapin and Schaffer, *Leviathan and the Air-Pump*; also Thomas A. Spragens, *The Politics of Motion: the world of Thomas Hobbes* (Lexington, KY: University of Kentucky Press, 1973).

6 See, for instance, Spragens, *The Politics of Motion* and D. P. Gauthier, *The Logic of Leviathan* (Oxford: Clarendon Press, 1969).

7 See, for instance, Michel Foucault, *Power/Knowledge: selected interviews and other writings* (Brighton: Harvester, 1980); *Language, Counter-Memory, Practice*, ed. and trans. D. F. Bouchard and S. Weber (Oxford: Blackwell, 1977); *The Foucault Reader*, ed. P. Rabinow (Harmondsworth: Penguin, 1986).

8 See entries under note 2 above; also Thomas S. Kuhn, *The Structure of Scientific Revolutions*, 2nd ed. (Chicago: University of Chicago Press, 1970); W. V. Quine, 'Two Dogmas of Empiricism', in *From a Logical Point of View*, 2nd edn (Cambridge, MA: Harvard University Press, 1961), pp. 20–46.

9 Ludwig Wittgenstein, *Philosophical Investigations*, trans. G. E. M. Anscombe (Oxford: Blackwell, 1953).

10 On the 'Principle of Symmetry', see Barnes, *Scientific Knowledge and Sociological Theory* (London: Routledge and Kegan Paul, 1974) and Bloor, *Knowledge and Social Imagery*; also – for some strong counter-arguments – W. H. Newton-Smith, *The Rationality of Science* (London: Routledge and Kegan Paul, 1981), pp. 247–65.

11 See, for instance, Ludwig Flieck, *Genesis and Development of a Scientific Fact* (Chicago: University of Chicago Press, 1979); K. Knorr-Cetina and M. Mulkay (eds), *Science Observed*; Bruno Latour, *Science in Action* (Milton Keynes: Open University Press, 1987); Andrew Pickering (ed.), *Science as Practice and Culture*.

12 See especially Roderick M. Chisholm, *Theory of Knowledge*, 2nd edn (Englewood Cliffs, NJ: Prentice-Hall, 1977).

13 Ernest Gellner, 'Concepts and Society', in Bryan R. Wilson (ed.), *Rationality* (Oxford: Blackwell, 1970), pp. 18–49.

14 Donald Davidson, *Inquiries into Truth and Interpretation* (Oxford: Clarendon Press, 1984).

15 Christopher Norris, *Resources of Realism: prospects for 'post-analytic' philosophy* (London: Macmillan, 1997).

16 See David Papineau, *Philosophical Naturalism* (Blackwell, 1993).

17 See entries under note 7 above.

18 On these Hobbesian elements in Foucault's thinking, see Michael Walzer, 'The Politics of Michel Foucault', in David C. Hoy (ed.), *Foucault: a critical reader* (Oxford: Blackwell, 1986), pp. 51–68.

19 William Empson, review of E. A. Burtt's *The Metaphysical Foundations of Modern Physical Science*, in *Argufying: essays on literature and culture*, ed. John Haffenden (London: Chatto and Windus, 1987), pp. 530–3; p. 531.

20 See especially Paul Feyerabend, *Against Method* (London: New Left Books, 1975).

21 See entries under note 8 above.

22 Kuhn, *The Structure of Scientific Revolutions*, p. 94.

23 Kuhn, 'Logic of Discovery or Psychology of Research?', in I. Lakatos and A. Musgrave (eds), *Criticism and the Growth of Knowledge* (Cambridge: Cambridge University Press, 1974), pp. 1–23; p. 21.

24 Kuhn, *The Structure of Scientific Revolutions*, p. 210.

11

Leviathan and the Turbojet: A Critique of Sociological Unreason

I

In this chapter I shall offer a constrastive account of two books that address central issues in the history and philosophy of science. One – Shapin and Schaffer's *Leviathan and the Air-Pump* – is already well established as a classic of its genre, namely the 'strong' programme in sociology of knowledge.[1] The other – Edward Constant's *The Turbojet Revolution* – has attracted nothing like the same degree of interest (reviews, citations, statements of methodological indebtedness) as measured by the usual academic indicators.[2] Perhaps this is not so surprising, given its relatively specialized field of concern and the fact that it offers a detailed reconstruction of one particular episode in the history of applied techno-scientific knowledge. But there are, I shall argue, some additional factors that have worked to promote *Leviathan and the Air-Pump* as a reference text among cultural theorists and socio-historians of science, and worked just as strongly to limit the reception of Constant's book and others like it. These have to do with the current predominance of anti-realist and cultural-relativist thinking in disciplines whose chief objective is to cast doubt on the truth-claims, methods and evaluative procedures of the natural sciences.[3] Above all, they reject what Constant adopts as a basic working premise: that there exists a real-world (mind- and theory-independent) physical domain whose properties are the object of scientific knowledge and whose better understanding is the prime source of technological advance.

Shapin and Schaffer challenge this view as just another instance of the standard collusion between 'mainstream' science, philosophy of science, and orthodox historiography. On their account, it is a preferred (largely mythical) self-image of scientific method whose origins can be dated to the seventeenth century and the emergence of an 'experimental culture' centred upon the physicist Robert Boyle and his famous series of air-pump trials to demonstrate the properties of vacuum phenomena.[4] More specifically, they challenge the received idea that this method established a space for the conduct of scientific debate that so far as possible removed it from the sphere of conflicting ideologies and sociopolitical interests.

Such was indeed the professed aim of Boyle and his Royal Society colleagues: to ensure 'a strict boundary between natural philosophy and political discussion', so that experiments could carried out (and 'matters of fact' established) in a spirit of free and open enquiry. In short, 'it was proper to speak of matters of fact because they were not of one's own making: they were, in the empiricist language-game, discovered rather than invented.'[5] However – Shapin and Schaffer argue – we should not be taken in by this rhetoric of *wertfrei* (value-neutral) scientific knowledge or pure, disinterested seeking after truth. Rather, we should see it as an ideological project designed to mobilize public assent for values and beliefs that were just as much political as scientific. Thus they aim to deconstruct Boyle's cardinal distinction between 'discovered' and 'invented' matters of fact, or those that refer to an objective reality beyond such values and beliefs, and those that might be viewed as social constructs reflecting some particular partisan interest. For if this distinction turns out to be untenable – or itself just a construct of one such dominant interest group – then the way is clearly open to a wholesale (strong sociological) assault on the ideas of scientific truth, reason, method, objectivity and progress.

'Experimenter's regress' is among their chief weapons in arguing against the claims of scientific realism. Thus: 'before any experimenter could judge whether his machine was working well, he would have to accept Boyle's phenomena as matters of fact. And before he could accept those phenomena as matters of fact, he would have to know that his machine would work well' (*LAP*, p. 226). But this circularity argument will appear convincing only if one maintains, like Shapin and Schaffer, that the 'phenomena' were products of interpretation through and through; that the air-pump apparatus was likewise a figment of the various discourses surrounding it; and that the (notional) vacuum was itself nothing more than a socially constructed 'space' wherein those discourses vied for control over the production of scientific 'knowledge'. For otherwise – on the principle of inference to the best explanation – it would seem much more rational to suppose that the air-pump apparatus was real enough (constructed on principles and with design features that Boyle was at pains to specify); that it produced phenomena subject to experimental testing and detailed observation; and moreover, that sometimes the results turned out to conflict with Boyle's expectations.

This latter is a crucial point since it goes clean against the anti-realist (and strong sociological) claim that experiments or research programmes are always set up so as to decide in advance – by convention, agreement or stipulative warrant – what shall count as a valid or acceptable result. However, it receives a quite different interpretation from Shapin and Schaffer. On their account, Boyle's writing-up of his 'failed' experiments was a credibilizing strategy that served two purposes: 'first, it allayed anxieties in those neophyte experimentalists whose expectations of success were not immediately fulfilled; second, it assured the reader that the relator was not wilfully suppressing inconvenient evidence, that he was in fact being faithful to reality' (*LAP*, p. 64). And again: 'a man who recounted unsuccessful experiments was such a man whose objectivity was not distorted by his interests' (p. 65). In which case the realist can only be deluded in arguing from the 'fact' of those failed experiments to the objectivity of Boyle's procedures, the substantive properties of his air-pump apparatus, and the existence of vacuum effects quite apart from the various interpretations placed upon them.

One is tempted to ask whether Shapin and Schaffer would prefer Boyle to have suppressed those discrepant results or written them up in such a way as to conceal the conflict between theory and observation. If so, then he would be doing no more than act on the standard ontological-relativist principle (as enounced by Quine, Kuhn and others) that anomalous data can always be finessed by 'redistributing predicates' over the range of existing beliefs or reinterpreting results so as to fit in with the prevailing paradigm or consensus view.[6] Such, according to Shapin and Schaffer, was the process of strategic 'negotiation' whereby Boyle and his colleagues contrived to gain assent for their experimental programme. It is evident in the various judgements – themselves a matter of 'social convention' – that decided when or whether an air-pump experiment had been 'replicated' in such a way as to confirm the original result and thus claim warrant as a genuine instance of 'virtual witnessing'. These judgements concerned, among other things,

> the moment when skill in making pumps had been transmitted, when a replica of the pump could be said to have been produced, when that replica had produced the same phenomenon as that reported by Boyle, and when a phenomenon could count as a challenge to Boyle's own claims. A range of commitments and investments bore on judgements whether replication had or had not been achieved, of whether or not a claimed phenomenon existed as a fact of nature. (*LAP*, p. 226)

Once again this passage exhibits a kind of reverse ('deconstructive') logic whereby the reality of Boyle's experiments or the veridical status of his scientific claims is thrown into doubt by treating them as purely conventional or social artefacts. For there is no difference, in Shapin and Schaffer's view, between the sorts of problem that arose in judging replications of Boyle's experimental procedure and the sorts of problem that arose in judging any results (or phenomena) produced by the 'original' air-pump. Indeed, they take a certain mischievous delight in tracing – or failing to trace – the various 'originals' and 'replicas' that were constructed in the wake of Boyle's early demonstrations. For it was *always* a matter of 'virtual witnessing' in the sense that any reality ascribed to vacuum phenomena – and even to the air-pump itself as a construct designed with certain purposes in view – was dependent on (or relative to) the interests of its observers.

Thus their point about the various 'technologies' involved is that these cannot be ranged on a scale of diminishing reality, so speak, from the physical apparatus at the one end to its social, cultural and 'literary' replications at the other. After all, 'matters of fact in Boyle's new pneumatics were machine-made,' and the machines that made them were themselves products of a discourse (that of experimental science) whose claims were legitimized by the newly emergent Restoration consensus (*LAP*, p. 26). In which case – following this argument through to its 'logical' conclusion – the space in Boyle's air-pump chamber was not so much the site of a plenum or a vacuum (the issue supposedly at stake between Hobbes and Boyle) as the site of an ideological struggle between different conceptions of politics, society and the means by which civil order could best be secured. For Hobbes,

> civil war flowed from any programme which failed to secure absolute compulsion. What was a judicious and liberal bracketing strategy to the Greshamites was, to Hobbes, a wedge opening the door which looked out on the war of each against each. Any working solution to the problem of knowledge was a solution to the problem of order. That solution had to be absolute. . . . All men make and sustain society, because all men that have natural reason can be made to see that it is in their interests that Leviathan be created and maintained. Having made civil society to protect themselves, the obligation to submit is total. The force by which submission is exacted is the delegated force of all those that enter into society and live as social beings. The intellectual enterprise which rationally demonstrates this to all men possesses an absolutely compulsory character. It is in philosophy what Leviathan is in society. (*LAP*, p. 152)

This passage provides a good brief statement of Hobbes's political and philosophic views as well as condensing the main points at issue in his controversy with Boyle.[7] Thus it explains (1) why Hobbes rejected any science based on observation and consensus, (2) why he adopted an 'absolutist' (demonstrative or a priori) approach, and also (3) why he espoused a plenist ontology which left no room for a vacuum in nature or for just those kinds of dissenting – potentially discordant – belief that the vacuum hypothesis seemed to invite. In other words, it offers as neat and convincing a piece of *sociological* explanation as any to be found in the recent literature. Still one may not wish to follow Shapin and Schaffer when they push their case beyond the realm of sociology (or cultural history) and aspire to 'deconstruct' some of the basic tenets of scientific method and 'mainstream' history/philosophy of science. For they are then committed to the much more dubious – indeed quite incredible – claim that science might have taken a different course (and come up with a whole alternative range of theories, observations, experimentally induced phenomena, applied technologies, etc.) had the sociopolitical climate of the time not favoured Boyle over Hobbes.

This is what Shapin and Schaffer have in mind when they recommend the 'stranger's perspective' as a useful means of defamiliarizing our standard (acculturated) habits of judgement and belief. A controversy such as that between Hobbes and Boyle is of particular interest in this respect since it enables us to take a fresh look at arguments which have long been written off – by mainstream opinion – as scarcely worth the effort of detailed reconstruction. Thus Hobbes occupies the role of a 'pretend stranger', one who 'attempt[s] to deconstruct the taken-for-granted quality of [his] antagonists' preferred beliefs and practices', and who does so moreover 'by trying to display the artifactual and conventional status of those beliefs and practices' (*LAP*, p. 7). Since this is precisely what Shapin and Schaffer are trying to do, one might expect them to show a certain methodological bias towards the Hobbesian side of the argument. For it is not just an issue – as they see it – of explaining why Hobbes should have held such views on account of his ideological commitments or motivating beliefs. More than that, it is a question of *justifying* Hobbes as a scientist (or natural philosopher) whose ideas happen to have fallen foul of a then newly emergent and thereafter predominant consensus view on these matters. 'Obviously,' the authors remark, 'many aspects of the programme [Boyle] recommended continue to characterize modern scientific activity and philosophies of scientific method' (p. 11). But rather than seek an explanation for Boyle's 'success' in the fact that his programme opened the way to so many scientific discoveries, Shapin and Schaffer turn this argument right around and treat the very notion of 'discovery' in science as a product of that same experimental ethos whose success was guaranteed – or at any rate solidly underwritten – by a prevailing social consensus. Thus 'an answer to the question . . . begins to emerge, and it takes a satisfying historical form,' namely that '[the] experimental form of life achieved local success to the extent that the Restoration settlement was secured' (p. 341).

It seems to me that Shapin and Schaffer exhibit a quite heroic determination to hold this puzzle very firmly the wrong way up. The strain comes out in the many passages where they cite Hobbes's counter-arguments to Boyle – his plenist ontology, his aether hypothesis, his theory concerning the 'congenital circular motion' of air particles – without so much as a hint that they might have been based on a false understanding of the physical phenomena involved.[8] Of course, it may be said that those 'phenomena' are just what Hobbes refused to accept, not only with regard to Boyle's interpretation of them but also – more radically – on the grounds that true science should operate on strictly demonstrative principles rather than rest content with matters of belief concerning phenomenal appearances. So it was that he arrived at the above set of a priori necessary ('absolute and compulsory') truths. Yet if we suppose Hobbes to have been right in all this then we should also perforce have to conclude that much of modern science (including the most basic principles of pneumatics, hydraulics, fluid mechanics and aerodynamics) is massively in error. More than that, we should be in the position of denying – like Hobbes – that experiments have any decisive (or even useful) role to play in the production of scientific knowledge, based as they are on such fallible resources as mere observation and the achievement of rational consensus through informed criticism and debate. Nor could we have recourse to arguments – such as inference to the best explanation – which take theories to be justified in so far as they account most convincingly for our knowledge of the growth of scientific knowledge as manifest in various fields of enquiry.[9] For this would once again be to beg the question of just what should count as 'knowledge', or 'progress', given the socially constructed character of scientific truth-claims and the lack of any neutral (non-interest-relative) criteria for deciding the issue between rival claims. In which case there could be no reason – conventional prejudice aside – for thinking Boyle to have established certain facts concerning vacuum phenomena which Hobbes was predisposed to reject on account of his deep-laid ideological commitments.

Of course, such an argument will count for nothing if one believes (as apparently do Shapin and Schaffer) that long-run judgements of scientific fact are no more reliable – or no less subject to pressures of circumstance – than those produced under far from ideal conditions at an early stage in some particular branch of enquiry. Thus the authors operate on two main assumptions: first, that the Restoration 'settlement' set a pattern for the liberal consensus view of good scientific method; and second, that present-day attitudes to the history of science have been largely shaped by that same still dominant consensus view. From which they conclude that Hobbes's theories have suffered a great injustice merely on account of Boyle's having captured the ideological high ground then and since. But this leaves them with the hard task of explaining why Boyle's results contributed to some of the more impressive achievements of modern science – such as aerodynamics (or airfoil theory) as a means of accounting for the possibility of heavier-than-air flight – while Hobbes's ideas, if true, would have rendered such achievements a priori impossible.[10]

As for the Hobbesian aether hypothesis, Boyle was quite prepared to let that one go by default, since the question of the aether's presence or absence was a purely metaphysical matter which could not be decided by any experimental (or properly scientific) method. Thus: 'the plenist could maintain belief in the aether, even its presence in the evacuated receiver, but he must not deploy the aether as a physical explanation of the pump's phenomena: this aether "is such a body as will not be made sensibly to move a light feather"' (*LAP*, p. 184). And again: 'Boyle permitted plenists to retain their commitment to an all-pervading subtle matter, so long as it did not figure in the interpretation of the air-pump experiments' (p. 184). Shapin and Schaffer incline to treat this as yet another instance of Boyle's liberal rhetoric disguising what was in fact a highly prescriptive and far from disinterested method of enquiry. Once again they take Hobbes's side, in effect, by demanding *why* this particular issue (the existence or non-existence of an aether) should be thus ruled out as beyond the remit of genuine scientific research. And indeed it is an issue that cannot be lightly dismissed if one judges by its continued prominence in recent – mostly strong sociological – writings.

Thus a similar case has been made with regard to the famous experiment by Michelson and Morley which attempted to settle this question through measurements of the time taken by two beams of light to traverse the same distance when projected at right angles to each other.[11] That is to say, if these measurements turned out to differ then the best explanation would run as follows: that the speed of light was relative to the earth's velocity and direction of travel through an aether – an all-pervasive fluid medium – which itself provided a fixed point of reference or a stable coordinate system. The experiments were especially crucial at this time, since any positive result (any proven difference in the measurements) would have cast serious doubt on the hypotheses advanced in Einstein's Special Theory of Relativity. For if the aether could indeed be shown to exist then it – and not the speed of light – was the ultimate determinant of space-time locations and velocities both earthly and extra-terrestrial. In other words, there would be no need for a theory, such as Einstein's, which entailed so radical a break with the assumptions of classical (post-Newtonian) physics.[12]

What makes this episode all the more interesting from the strong sociological viewpoint is that some of the early Michelson/Morley results seemed to indicate that there was in fact a marked discrepancy between the two sets of measurements. That is, it appeared that the speed of light could be shown to differ – albeit by a scarcely detectable margin – according to whether it travelled with or against the direction of the earth's movement relative to the aether. Einstein none the less continued to maintain his theory, partly on account of its ability to explain various problematic phenomena, and partly on the grounds of its elegance – or strong intuitive appeal – along with its power to unite otherwise discrete fields of scientific enquiry. In the event, his conviction was borne out by repeated trials of the Michelson–Morley experiment, which produced a result in keeping with Einstein's hypothesis, i.e., that the velocities would not differ since the speed of light was a universal constant whose value was itself the fixed point of reference relative to which all other variables (observational standpoints or space-time locations) must henceforth be determined. That the early results had appeared to falsify the Special Theory was now put down to problems with the first experimental set-up or excessive margins of error in carrying out so immensely delicate a series of measurements. And in the long-run perspective it can surely be claimed that Einstein's hypothesis has proven its worth through a range of subsequent discoveries – from the red-shift phenomenon in astrophysics to the entire development of nuclear technology and sub-atomic particle research – which would otherwise lack any remotely plausible explanation.

However, such arguments are quite beside the point if one assumes, like the strong sociologists, that 'mainstream' history of science is always written by the victors and embodies nothing more than a partisan view of scientific progress and method. For it is then possible to argue, concerning the Michelson–Morley experiments, that the aether issue was by no means definitively settled; that the results (both early and late) were open to doubt on various observational and interpretative grounds; and hence that historians of science have applied their usual techniques of selective hindsight in making the story come out right from an orthodox (post-Einsteinian) perspective. Nor is it in any way coincidental that the aether figures here – as likewise in *Leviathan and the Air-Pump* – as a locus for disputes having to do with the status of truth-claims in the physical sciences and with the issue of realism versus anti-realism that is often raised in this context. Thus, according to Shapin and Schaffer,

> Boyle's most consequential stipulation about Hobbes's views on leakage and the constitution of the air concerned the role of an aether. Boyle agreed with his adversary that the air might be regarded as a heterogeneous mixture. He allowed that one of its fractions might be an aether, and even that this aether might intrude itself (or be always present) in the receiver. But he took exception to Hobbes's alleged use of the aether to impugn the findings of the pump. . . . This aether must either be demonstrated by experiment to exist or it was to be regarded as a *metaphysical* entity. Plainly, Hobbes's introduction of an aether in the context of the air-pump trials was, for Boyle, based on no such experimental evidence. (*LAP*, p. 181)

Once again the authors incline to view this as a stipulative fiat – or an authoritarian prescription – imposed by Boyle for no better reason than to render his programme proof against challenge from an adversary quarter. But then it must asked just where – if anywhere – they would draw the line between hypotheses admissible in a scientific context and hypotheses that lack any possible means of verification (or falsification) and are therefore simply not candidates for the status of scientific knowledge. For the trouble with Hobbes's aether-based theory was that it *made no difference*, in practical terms, whether or not one took the receiver to be filled at all times with a 'subtle fluid' whose insensible effects were *by very definition* beyond reach of experimental witness. In other words, the question of its existence or non-existence was a matter of no scientific interest, whatever its crucial importance (for Hobbes and his disciples) as a matter of a priori metaphysical belief.

In this respect, Hobbes's theory of the aether – and his entire plenist ontology – were hypotheses on a par (say) with Aristotle's theory of the four elements or the four humours. That is to say, they were compatible with *all and any* results that might turn up in the course of enquiry, since nothing could count – scientifically speaking – as counter-evidence to claims that were wholly devoid of substantive content, or were framed in such a way as to elude the best efforts of experimental proof. Of course, this criticism cannot apply to the Michelson–Morley experiments. For here it was precisely a matter of testing a well-defined scientific hypothesis – i.e., that the speed of light would be affected by its directionality relative to the aether – through application of the best (most refined and technologically advanced) methods to hand. This can scarcely be said of Hobbes's arguments, compounded as they were of dogmatic certitude, deep-laid ideological bias, and an absolute refusal to accept any evidence – least of all any mere experimental results – which came into conflict with his own fixed beliefs. Nevertheless, as Shapin and Schaffer note, 'Boyle permitted plenists to retain their commitment to an all-pervading subtle matter, so long as it did not figure in the interpretation of the air-pump experiments' (*LAP*, p. 184). He could well afford to do so since the aether hypothesis – together with the entire plenist metaphysic – was an issue which could have no possible bearing on the validity of Boyle's experimental results or of his scientific programme as a whole. Nor should it give much comfort to the strong sociologists and other sceptics that the Michelson–Morley results were at first very far from proving the validity of Einstein's conjecture. For in this case the final outcome – discounting for experimental errors – was such as to refute the aether hypothesis and leave no room for the kind of accommodating gesture that Boyle extended to his adversaries.

However, the main point, in this context, is not so much that Einstein's theory was eventually vindicated but that the strong sociologists are prone to rest their case on a false understanding of experimental method and its role in the process of scientific theory formation. This is the idea – exploited to the full by Shapin and Schaffer – that in showing some *particular* experiment not to have attained the highest possible standards of demonstrative rigour, technical refinement, observational accuracy and so forth, one has thereby knocked a sizeable hole in the *entire enterprise or research programme* of which that experiment formed a part. In *Leviathan and the Air-Pump* this approach works out as a systematic bias towards Hobbes's side of the argument, that is to say, an implicit demand that Boyle's methods match up to the standard – of 'absolute', 'coercive' or 'compulsory' assent – laid down by Hobbes for the proper regulation of natural and civic philosophy alike. No doubt the authors would reject such a claim, committed as they are by overt profession to a Principle of Symmetry that eschews all judgements of scientific truth or falsehood in the interests of a purely sociohistorical approach. But the effect of this approach – as I have argued – is to subtly weigh the scales against any method (such as Boyle's) which appeals to established scientific criteria of truth, validity or rational debate, and in favour of any philosophy (such as Hobbes's) that takes those values to be merely the instrument of some well-placed special interest group.

II

Thus there can be little doubt of Shapin's and Schaffer's scepticism when they describe Boyle's 'solution' to the problem of knowledge as one that involved 'set[ting] the house of natural philosophy in order by remedying its divisions and withdrawing it from contentious links with civic philosophy' (*LAP*, p. 21). For their entire line of argument in *Leviathan and the Air-Pump* is such as to deny that issues of scientific method can ever be 'withdrawn' – or held safely apart – from issues of political power. It is perhaps not so obvious that their approach tends to favour the alternative (Hobbesian) 'solution', namely 'that order was only to be ensured by erecting a demonstrative philosophy that allowed no boundaries between the natural, the human, and the social, and which alllowed for no dissent within it' (p. 21). But there is, none the less, a very marked elective affinity between Shapin and Schaffer's methodological premises – widely shared among strong sociologists – and Hobbes's absolutist conception of the power/knowledge nexus. It comes out clearly in their repeated statements to the effect that 'the problem of generating and protecting knowledge is a problem in politics, and, conversely . . . the problem of political order always involves solutions to the problem of knowledge' (p. 21). For this was exactly Hobbes's belief and his chief reason for opposing what he saw as the emergence of a discourse – that of experimental science – which opened a space for dissenting views on matters (scientific and political) that should properly stand beyond reach of dispute.

Of course, Shapin and Schaffer are as far from endorsing Hobbes's politics as they are from claiming that his natural philosophy was in any way superior to Boyle's experimental method. Such matters, after all, can be none of their concern as historians committed to the strong sociological premise that judgements of truth – whether short-run or long-run – always reflect some prevailing cultural or ideological bias. However, it is just this presumption, I suggest, that constitutes their own prevailing bias towards a Hobbesian 'solution' to the 'problem of knowledge'. Thus when they remark that '[i]n this book [*Leviathan*] Hobbes took away vacuum on definitional, historical, and, ultimately, on political grounds' (p. 91) the description could as well apply to the procedures adopted in *Leviathan and the Air-Pump*. That is to say, Shapin and Schaffer 'take away vacuum' as a natural phenomenon – a 'matter of fact' – the existence of which (and its determinate effects) could best be established through the methods of experimental science. They do so, moreover, by a studious application of those three main argumentative techniques – 'definitional' 'historical' and 'political' – that Hobbes brought to bear in his critique of Boyle. For their argument takes hold precisely through a stipulative redefinition (in this case a defining-away) of what Boyle took to be the proper boundary between scientific method and matters of sociopolitical dispute. Thus: '[t]he vacuism Hobbes attacked was not merely absurd and wrong, as it was in his physical texts; it was *dangerous*. Speech of a vacuum was associated with cultural resources that had been illegitimately used to subvert proper authority in the state' (*LAP*, p. 91). Again, it would be wrong – naïve in the extreme – to take such passages of *oratio obliqua* (or indirect style) as suggesting that the authors are somehow in agreement with Hobbes on his substantive scientific, social or political claims. But they do make it possible for Shapin and

Schaffer to suspend all judgements of scientific truth and falsehood in favour of a strong sociological approach that treats such judgements – after the manner of Nietzsche and Foucault – as products of an epistemic will-to-power that pervades every 'discourse' of accredited knowledge in the natural and social sciences.[13] And from here it is but a short step to the notion that Hobbes's plenist ontology was perfectly valid on its own terms, that is to say, as a discourse which – unlike Boyle's – acknowledged the omnipresent workings of power and the impossibility of severing the link between questions of truth or valid method in the scientific domain and issues of legitimacy in the sociopolitical sphere.

Thus where Boyle 'attempted to appropriate this paradigmatic phenomenon [the vacuum] for the discursive practices of the new experimental programme', Hobbes conversely 'addressed cohesion [of two flat-surfaced objects inside the air-pump receiver] in order to demonstrate that its correct physical explanation was incompatible with vacuism' (*LAP*, p. 90). His alternative explanations – the 'flexing' of those surfaces no matter how hard their material, the inrush of 'subtle' or 'purified' air, the presence of an aether (if all else failed) as a matter of 'absolute', 'demonstrative' warrant – are not to be asssessed by received standards of scientific truth and method. For, of course, those standards are derived in large part from the 'discourse' of experimental science which Boyle and his colleages managed to propagate so effectively and which has since then set the very terms for what counts as a good scientific practice. Thus the dispute between Boyle and Hobbes was not the kind of difference that could ever be resolved – then or now – on 'purely' scientific (least of all experimental) grounds. Rather, '[i]t was a difference in conceptions of proper speech about such phenomena, and, therefore, a difference in exemplifying how the natural philosopher was to go on' (p. 91). And if the very choice of term 'natural philosopher' suggests a certain Hobbesian bias – that is, the idea that scientific theories are always bound up with issues of a wider (sociopolitical) import – then this seems to follow directly from Shapin and Schaffer's practice as cultural historians. For there is nothing on their account that could possibly explain how the methods of scientific enquiry – conjecture, hypothesis, experimental testing, criticism, inference to the best explanation, etc. – could ever break free of this collusive imbrication with structures of socially legitimized power/knowledge.

There are passages in their book where Shapin and Shaffer do seem to acknowledge the value (even the truth-productive yield) of a 'liberal' approach, like Boyle's, that eschewed the a priori dogmatic method of Hobbesian natural philosophy. Thus: 'what the *New Experiments* did was to *exemplify* a *working* philosophy of scientific knowledge. . . . In a concrete experimental setting it showed the new natural philosopher how he was to proceed in practical matters of induction, hypothesizing, causal theorizing, and the relating of matters of fact to their explanations' (p. 49). And again: 'Boyle's epistemological armamentarium included matters of fact, hypotheses, conjectures, doctrines, speculations, and many other locutions serving to indicate causal explanations' (p. 49). All of which amounts – one might think – to a strong case for Boyle's approach as one that found room for a wide range of scientific procedures just so long as they were ultimately subject to experimental check in the sphere of informed participant debate. However, Shapin and Schaffer take a different view, treating it as yet further evidence of Boyle's and his colleagues' skill in conjuring assent for a scientific programme which imposed its own hegemonic values under cover of a liberal-sounding rhetoric. Such is the strong suggestion carried by their phrase 'epistemological armamentarium', working as it does to convey the idea that all this apparent liberality was a ruse in the service of power/knowledge interests with coercive sanctions attached. And their suspicion is reinforced by Boyle's appeal to experiment – or observational witness – as the distinguishing mark of any scientific programme that would seek to move beyond the stage of mere 'conjecture' or unsupported 'hypothesis'. For it is just this appeal which they (like Hobbes) set out to 'deconstruct' as an instrumental fiction adopted by one party – the new experimentalists – with a view to silencing their opponents.

As I have said, Shapin and Schaffer make much of the fact that some of Boyle's experiments were inconclusive and others conducted under laboratory conditions – with crude apparatus, possibilities of leakage, inaccurate means of measurement, wide margins of observational error, etc. – which cannot but raise questions concerning their scientific validity or probative force. Again they attempt to be even-handed about this. Thus: 'True, there were points at which Hobbes's criticisms were less than well-informed, just as there were aspects of Boyle's position that might be regarded as ill-informed and even sloppy.' In short, '[i]f the historian *wanted* to evaluate the actors by the standards of present-day scientific procedure, he would find both Hobbes and Boyle vulnerable' (*LAP*, p. 13). Of course, this can be no part of the authors' intention, given the strong sociological premise that all truth-claims should be viewed without prejudice regarding their 'scientific' status. All the same, it is clear that Hobbes most often gets the benefit of this methodological doubt while Boyle and the new experimentalists are under a constant pressure to justify claims which the authors – in company with Hobbes – tend to treat as mere products of conventional belief or figments of the social imaginary. Hence their alacrity in seizing on just those aspects of the air-pump experiments which can be argued to exemplify the failings or inadequacies of the new experimental programme.

However, there is a basic confusion here concerning the nature of scientific experiments and the way they contribute to the growth of knowledge in particular realms of enquiry. It is the error of supposing that *any* experiment – or sequence of experiments – could possibly achieve that 'absolute', 'demonstrative' power of securing rational assent which Hobbes took to be the mark of genuine science, and which seems (oddly enough) to furnish the unspoken criterion for Shapin and Schaffer's account of the air-pump controversy. For if one thing is clear from the history of science, it is the fact that experiments can lend strong support to hypotheses, conjectures or theories while often in some respect falling short of the strictest procedural requirements.[14] What counts is not so much their 'compulsory' force – their power to demonstrate a necessary or law-governed connection between particular events – as their role in the establishment of higher-level (causal-explanatory) theories to which they stand as confirmatory instances. In Roy Bhaskar's useful phrase, such theories have to do with laws of nature – and cognate forms of explanation – that are 'transfactually efficacious', that is to say, which possess a pertinence and scope beyond any particular context of experiment or local observation.[15] A causal-realist approach can thus be defended despite all those well-known problems (like the underdetermination of theory by evidence or the theory-laden character of observation statements) that plagued the logical-empiricist programme.[16] It was partly due to the collapse of that programme – brought about by its narrow conception of scientific method – that there emerged a whole range of reactive (anti-realist) doctrines which found no place for causal explanations except as constructs of this or that language game, paradigm, ontological scheme or whatever. Among the strong sociologists, likewise, it encouraged the idea that experimental 'findings' were always to be viewed as products of a given research culture (or 'interpretive community') whose interests, values or criteria were instrumental in deciding what counted as a 'matter of fact'.[17] For with the failure of logical empiricism – so it seemed – there could be no appeal to standards of scientific method or reasoning on the evidence which stood apart from those same acculturated habits of belief. And in so far as observations were theory-laden (or different possible theories supported by the evidence) it could further be claimed that sociocultural factors were most often what decided the issue.

However, this whole line of argument rests on a false understanding of the role that experiments typically play in the process of scientific theory construction or hypothesis testing. For Shapin and Schaffer – as for others of the strong sociological persuasion – any evidence (or suspicion) of less than conclusive findings, of ambiguous data, observational margins of error, methodological problems, leeway for interpretive doubt, etc., is sufficient to subvert not only the result of some particular experiment but the also the entire research programme of which that experiment formed a part. Moreover, it extends to the whole experimental 'culture' of modern science in so far as that culture still relies heavily on those same methods that Boyle and his colleagues were among the first to establish. However, there is no reason to accept this extreme cultural-relativist conclusion if one takes Bhaskar's point about the 'transfactual' dimension of scientific knowledge and the way that experiments may contribute to knowledge without necessarily – or indeed while never – affording demonstrative or absolute proof.

Alan Chalmers puts the case as follows in a style more succinct than Bhaskar's:

> scientific laws and theories cannot be construed as expressing relationships between sets of events, as many empiricists would have it. Laws in science cannot be appropriately interpreted as expressing constant conjunctions of events of the form 'events of type A are invariably accompanied or followed by events of type B'. . . . Human agents devise and assemble experimental set-ups, which constitute the approximately closed systems appropriate for testing scientific laws and theories. The events that occur during the performance of an experiment, the flashes on screen, positions of pointers on scales and so on, are in one sense brought about by human agents. They would not occur were it not for the intervention of human agents. Whilst it is the case that, in this sense, the conjunctions of events relevant to the testing of laws are brought about by humans, the laws that the experiments make it possible to test are not brought about by humans. (I can easily upset the workings of an experiment by some clumsy intervention and thereby upset the sought-for conjunction of events. In so doing I do not upset the laws of nature.) Consequently there must be a distinction between the laws of physics and the sequences of events that are typically produced in experimental activity and which constitute the evidence for those laws.[18]

I have cited this passage at length because it pinpoints some of the major confusions in the strong sociological approach. One is the idea that scientific theories stand or fall on the outcome (and probative force) of particular experiments whose every last detail must be fully secure against challenge if their results are to qualify as genuine contributions to the advancement of knowledge. This in turn gives the sceptic a hold for arguing (1) that no experiment could *ever* pass such a test (since observations are theory-laden, theories underdetermined by the evidence, etc.), and (2) that a great many 'classic' experiments can be shown to have fallen short of that ideal on more specific (circumstantially documented) grounds. But again, this can be seen as the direct result of applying a wholly inappropriate standard – like Hobbes in his critique of Boyle – and thus reaching a negative verdict on the methods and findings involved.

Such has often been the case with those sceptical arguments, from Hume on down, which purport to demonstrate the lack of adequate (philosophical or logical) grounds for accepting some otherwise well-tested item of scientific knowledge. Thus in Hume – as in latter-day sceptics such as Nelson Goodman – inductive procedures are subject to challenge not so much on their own self-explanatory grounds but according to strictly *de*ductive criteria of truth, consistency and rigour.[19] For it should otherwise be clear that our inherited body of scientific knowledge – along with its cognate methods, procedures, validity conditions, etc. – gives adequate reason to reject any argument (no matter how 'logically' cogent) which views induction as a flawed methodology or mere product of ingrained cultural belief. What one needs to do, rather, is turn this argument around and ask why anti-realist and cultural-relativist doctrines have enjoyed such prominence in recent debate. One reason – first pointed out by

Aristotle – is the fallacy of placing undue weight on criteria which have only a limited role in matters of inductive reasoning or experimental warrant. Another (closely related to this) is the notion that there must exist a perfect correspondence between theories, laws of nature and experimental findings if those theories and findings are to carry any kind of causal-explanatory force. But again this ignores the transfactual character of laws and explanations, that is to say, their holding good despite any localized factors (disturbances, anomalies, interference effects and so forth) that may often – or always – prevent such a perfect correspondence.

Chalmers makes the point with reference to Newton's first law of motion. 'Certainly,' he writes,

> no body has ever moved in a way that perfectly exemplifies that law. Nevertheless, if the law is correct, all bodies obey it, although they rarely get a chance to show it. The purpose of experimentation is to give them a chance to show it. If Newton's laws are 'true' they are always 'true'. They are not true only under experimentally controlled conditions. If that were so we would not be justified in applying them outside of experimental conditions. If Newton's laws are true they are always true, but are usually accompanied by the simultaneous action of other tendencies. If Newton's laws correspond to anything it is to transfactual tendencies, which are very different from localized states of affairs such as cats being on mats.[20]

The same point could be made with regard to any scientific discipline where the process of theory construction starts out from idealized premises that cannot be met – only approximated – under real-world experimental conditions. Thus, for instance, the science of aerodynamics has involved a good deal of advanced theoretical work which adopts the principles of classic fluid mechanics and discounts for all the complicating factors that arise even in well-conducted wind-tunnel experiments. That is to say, it assumes the presence of a fluid medium that is frictionless, irrotational and non-viscous and which yields precise theoretical results with regard to the performance of some given airfoil section at some given velocity. If this were really the case – if actual conditions matched this idealized picture – then no heavier-than-air machine could ever have got off the ground, since aerodynamic lift depends in part on precisely those factors (friction, circular airflow and viscosity) which find no place within the classic theory.[21] Yet here again, as with Newton's first law, it is wrong to suppose that the theory is discredited – or the phenomenon of flight placed beyond theoretical grasp – by the lack of any perfect correspondence between them.

Such might very well appear to be the case if one adopted a narrowly positivist (or logical-empiricist) view of what counts as an adequate covering-law approach in matters of scientific explanation. However, that approach has the twofold disadvantage of failing to *explain* anything – since it stops short at the stage of merely correlating observed phenomena with law-like generalizations – and casting doubt (in methodological terms) on a large portion of the knowledge arrived at by well-established inductive and causal-explanatory procedures.[22] Also it fails to account convincingly for instances where a theory manages to predict some otherwise unexpected phenomenon, or where it first runs up against apparent problems (observational discrepancies, anomalous results, experimental disconfirmation, etc.) and then turns out – at a later stage of enquiry – to have explained the cause of just those problematic data. The classic case here is the trouble created for Newton's theory of celestial motion by observed irregularities in the orbit of the planet Uranus, deviations brought about (as it subsequently proved) by the existence and gravitational effect of the hitherto unobserved planet Neptune. But there are many such cases that might be adduced in support of the causal-realist claim that scientific theories receive their most impressive confirmation from explanatory hypotheses whose predictive power may sometimes go beyond our current best methods of experimental testing or verification. For they demonstrate not only the limits imposed by a reductive (logical-empiricist) model of scientific theory construction, but also the fallacy of anti-realist or cultural-relativist doctrines which exploit those limits in order to assert the socially constructed or conventional character of all scientific knowledge.

Where these arguments go wrong is in confining their focus to the original context of discovery, thus taking no account – or minimal account – of the longer-run perspective which often provides a quite different (methodologically more adequate) grasp of the issues involved. The result of such social-constructionist approaches, as Ian Hacking writes, is that 'we are left with a feeling of absolute contingency' since 'they give us little sense of what holds the constructions together beyond the networks of the moment, abetted by human complacency.'[23] In other words, they fail to explain our knowledge of the growth of knowledge as evidenced by those many well-documented cases where scientific theories have eventually proven their worth despite going against the dominant paradigm, the received theoretical wisdom, and even – on occasion – the weight of accepted observational or experimental 'proof'. For to the realist way of thinking, it must surely be an error to 'explain' discoveries in terms of pre-existent beliefs, values, ontological commitments, motivating interests and so forth, when it is precisely the nature of genuine discoveries to venture beyond such fixed habits of thought.

III

Edward Constant offers some interesting examples in his book *The Turbojet Revolution*, a study of the various lines of research that led to the achievement of jet-powered flight.[24] Constant is far from rejecting the idea that social factors played a large – sometimes decisive – role in determining particular programmes of research in different national or geopolitical contexts and among different techno-scientific communities. Thus, for instance, there were strong military and government pressures to develop a faster, more efficient means of propulsion in those countries (chiefly Britain and Germany) where air superiority was quickly seen as a crucial part of the strategy for gaining victory in World War Two. These incentives counted for less in the US context, partly by reason of America's later entry into the war, and partly because the geography of that country required that more resources be put into building larger, longer-haul passenger and cargo machines where speed was not of the essence and where propeller-driven aircraft still seemed to offer the best solution. 'American technology tended to be functional, cheap, rugged, and energy-intensive. . . . the quest for technical excellence or extraordinary performance at the expense of utility or reasonable cost was quite foreign to most American technological practice' (*TJR*, p. 176). Thus Constant takes full account of those factors – social, political, cultural, geographical, military-strategic and so forth – which are standardly adduced by strong sociologists against the idea of scientific 'progress' as a product of disinterested truth-seeking enquiry. Moreover, he is sufficiently in sympathy with their approach to invoke Kuhn on paradigm change (along with a variety of kindred arguments) and thus make the point that differing ideas about the feasibility of jet propulsion were very largely shaped by differences of view as to what should count as a viable programme of research and development. 'It would appear,' he writes, 'that national scientific traditions and national political, social, and geographical exigencies shaped both the distribution and nature of major effort in aerodynamic science (and thus the distribution and nature of major achievements therein), and the scope and particular form of application of that science to aeronautical technology' (*TJR*, p. 100). In short, Constant's book might conceivably be read – albeit in a selective fashion – as a useful source-text for the strong sociological and cultural-relativist case.

However, this ignores everything that he has to say about other main aspects of the turbojet revolution. These were (1) its dependence on a range of existing technologies that were now harnessed in a novel and indeed 'revolutionary' way, and (2) the fact that the turbojet engine would not have been invented – or its operative principles discovered – were it not for certain strong theoretical conjectures put forward by Frank Whittle and other pioneering researchers. Those existing technologies (some going back many centuries, others comparatively recent) included, for instance, the water-turbine, the turbine-pump, turbo-air compressors, steam- and gas-turbines, and the use of turbo-superchargers to boost the performance of conventional (piston or reciprocating) engines. All of these played a crucial contributory role, along with various principles derived from the allied technologies of fluid mechanics, hydraulics and aerodynamics. Thus the jet engine could not have been conceived – let alone brought to practical fruition –

had it not been for the knowledge acquired (through theory and experiment) of airflow characteristics, turbine-blade performance, differentials of gas pressure and velocity at various stages of the combustion process, etc. To this extent the way was already prepared by a long preceding history of specialized research in these and related fields. But yet more crucial was the realization – vividly evoked by Whittle – that there existed the scope for a radically new form of propulsive mechanism that drew upon various extant technologies but which promised performance levels far beyond anything hitherto attained.[25]

Constant uses the phrase 'presumptive anomaly' to describe this sense that certain aspects of a given techno-scientific conjuncture give reason to envisage some future advance that is blocked at present by limited resources or by habits of thought trained up on more immediate practical concerns. At that time the main obstacle to faster flight had to do with the sharply declining efficiency of propeller-blade airfoils above a certain speed, i.e., the resultant of combined forward motion and rotational velocity. This was the phenomenon of so-called 'compressibility burble', the 'sudden large increase in drag and corresponding loss of lift characteristic of shock wave formation' (*TJR*, p. 109). It thus became clear that there existed an absolute limit to what could be achieved in the way of higher speeds with conventional (piston-engine and propeller-driven) aircraft. Yet it also seemed clear – with the advent of lighter airframes, stronger materials, more powerful engines, higher degrees of streamlining, etc. – that this was an isolated factor and that in principle much greater (supersonic) velocities ought to be attainable. Such was the 'presumptive anomaly' that confronted Whittle and others who were working at the limits of conventional design and engineering practice. What made this problem all the more challenging was the inverse trade-off between greater airspeed (which required flying at higher altitudes to lessen the effects of air-resistance or drag) and the fact that piston-engines and propellers suffered a drastic fall-off in performance at such altitudes owing to decreased volumetric efficiency and increased blade-tip turbulence.

Yet, as Constant shows, it was precisely these factors that led Whittle to reformulate the problem and thus see beyond the limits imposed by existing ideas and technologies. 'By early 1929,' he writes,

> Whittle seems to have been in somewhat of a quandary. Initially, he had sought high speed not through airframe streamlining but rather through lessened drag at very high altitudes. For flight under such conditions, he required a light, reasonably efficient propulsion system which was altitude compensating (as a turbo-charged system or a gas turbine was) and which dispensed with the propeller (because of insufficient propeller 'bite' in less dense air, and because of the earlier onset of blade-tip compressibility at high altitudes). A pure rocket could meet neither of the first two conditions, a piston engine with ducted fan none but the last, and a turbo-prop all but the last. (*TJR*, p. 183)

So it was that Whittle achieved his conceptual breakthrough: one that combined elements of all these technologies with an understanding of how they might be reconfigured in such a way as to transform their potential for high-altitude and high-velocity performance. In his own words,

> it suddenly occurred to me to substitute a turbine for the piston engine [as used in the ducted fan system]. This change meant that the compressor would have to have a much higher pressure ratio than the one I had visualized for the piston-engined scheme. In short, I was back to the gas turbine, but this time of a type which produced a propelling jet instead of driving a propeller. Once the idea had taken shape, it seemed rather odd that I had taken so long to arrive at a concept which had become very obvious and of extraordinary simplicity. My calculations satisfied me that it was far superior to my earlier proposals.[26]

It seems to me that this passage – along with its copiously detailed background history – helps to sharpen the focus on some central issues in philosophy of science and the sociology of knowledge. Most important, it pinpoints a basic confusion in the strong sociological 'principle of symmetry' according to which *all* theories, methods, research programmes, discovery procedures, etc., are equally subject to explanatory treatment in cultural or sociopolitical terms, no matter how great their apparent contribution to the growth of scientific knowledge.[27] For it would take a quite remarkable strength of anti-realist prejudice to maintain that Whittle's decisive insight came about *not at all* through combined theoretical insight and grasp of transfactually efficacious scientific laws but *entirely* through those various localized pressures and incentives that Constant attributes to the social context of discovery.

From this latter point of view, he concedes, 'the historical explanation for the particular spatial and temporal configuration of the turbojet revolution may lie at the fairly abstruse level of national cultural differences' (*TJR*, p. 244). And again, more specifically: 'the information field in which American practitioners worked, in sharp contrast to that of the Europeans, was clearly biased against recognition of the theoretically founded presumptive anomaly so critical to the turbojet revolution' (p. 244). But to make this claim is, of course, not to say – in Kuhnian fashion – that the various geographically divided research communities somehow quite literally 'lived in different worlds', or that they shared so little in the way of agreed-upon knowledge that nothing could count as a common measure of technological achievement. For the point about national cultural difference is that some contexts are apt to encourage certain kinds of genuinely knowledge-productive interest, while others – for various socially explicable reasons – tend to safeguard existing ideas and thus put a brake on innovatory thinking. After all, it could make no sense to talk about 'theoretically founded presumptive anomaly' if the anomaly in question were merely a product of local preconceptions or prejudices. (Indeed, it is hard to see how anomalies of *any* sort could arise in that situation.) And if cultural differences belong, in contrast, to a 'fairly abstruse level' of explanatory treatment, then this also points to a crucial weakness in the strong sociologists' programme. For that programme rests its case very heavily on the idea that 'social facts' – items of knowledge concerning the historical and cultural contexts of scientific discovery – are somehow more reliable (less subject to effects of partial or distorted hindsight) than the methods of science itself.[28]

In response to such claims it is tempting to adopt the exasperated tone of Richard Dawkins and declare flatly: 'show me a cultural relativist at 30,000 feet and I'll show you a hypocrite.'[29] That is to say, any inference to the best explanation with regard to the capacity of jet-powered aircraft to remain (for the most part) safely aloft is one that must acknowledge several points which the relativists routinely deny. These include (1) the existence of a real-world (mind- and belief-independent) physical domain; (2) the existence of law-governed (e.g., causal) regularities within that domain; (3) their transfactual efficacy or character of holding good from one particular context to another; and (4), following directly from this, their invariance across cultures – likewise across languages, paradigms or sociohistorical locales – regardless of whether they are known or accepted by members of some given community. Thus the various kinds of knowledge that came together in the making of the turbojet 'revolution' were valid not only by the cultural lights of those – like Whittle – who perceived their significance at the time but also in virtue of their proven capacity to *produce and explain* certain otherwise inexplicable results, among them the achievement of supersonic flight at high altitudes. To put it bluntly: there was nothing in Aristotle's, in Leonardo's, or in Newton's 'world' that could somehow have prevented that achievement had only the relevant principles then been discovered, along with the required materials technology, guidance systems, engineering skills and so forth. As it happens, two of these thinkers (Aristotle and Leonardo) came up with some intriguing but ultimately wrong or dead-end proposals, while the third (Newton) advanced a theory which, if valid, would have ruled out the very possibility of heavier-than-air flight.[30] But this detracts not a whit from the realist's point: that such possibilities – pertaining as they do to every world physically compatible with ours – have always existed whatever the state of contemporary knowledge regarding them.

Thus one might suggest, loosely speaking, that aerodynamics can be dated to the year 1738, since it was then that Bernoulli propounded his theorem to the effect that fluid pressure in a pipe varies inversely with flow velocity. This theorem established a working basis for all those later developments – in airfoil theory, propeller design, turbine technology, etc. – which in turn laid the ground for turbojet propulsion as a means of exploiting their combined potential. Yet, of course, it is not so much aerodynamics as the modern *science* of aerodynamics that has a datable history and can be shown to have progressed through stages of development marked not only by signal advances but also by various contingent social and cultural factors. For Bernoulli's theorem has this much in common with every other law in the natural sciences: that it holds transfactually – regardless of those factors – wherever the relevant physical conditions obtain.

Constant amplifies this point as follows in discussing the contribution of the Russian aerodynamicists Kutta and Joukowski. Their theorem showed

> that for a two-dimensional wing (of infinite span) lift could be computed, or predicted, by computing the amount of theoretical circulation about the wing necessary to have equal velocities of flow off the upper and lower trailing edges. For the first time, generation of lift became subject to rigorous mathematical treatment, and students of aerodynamics realised for the first time that lift was mainly the result of lessened pressure above the wing due to accelerated flow (according to Bernoulli's theorem) rather than the result of air particles striking the lower surface of the inclined plane. The Kutta–Joukowski theorem was corroborated by wind tunnel experiments. Aerodynamics had reached the level of mathematical rigor and experimental testability essential to true science. (*TJR*, p. 104)

The same applies to Sir George Cayley's cardinal insight that aerodynamic lift could be achieved not only, as many previous theorists had believed, through the deployment of upward rotational thrust – i.e., on the helicopter principle – but also by utilizing forward (propeller-driven) movement as applied to a plane surface (a wing) suitably shaped and angled to exploit the pattern of airflow over it. The historian of science will, of course, place a date (1809) on Cayley's lapidary statement and treat it as one – albeit crucial – stage in the progress of aerodynamic theory. The sociologist of knowledge will further wish to claim that this insight could not have been achieved in the absence of certain sociohistorical factors – incentives to research and development – which set the conditions for Cayley's break with established habits of thought. But it is also the case, as with Bernoulli's theorem, that the insight thus achieved was one that transcended the circumstances of its own time and place in so far as it stated a principle of flight whose truth would have held – counterfactually speaking – in any locale or at any historical period where the same laws of nature (or physical constants) applied. Moreover, it marked a decisive advance not only by establishing the theoretical possibility of heavier-than-air flight but also by explaining why earlier proposals had appeared to foreclose that possibility through their failure to grasp the relevant point with regard to lift as a joint resultant of forward motion and decreasing pressure on the upper surface of a wing.

Thus to think of Cayley's theory as a product of its times – or as taking rise within a particular sociohistorical locale – is a valid approach so long as it is applied to discussions of the context of discovery and not then extended (in strong sociological fashion) to the context of justification.[31] For the principle he established is one that must hold true *across and despite* all the varying cultures, scientific interests, prevailing beliefs, metaphysical convictions, and so forth, which make up the long history of thought about heavier-than-air aviation. In other words – to repeat – that principle is valid for any 'world' that is physically compossible with ours in respect of certain laws or constants (i.e., those of dynamics, fluid mechanics, energy conservation, etc.) whose validity depends not at all on any beliefs that we or others may hold concerning them. Which is also to say – contra Kuhn and the cultural relativists – that talk of radically different ('incommensurable') worlds, paradigms or discourses is talk that systematically confuses the issue with regard to these questions in history and philosophy of science. For the reason that we – unlike the ancient Greeks – inhabit a world which has in some sense been transformed by the achievement (among other things) of supersonic flight is *not* that there has occurred some fundamental change in the laws of nature which would justify Kuhn's avowedly Quinean ontological-relativist position.[32] Rather, it is the fact that we now know a great deal more about just those physical phenomena that were quite unknown to the ancient Greeks, or whose nature and workings remained obscure for lack of the requisite scientific understanding. Thus there is no reason to suppose that a Boeing 747 built (*per impossibile*) in fourth-century BC Athens would have failed to fly or indeed to exhibit all the same performance characteristics – take-off velocity for a given payload, steady cruising altitude, maximum fuel efficiency, induced drag, stalling speed as a function of its critical Reynolds number, etc. – had the ancient Greeks only possessed the kind of knowledge required to design and construct such an aircraft. For it would then have been subject to just the same physical conditions – both enabling and restrictive – as those that apply in the present-day (real-world) context and whose discovery constitutes the record of advance in aerodynamics and related fields.

IV

I should perhaps say again – having stressed those aspects of his argument which count heavily against the strong sociological approach – that Constant is very far from ignoring the extent to which cultural differences do play a role in the choice of particular scientific priorities or directions for future research. Thus:

> The development of commercial aviation in the United States, in the British Empire, and in Germany represents the differential utilization of the same fundamental scientific and technological foundation. Subsonic technological capabilities in all advanced countries were or were potentially about the same. But a variety of geographical, political, social, and economic factors intervened to produce sharply divergent choices in the development of commercial aviation. (*TJR*, p. 169)

Out of context, this passage might just be taken as arguing the case for a 'strong' anti-realist and social-constructionist approach. However, this would be to miss the point of Constant's clearly drawn distinction between (1) the fact that different cultures 'utilize' potential knowledge resources in different ways, and (2) the fact that those knowledge resources derive from 'the same scientific and technological foundation'. It also ignores his argument from 'presumptive anomaly', that is to say, his claim that although varying cultural and national contexts were such as to place greater or fewer obstacles in the way of any major scientific advance, nevertheless it was a matter of objective possibility – of transfactual and non-culturally-emergent truth – that there existed the scope for just such a breakthrough.

Thus '[p]rior technology (both structural and functional antecedents) and fundamental science (aerodynamics, operating through the mechanism of presumptive anomaly) were the intellectual substance of the turbojet revolution' (p. 152). And again: 'Without presumptive anomaly, without prior conceptualization and development of the turbojet, no plausible amount of money spent on aeronautics for any purpose would seem likely to have led quickly to the turbojet' (p. 243). For it is Constant's point that the anomaly in question – the limits of existing aircraft/engine performance as compared with what ought to be possible given developments in various (as yet uncoordinated) fields of research – was by no means restricted to certain cultural contexts. Rather, it emerged from a complex conjuncture of theories, principles, technologies and (relatively localized) research interests which acquired a different salience from one context to another, but whose level of progress – or rate of advance – was ultimately set by their capacity to register the existence of that same anomaly. Moreover, its emergence cannot be thought of – in strong sociological terms – as a product of certain predominant interests characteristic of this or that research community. For the discrepancy between actual and potential airspeed (or engine performance) was a conclusion arrived at precisely by extrapolating from known principles of aerodynamics, thermodynamics, and other established sciences. And those principles in turn derived their validity – in this case their manifest problem-solving power – from transfactually efficacious laws of physics which depended not at all on their investigators' interests or beliefs.

Constant is equally clear about the limits of Kuhnian paradigm-talk as applied to 'revolutions' like that which occurred with the advent of turbojet propulsion. Up to a point, he concedes, this did mark a break with the 'normal science' of that time, and a break whose effects were so dramatic and far-reaching as almost to justify such talk. Thus: 'The turbojet would overthrow and very nearly extinguish the technological tradition founded upon piston engines and propellers, and while fully exploiting the production technology associated with steam turbines, the turbojet would revolutionize their design as well' (*TJR*, p. 241). But he also stresses a point often missed by the proponents of radical paradigm change or those – the strong sociologists among them – who routinely endorse Kuhn's extravagant claim that scientists before and after a revolution quite literally 'live in different worlds'. For if this were the case, Constant argues, then we should have no means of explaining how major advances typically come about through the convergence of various existing paradigms – theories, technologies, forms of acquired expertise – and the perception of anomalies that then give rise to some decisive transformation in the techno-scientific sphere. Thus '[t]he protagonists of the turbojet revolution, no less than their adversaries who were committed to normal technology, fully understood and fully subscribed to community-defined requirements of testability and replicability' (p. 241). And again: 'the turbojet pioneers, no less than their opponents, recognized the necessity of adapting their ideas to meet total-systems interface constraints imposed by the whole community of aeronautical practitioners' (ibid.). But there is no support here for strong sociologists – or followers of Kuhn – who would seize upon these passages as evidence that *all* such 'requirements' and 'constraints' are 'community-defined', and hence (with reference to Wittgenstein) incapable of justification beyond the norms of some particular language game or form of life.[33] For in that case, again, there could be no explaining why scientists should ever change their ideas, and least of all – to adopt Kuhn's own terminology – why 'normal' science should ever give way to those periods of 'revolutionary' turmoil when orthodox ideas are subject to radical challenge.

What makes this, strictly speaking, an unreal problem is the fact that some standards may indeed be 'community-defined' (in the sense that they embody the interests, values and priorities of some given scientific discipline or research programme) while at the same time possessing a validity beyond that particular cultural locale. Such, as I have argued, is the case with aerodynamics: a science whose various stages of advance can be dated (and located) by the methods of sociohistorical enquiry, but which also lays claim to a record of achievement – theoretical, practical, problem-solving, explanatory, predictive and so forth – that manifestly transcends any localized cultural horizon. Thus when Constant refers to 'community-defined standards of testability and replicability' he can hardly mean to say – against all the evidence assembled in his book – that those standards are based on *nothing more* than the assent of some particular scientific community. For he would then be adopting the same approach as strong sociologists like Shapin and Schaffer, that is, the idea of 'experimental culture' as a product of sociopolitical forces which can best be explained in terms not beholden to scientific notions of truth, method or good practice. However, as Constant very clearly demonstrates, this approach can come nowhere near accounting for the advent of the turbojet 'revolution'. For this occurred *despite and against* the preponderance of established (communally sanctioned) views concerning the best means of achieving greater propulsive efficiency and higher airspeeds. After all, '[t]he internal combustion engine was then the preeminent form of aircraft power, and the natural first step towards replacing propeller drive with reaction propulsion seemed to be the adaptation of the piston engine to the new system' (*TJR*, p. 142). Hence what must now appear the various forms of half-way or compromise solution. These latter ranged from the turbo-supercharger (which utilized elements of turbine technology to boost the gas-flow performance of piston engines) to the turbo-prop arrangement which pre-empted the turbo-jet in many of its design features but still embodied the ruling conception that reactive energy could only be harnessed by converting it to rotary movement via shaft drive and propeller.

So it was, in Constant's well-chosen words, that '[t]he turbojet revolution became a community revolution fomented by outsiders' (p. 241). Its pioneers were 'outsiders' *not* in the sense that they rejected the entire body of knowledge – the theoretical assumptions, acquired expertise, experimental methods and so forth – that set the parameters for any advance within the field. Rather, they stood just sufficiently askew to those prevailing conceptions for the fact to register that a decisive advance was conceivable and that its achievement was prevented only by the hold of existing (community-relative) habits of belief and expectation. Thus '[e]ach would recognize the urgent need to redefine critical performance parameters. . . . [y]et each would work to overthrow only a technological tradition, not to violate the norms of a scientific or technological culture' (p. 179). And this is not just a point – as Kuhnians might argue – about the staying power of 'normal' science or the resistance to change naturally encountered in expert communities with every motive for maintaining their hold on the discourse of accredited scientific knowledge. For it would then be quite impossible to explain why some – and not other – members of the relevant research community saw fit to question some (and not other) of the presuppositions that effectively defined their communal enterprise. As Constant puts it:

> The creation of the first turbojets, while clearly linked to prior turbine technology, to previous aeronautical practice, and to the traditions and values of scientific technology, depended most critically upon radical assumptions about airframe and gas turbine performance deduced from contemporary advances in aerodynamics. The perception of presumptive anomaly derived from science demarcates successful from unsuccessful proponents of revolutionary alternative aircraft propulsion systems. Only the complete recognition of presumptive anomaly separates the four men who created turbojets from the two . . . who came so close or from proponents of turboprops or other reaction propulsion systems. (*TJR*, p. 242)

Moreover, such anomalies cannot be explained – unless as the result of mere random cultural drift – if one takes them to exist only relative (or internally) to this or that Kuhnian paradigm, Wittgensteinian language game, Foucauldian discourse, Quinean ontological scheme, etc. For then the question arises: 'anomalous' by what standards or with reference to what possibilities of further advance? Clearly an answer cannot be sought through the appeal to existing ('community-sanctioned') criteria. Nor can the anomaly be thought of – in Foucauldian or Kuhnian terms – as arising somehow from intra-discursive (or intra-paradigm) conflicts and tensions which eventually produce a full-scale 'epistemological break' or radical paradigm change.[34] On this account it remains a mystery just *why* such changes should ever occur, sealed off as they are within a realm of discourse – of knowledge-constitutive practices or conventions – which is taken to define the very horizon of enquiry from one paradigm to the next. That is to say, any conflicts or presumptive anomalies could emerge only within and against that same all-encompassing horizon. Thus the radical-sounding language of 'breaks' and 'revolutions' does nothing to explain – and indeed works chiefly to obscure – the processes involved in real-world examples of scientific theory change.[35] For these latter require not only the revision of established community norms but also, more decisively, a new understanding of causal powers or transfactual possibilities that could have found no place in the existing range of accredited scientific knowledge.

Constant makes this point very strongly with regard to Whittle and the other pioneers of turbojet propulsion. I shall quote one further passage at length since it offers a clear statement of his case for 'presumptive anomaly' as the motor of techno-scientific advance and also as one of the strongest arguments for causal realism via inference to the best explanation. The anomaly in question – i.e., the discrepancy between actual and potential performance levels – was one that quite simply could not have emerged except through a process of creative extrapolation from known theoretical principles and acquired expertise in various fields (most notably those of aerodynamics and advanced materials technology). These in turn gave rise to some strong predictive insights based on the assumption that certain laws of nature held good – were transfactually efficacious – across a range of potential applications extending well beyond their original context of discovery. In Constant's words:

> the presumptive anomaly on which [the inventors of the turbojet] predicated their actions derived solely and directly from advances in aerodynamics and comprised the conjunction of radical assumptions about total airframe performance with radical assumptions about internal combustion gas turbine component efficiencies. [They] saw what mature subsonic aerodynamics and the first insights of supersonic aerodynamics implied: that but for the propeller the new stressed-skin, well-streamlined airframes were perfectly capable of flying at near-sonic speeds. [They] saw that the application of mature subsonic theory to gas turbine component design could produce internal combustion gas turbines of unprecedented power, efficiency, simplicity, and lightness. [They] saw that such a gas turbine, when used for reaction propulsion at high speeds, could be propulsively efficient even though thermally less efficient than existent piston aero-engines or proposed gas turbines used for driving propellers. (*TJR*, p. 178)

Thus cultural factors may go some way towards explaining why different research communities displayed a greater or lesser willingness to break with existing (paradigm-governed) normative conventions or constraints. But they cannot explain what it was about the turbojet revolution that acted – so to speak – as a propulsive force towards new and hitherto unthought-of performance parameters. For at this stage the story has to take account of non-social, non-historical and non-culturally-emergent factors which alone made it possible for aviation science to achieve that particular great leap forward. Those factors were embodied in the various pre-existent technologies that Constant describes, along with the various known physical constants – gas-flow properties, lift/drag ratios, airfoil efficiency levels and so forth – which pointed the way to a signal advance through the presence of presumptive anomaly. But again this advance could never have come about – either in theory or in practice – had it not been for (1) the transfactual validity of aerodynamic and thermodynamic principles, and (2) the fact that those principles expressed certain real-world operative causal powers that transcended (or in no way depended upon) the conventions or the knowledge-constitutive interests of this or that research community.

Constant's book has the great virtue of approaching these issues through a detailed reconstructive treatment of one particular episode. It is also indicative of a wider – and I think beneficial – shift towards the study of technological innovations as a test-ground for claims in philosophy of science that are often advanced with insufficient regard for the real-world, practical and material constraints upon the process of scientific theory change.[36] For the trouble with the strong sociological approach exemplified by Shapin and Schaffer's *Leviathan and the Air-Pump* is that it fails to offer a convincing account – an inference to the best explanation – as to how and why the methods of the physical sciences should have produced such a range of otherwise inexplicable discoveries, inventions and stages of advance in our better understanding of the physical world. At this point, of course, the anti-realist will protest that each of these terms – 'discoveries', 'inventions', 'advances', 'better understanding' – is a product of that mainstream cultural bias that takes for granted the pre-eminence of scientific method as a source of reliable knowledge and which thus fails to notice the inherent circularity of its own evaluative criteria. Moreover, so it is held, such an argument ignores the various well-known problems (of ontological relativity, meaning variance, paradigm incommensurability and so forth) which leave no room for any notion of scientific 'progress' except as construed from the limiting perspective of our own cultural time and place. But these objections must appear ill-founded if one considers how shaky are their own premises – a range of highly debatable ideas about semantics, discourse, representation, the 'social construction of reality', etc. – as compared with the cumulative warrant for our trust in the methods and procedures of the physical sciences. For in the latter case there is nothing, hyper-cultivated scepticism aside, that could give serious reason to doubt the evidence of scientific progress in various fields of enquiry. If this book has achieved anything, then I hope it will have convinced at least a few cultural and social theorists that anti-realism is far from having won all the arguments except on its own (decidedly partisan) terms of reference.

References

1 Steven Shapin and Simon Schaffer, *Leviathan and the Air-Pump: Hobbes, Boyle, and the experimental life* (Princeton, NJ: Princeton University Press, 1985).

2 Edward W. Constant, *The Origins of the Turbojet Revolution* (Baltimore: Johns Hopkins University Press, 1981).

3 See, for instance, Barry Barnes, *About Science* (Oxford: Blackwell, 1985); David Bloor, *Knowledge and Social Imagery* (London: Routledge and Kegan Paul, 1976) and *Wittgenstein: a social theory of knowledge* (New York: Columbia University Press, 1983); Harry Collins, *Changing Order: replication and induction in scientific practice* (Chicago: University Chicago Press, 1985); Steve Fuller, *Social Epistemology* (Bloomington: Indiana University Press, 1988) and *Philosophy of Science and its Discontents* (Boulder, CO: Westview Press, 1989); K. Knorr-Cetina and M. Mulkay (eds), *Science Observed* (London: Sage, 1983); Bruno Latour and Steve Woolgar, *Laboratory Life: the social construction of scientific facts* (London: Sage, 1979); Andrew Pickering, *Constructing Quarks: a sociological history of particle physics* (Edinburgh: Edinburgh University Press, 1984); Andrew Pickering (ed.), *Science as Practice and Culture* (Chicago: University of Chicago Press, 1992); Andrew Ross, *Strange Weather: culture, science and technology in the age of limits* (London: Verso, 1991); Steve Woolgar, *Science: the very idea* (London: Tavistock, 1988).

4 See Robert Boyle, *The Sceptical Chymist* [1661] (London: Dent, 1911); *The Works of the Honourable Robert Boyle*, ed. Thomas Birch, 2nd ed. 6 vols. (London: J. and F. Rivington, 1772); also John F. Fulton, *A Bibliography of the Honourable Robert Boyle*, 2nd edn (Oxford: Clarendon Press, 1961).

5 Shapin and Schaffer, *Leviathan and the Air-Pump*, p. 67. All further references indicated by *LAP* and page number in the text.

6 See W. V. Quine, 'Two Dogmas of Empricism', in *From a Logical Point of View*, 2nd edn (Cambridge, MA: Harvard University Press, 1961), pp. 20–46 and Thomas S. Kuhn, *The Structure of Scientific Revolutions*, 2nd edn (Chicago: University of Chicago Press, 1970).

7 See the detailed bibliography of Hobbes's English and Latin works in Shapin and Schaffer, *Leviathan and the Air-Pump*, pp. 395–7.

8 See also Thomas A. Spragens, *The Politics of Motion: the world of Thomas Hobbes* (Lexington, KY: University of Kentucky Press, 1973).

9 See, for instance, Peter Lipton, *Inference to the Best Explanation* (London: Routledge, 1993) and Wesley C. Salmon, *The Foundations of Scientific Inference* (Pittsburgh: University of Pittsburgh Press, 1967); also G. Harman, 'Inference to the best explanation', *Philosophical Review*, 74 (1965), pp. 88–95.

10 For an illuminating account of this history see Norwood Russell Hanson, 'A Picture Theory of Theory-Meaning' and 'The Theory of Flight', in *What I Do Not Believe, and Other Essays* (Dordrecht: D. Reidel, 1971), pp. 4–49 and 333–90; also Clive Hart,

The Prehistory of Flight (Berkeley and Los Angeles: University of California Press, 1985).

11 See, for instance, Harry Collins and Trevor Pinch, *The Golem: what everyone should know about science* (Cambridge: Cambridge University Press, 1993).

12 For a good brief account of the Michelson–Morley episode see Rom Harré, *Great Scientific Experiments* (London: Oxford University Press, 1983).

13 See Michel Foucault, *The Order of Things: an archaeology of the human sciences* (London: Tavistock, 1973) and *Language, Counter-Memory, Practice*, ed. and trans. D. F. Bouchard and S. Weber (Oxford: Blackwell, 1977); also Gary Gutting, *Michel Foucault's Archaeology of Scientific Reason* (Cambridge: Cambridge University Press, 1989).

14 See, for instance, Rom Harré, *Great Scientific Experiments* and – from a more sceptical viewpoint – Collins and Pinch, *The Golem* and Peter Gallison, *How Experiments End* (Chicago: University of Chicago Press, 1987).

15 Roy Bhaskar, *A Realist Theory of Science* (Leeds: Leeds Books, 1975); *Scientific Realism and Human Emancipation* (London: Verso, 1986); *Reclaiming Reality: a critical introduction to contemporary philosophy* (London: Verso, 1989).

16 See Kuhn, *The Structure of Scientific Revolutions*; also Wesley C. Salmon, *Four Decades of Scientific Explanation* (Minneapolis: University of Minnesota Press, 1989).

17 See entries under note 3 above.

18 Alan F. Chalmers, *What Is This Thing Called Science?*, 2nd edn (Queensland: University of Queensland Press, 1982), pp. 154–5.

19 Nelson Goodman, *Fact, Fiction, and Forecast* (Cambridge, MA: Harvard University Press, 1955).

20 Chalmers, *What Is This Thing Called Science?*, p. 155.

21 See entries under note 10 above.

22 See Salmon, *Four Decades of Scientific Explanation*; also Adolf Grunbaum and W. C. Salmon (eds.), *The Limitations of Deductivism* (Berkeley and Los Angeles: University of California Press, 1988).

23 Ian Hacking, 'Statistical Language, Statistical Truth, and Statistical Reason: the self-identification of a state of scientific reasoning', in Ernan McMullin (ed.), *The Social Dimensions of Science* (Notre Dame, IN: University of Notre Dame Press, 1992), p. 131.

24 Edward W. Constant, *The Origins of the Turbojet Revolution* (Baltimore: Johns Hopkins University Press, 1981). All further references indicated by *TJR* and page number in the text.

25 Frank Whittle, *Jet: the story of a pioneer* (London: Frederick Muller, 1953).

26 Ibid., pp. 24–5.

27 See entries under note 3 above.

28 For further discussion see Margaret Gilbert, *On Social Facts* (London: Routledge, 1989).

29 Richard Dawkins, *River Out of Eden* (London: Weidenfeld and Nicolson, 1995), p. 32.

30 See entries under note 10 above.

31 For the classic statement of this 'two contexts' principle, see Hans Reichenbach, *Experience and Prediction* (Chicago: University of Chicago Press, 1938). The distinc-

tion is attacked by various proponents of the 'strong sociological' line, among them Karin Knorr-Cetina, *The Manufacture of Knowledge: an essay on the constructivist and contextual nature of knowledge* (London: Oxford University Press, 1981). See also entries under note 3 above.

32 See especially Kuhn's 1969 'Postscript' to *The Structure of Scientific Revolutions.*

33 See Bloor, *Wittgenstein: a social theory of knowledge* and Derek L. Phillips, *Wittgenstein and Scientific Knowledge: a sociological perspective* (London: Macmillan, 1977).

34 See Foucault, *The Order of Things.*

35 For further discussion see Gary Gutting (ed.), *Paradigms and Revolutions* (Notre Dame, IN: University of Notre Dame Press, 1980); Ian Hacking (ed.), *Scientific Revolutions* (London: Oxford University Press, 1981); Mary M. Hesse, *Revolutions and Reconstructions in the Philosophy of Science* (Brighton: Harvester, 1980); John Krige, *Science, Revolution and Discontinuity* (Brighton: Harvester, 1980).

36 See, for instance, Walter G. Vincenti, *What Engineers Know and How They Know It: analytical studies from aeronautical history* (Baltimore: Johns Hopkins University Press, 1990); also Jon Elster, *Explaining Technical Change: a case study in the philosophy of science* (Cambridge: Cambridge University Press, 1983); Andrew Feenberg, *Critical Theory of Technology* (New York and London: Oxford University Press, 1991); Don Ihde, *Instrumental Realism: the interface between philosophy of science and philosophy of technology* (Bloomington, IN: Indiana University Press, 1991); Carl Mitcham and Robert Mackey (eds), *Philosophy and Technology* (New York: Free Press, 1972).

Index of Names